China in Africa

China's expansion and growing influence in Africa is arguably the most remarkable global political and economic development in the twenty-first century. China's foray into Africa started in the late 1990s, propelled by its desire to obtain new sources of raw materials and energy for its economic growth, as well as new markets for its manufactured goods. While China's 'no political strings attached' policy proves attractive to many African leaders, China has been criticized as neo-colonialist, interested solely in stripping Africa of its mineral wealth without proper environmental or social precautions.

This book addresses the controversy by exploring the motivations and practices of China's African engagement, providing a comprehensive account of the intensified interactions between China and African states. The first part examines the debate surrounding whether China has pursued a neo-colonialist path in Africa, by looking at the perception of China by the locals and the challenges that the intensified relationship has posed for African states. The second part analyses China's strategic motivations to see if Beijing has acquired sustaining power and influence in Africa in competition with the West. The third part focuses on economic and business practices of Chinese companies in Africa, as well as China–Africa trade patterns.

The articles in this book were originally published in special issues of the *Journal of Contemporary China*.

Suisheng Zhao is Professor and Director of the Centre for China–US Cooperation at the Josef Korbel School of International Studies, University of Denver, Colorado, USA, and founding editor of the *Journal of Contemporary China*.

China in Africa

Strategic motives and economic interests

Edited by
Suisheng Zhao

LONDON AND NEW YORK

First published 2015
by Routledge
2 Park Square, Milton Park, Abingdon, Oxon, OX14 4RN, UK

and by Routledge
711 Third Avenue, New York, NY 10017, USA

Routledge is an imprint of the Taylor & Francis Group, an informa business

© 2015 Taylor & Francis

All rights reserved. No part of this book may be reprinted or reproduced or utilised in any form or by any electronic, mechanical, or other means, now known or hereafter invented, including photocopying and recording, or in any information storage or retrieval system, without permission in writing from the publishers.

Trademark notice: Product or corporate names may be trademarks or registered trademarks, and are used only for identification and explanation without intent to infringe.

British Library Cataloguing in Publication Data
A catalogue record for this book is available from the British Library

ISBN 13: 978-1-138-89900-1

Typeset in Times New Roman
by RefineCatch Limited, Bungay, Suffolk

Publisher's Note
The publisher accepts responsibility for any inconsistencies that may have arisen during the conversion of this book from journal articles to book chapters, namely the possible inclusion of journal terminology.

Disclaimer
Every effort has been made to contact copyright holders for their permission to reprint material in this book. The publishers would be grateful to hear from any copyright holder who is not here acknowledged and will undertake to rectify any errors or omissions in future editions of this book.

Contents

Citation Information   vii

Part I: The Debate on China in Africa

1. A Neo-Colonialist Predator or Development Partner? China's engagement and rebalance in Africa
   Suisheng Zhao   1

2. Why Do We Need 'Myth-Busting' in the Study of Sino–African Relations?
   Miwa Hirono and Shogo Suzuki   21

3. China in Africa: presence, perceptions and prospects
   Fei-Ling Wang and Esi A. Elliot   40

Part II: Strategic Interactions and Motivations

4. China–Africa Cooperation: promises, practice and prospects
   Sven Grimm   61

5. China Goes to Africa: a strategic move?
   Jianwei Wang and Jing Zou   80

6. China's Libya Evacuation Operation: a new diplomatic imperative—overseas citizen protection
   Shaio H. Zerba   100

7. China's Exceptionalism and the Challenges of Delivering Difference in Africa
   Chris Alden and Daniel Large   120

Part III: Business Practices and Economic Relations

8. Bashing 'the Chinese': contextualizing Zambia's Collum Coal Mine shooting
   Barry Sautman and Yan Hairong   138

9. Workforce Localization among Chinese State-Owned Enterprises (SOEs) in Ghana
   Antoine Kernen and Katy N. Lam   158

CONTENTS

10. Chinese State-owned Enterprises in Africa: ambassadors or freebooters 178
    Xu Yi-Chong

11. China–Africa Trade Patterns: causes and consequences 197
    Joshua Eisenman

    Index 215

Citation Information

The following chapters were originally published in various issues of the *Journal of Contemporary China*. When citing this material, please use the original page numbering for each article, as follows:

Chapter 1
A Neo-Colonialist Predator or Development Partner? China's engagement and rebalance in Africa
Suisheng Zhao
Journal of Contemporary China, volume 23, issue 90 (November 2014) pp. 1033–1052

Chapter 2
Why Do We Need 'Myth-Busting' in the Study of Sino–African Relations?
Miwa Hirono and Shogo Suzuki
Journal of Contemporary China, volume 23, issue 87 (May 2014) pp. 443–461

Chapter 3
China in Africa: presence, perceptions and prospects
Fei-Ling Wang and Esi A. Elliot
Journal of Contemporary China, volume 23, issue 90 (November 2014) pp. 1012–1032

Chapter 4
China–Africa Cooperation: promises, practice and prospects
Sven Grimm
Journal of Contemporary China, volume 23, issue 90 (November 2014) pp. 993–1011

Chapter 5
China Goes to Africa: a strategic move?
Jianwei Wang and Jing Zou
Journal of Contemporary China, volume 23, issue 90 (November 2014) pp. 1113–1132

CITATION INFORMATION

Chapter 6
China's Libya Evacuation Operation: a new diplomatic imperative—overseas citizen protection
Shaio H. Zerba
Journal of Contemporary China, volume 23, issue 90 (November 2014) pp. 1093–1112

Chapter 7
China's Exceptionalism and the Challenges of Delivering Difference in Africa
Chris Alden and Daniel Large
Journal of Contemporary China, volume 20, issue 68 (January 2011) pp. 21–38

Chapter 8
Bashing 'the Chinese': contextualizing Zambia's Collum Coal Mine shooting
Barry Sautman and Yan Hairong
Journal of Contemporary China, volume 23, issue 90 (November 2014) pp. 1073–1092

Chapter 9
Workforce Localization among Chinese State-Owned Enterprises (SOEs) in Ghana
Antoine Kernen and Katy N. Lam
Journal of Contemporary China, volume 23, issue 90 (November 2014) pp. 1053–1072

Chapter 10
Chinese State-owned Enterprises in Africa: ambassadors or freebooters?
Xu Yi-Chong
Journal of Contemporary China, volume 23, issue 89 (September 2014) pp. 822–840

Chapter 11
China–Africa Trade Patterns: causes and consequences
Joshua Eisenman
Journal of Contemporary China, volume 21, issue 77 (September 2012) pp. 793–810

Please direct any queries you may have about the citations to clsuk.permissions@cengage.com

A Neo-Colonialist Predator or Development Partner? China's engagement and rebalance in Africa

SUISHENG ZHAO

While China has found niches based on its comparative advantages and gained a solid ground in Africa, its almost single-minded way of pursuing business interests without regard to many issues of local and international concern has caused backlashes. Learning lessons the hard way, China has made efforts to adjust such insensitive business practices in recent years, motivated by its changing economic and strategic interests.

China's fast-growing African engagement has attracted wide attention and heated debate about whether China has pursued a colonialist path in the continent. During her visit to Zambia in June 2011, then US Secretary of State Hillary Clinton accused China of a creeping 'new colonialism' in Africa from foreign investors and governments interested only in extracting natural resources to enrich themselves. Secretary Clinton did not identify any alleged culprits, but a day earlier she had urged scrutiny of China's investments and business interests in Africa so that the African people were not taken advantage of.[1] One year later, at the 2012 Forum on China–Africa Cooperation (FOCAC), South African President Jacob Zuma warned an audience that included Chinese President Hu Jintao and UN Secretary General Ban Ki-moon that Africa's burgeoning economic ties with China were 'unsustainable in the long term'. Referring to the continent's colonial past, he said that 'Africa's past economic experience with Europe dictates a need to be cautious when entering into partnerships with other economies'.[2] Reflecting the fear that the continent's industrial sector was getting battered by cheap Chinese imports, Nigeria's central bank governor Lamido Sanusi wrote in 2013 that Africa must shake off its romantic view

---

* Suisheng Zhao is Professor and Director of the Center for China–US Cooperation at Josef Korbel School of International Studies, University of Denver, senior fellow at Chahar Institute, and Editor of the *Journal of Contemporary China*.
 1. 'Hillary Clinton warns Africa of "new colonialism"', *Huffington Post*, (11 July 2011), available at: http://www.huffingtonpost.com/2011/06/11/hillary-clinton-africa-new-colonialism_n_875318.html.
 2. Leslie Hook, 'Zuma warns on Africa's trade ties to China', *Washington Post*, (19 July 2012), available at: http://www.washingtonpost.com/world/asia_pacific/zuma-warns-on-africas-trade-ties-to-china/2012/07/19/gJQAFgd7vW_story.html.

of China and recognize that Beijing is a competitor as much as a partner and capable of the same exploitative practices as the old colonial powers.[3]

Taking a position of 'Chinese exceptionalism',[4] however, Beijing insists that it comes to Africa as equals without colonial intentions. A cross section of Africans has a favorable opinion of China because, while the Western investments and assistants failed to lift African countries out of poverty, China's way of doing business provides an alternative opportunity for Africans. As Chinese Vice-Foreign Minister Zhai Jun said, 'China's economic backing is giving African countries options they never had under a Western-led world order'.[5] Lashing out at Secretary Clinton's 'warning of the "new colonialism" looming on the continent in a veiled swipe at Beijing's efforts to forge closer ties with Africa', a *Xinhua* commentary accused Clinton as being 'vicious', 'rude' and taking a 'cheap shot' at Beijing, and was either 'ignorant of the facts on the ground or chose to disregard them'.[6] Beijing's view is echoed by some scholars with the 'primary purpose' to 'engage in "myth-busting" in the study of Sino–African relations'. Putting China's African engagement in context, these scholars point out that 'not only Chinese but also Western actors are responsible for their lack of political will to solve Africa's security and governance problems'.[7]

Some African leaders also share this view and describe the relationship as neutral and business-oriented to generate economic growth for both China and African countries. Because China wants to buy the minerals and Africa has them for sale, it is a win–win relationship. While African resources help feed China's economic boom, Africa's economic growth benefits from trade and infrastructure built and financed by China. New roads, railways, ports and airports have filled a critical gap that Western donors had been shy to provide and opened the door of many African countries to a future of real development. As Zambia's Deputy Finance Minister Miles Sampa said, 'The Chinese are not our relatives or friends, they are here for business and they are our partners'.[8] A Zambian born economist also suggested that

> China's motives for investing in Africa are actually quite pure. To satisfy China's population and prevent a crisis of legitimacy for their rule, leaders in Beijing need to keep economic growth rates high and continue to bring hundreds of millions of people out of poverty. And to do so, China needs arable land, oil and minerals. Pursuing imperial or colonial ambitions with masses of impoverished people at home would be wholly irrational and out of sync with China's current strategic thinking. Moreover, the evidence

---

3. Lamido Sanusi, 'Africa must get real about Chinese ties', *Financial Times*, (11 March 2013), available at: http://www.ft.com/intl/cms/s/0/562692b0-898c-11e2-ad3f-00144feabdc0.html#axzz2kergjtnD.

4. Chris Alden and Daniel Large, 'China's exceptionalism and the challenges of delivering difference in Africa', *Journal of Contemporary China* 20(68), (2011), p. 21.

5. 'Stronger China–Africa relations no damage to other countries' interests: Chinese vice FM', *Xinhua*, (12 July 2012), available at: http://news.xinhuanet.com/english/china/2012-07/12/c_131712107.htm.

6. Commentary, 'New colonialism targeting China–Africa cooperation untenable', *Xinhua*, (18 July 2012), available at: http://www.chinadaily.com.cn/xinhua/2012-07-18/content_6474149.html.

7. Miwa Hirono and Shogo Suzuki, 'Why do we need "myth-busting" in the study of Sino–African relations?', *Journal of Contemporary China* 23(87), (2014) p. 443–461.

8. Rachael Akidi, 'African nations lack vision on China investment', *South China Morning Post*, (23 June 2012), available at: http://www.scmp.com/portal/site/SCMP/menuitem.2af62ecb329d3d7733492d9253a0a0a0/?vgnextoid=8cf8719d7d318310VgnVCM100000360a0a0aRCRD&ss=China&s=News.

does not support a claim that Africans themselves feel exploited. To the contrary, China's role is broadly welcomed across the continent.[9]

Addressing the debate, this article argues that while it may be too innocent to describe China as a neutral business partner, China has found niches based on its comparative advantages and gained a solid ground in Africa. Almost single-mindedly pursuing its business interests, however, China's business practices have been controversial in many issue areas. Learning lessons the hard way, China has tried to rebalance its policy and adjust its insensitive business practices, motivated by its changing economic and strategic interests.

China's African engagement: comparative advantages and resources mobilization

China's engagement with Africa has come in three periods. China started venturing into the continent in the early 1950s, motivated by the strategic interest of breaking the diplomatic isolation imposed by the Western powers. To find diplomatic partners, Beijing built a costly railway to link Tanzania and Zambia and constructed stadiums for football matches and political rallies in many African countries that supported Mao's anti-imperialist struggle. These investments paid off as African countries played a decisive role in Beijing's 1971 entry into the UN.

The second period was the 1980s, known as the 'decade of neglect' of Africa in China's foreign policy.[10] Launching market-oriented economic reform, Beijing was not sure how it could benefit from relations with economically stagnated African countries. While the rhetoric of third world solidarity continued, China moved its foreign policy priority toward Western industrialized countries. The 'neglect' was to be corrected after the Tiananmen crackdown in 1989, motivated initially by the strategic interest of working with its third world allies to resist Western sanctions as well as the competition with Taiwan over diplomatic recognition by African countries. But the economic incentives of securing natural resources to sustain China's phenomenal growth started to take top priority in the wake of the twenty-first century. As rapid economic growth brought China to an unprecedented resource vulnerability that could threaten China's sustainable development and political stability, China developed a series of diplomatic measures to deepen political and commercial relationships with all resource-rich nations. Essentially a neo-mercantilist approach to the search for readily available and affordable natural resources supplies,[11] China rediscovered Africa, one of the only places in the world where so many resources are still up for grabs.

The rediscovery started the third period. China's renewed interest in Africa is more or less similar to those of the European powers several centuries earlier: to seek fortunes in Africa's natural resources and tap into the continent's unsaturated markets for its manufactured goods. While Africa's traditional business partners either

---

9. Dambisa Moyo, 'Beijing, a boon for Africa', *New York Times*, (27 June 2012), available at: http://www.nytimes.com/2012/06/28/opinion/beijing-a-boon-for-africa.html/.

10. Ian Taylor, 'China's foreign policy towards Africa in the 1990s', *Journal of Modern African Studies* 26(3), (1990), pp. 443–460.

11. Suisheng Zhao, 'China's global search for energy security: cooperation and competition in Asia–Pacific', *Journal of Contemporary China* 17(55), (2008), p. 207.

abandoned or found many problems operating in Africa, many Chinese companies saw opportunities because of the following comparative advantages. First, Chinese managers and workers are not only very diligent and disciplined but also normally do not ask for the comfort and expenses that Western expatriates often demand. Chinese equipment and materials are also cheaper. Although in some ways Chinese companies come to Africa at a disadvantage with language and other cultural barriers, their greatest asset is their willingness to take a leap into the unknown because China's economic reform created a frenzy of entrepreneurialism. Given their energies and talents working for a pittance to turn a profit, the Chinese have engaged in cutthroat competition, which creates a virtually unbeatable China price and efficiency. Africans watched in surprise as buildings were erected in weeks.

Second, compared with Western powers, China has no history of enslavement, colonization, financing coups against unfriendly African regimes or deploying military forces in support of its foreign policies. Promoting its efforts with flowery rhetoric touting China as a traditional friend of Africa to foster a harmonious world and invoking the memory of colonial aggression and their common history with Africans as the subjects of outside oppression, China designed a package of infrastructure construction for extraction rights by signing agreements to build infrastructure projects in exchange for minerals. Although much of the infrastructure was crucial to China's ability to operate effectively, they provided a much-needed stimulus to the local economy because the infrastructure deficit is a major impediment to growth and Western investors failed to fill this massive infrastructure investment gap: 'Since many African countries lack the indigenous engineering capability to construct these large-scale projects or the capital to undertake them, African governments with limited resources welcome Chinese investments enthusiastically'.[12]

Third, while Western companies have come under increasing pressure and oversight from shareholders and regulators to tell their African partners to abide by certain political and ethical conditions, the Chinese have followed a policy of 'non-interference' in domestic affairs. Constantly reminding its African partners that China will never impose its will on another country, Beijing requires no political conditions before signing business contracts. This is a welcome relief after years of Western investment and assistance offers inconveniently premised on high benchmarks about environmental damage, human rights, transparency and good governance. For most African leaders, the paramount task is how to eradicate poverty. While only three decades ago, China was as poor as some of the poorest African countries, China has lifted hundreds of millions of their own people out of poverty without democratization. China's model of a strong government and its focus on economic growth is looked upon by many African leaders as an example to follow.[13]

Cutting a swath across the entire continent, the Chinese government has mobilized diplomatic, cultural and financial resources in pursuing its African interests. No other major power has shown the same interest, muscle and sheer

---

12. Mamta Badkar, 'MAP: here are all of the big Chinese investments in Africa since 2010', *Business Insider*, (13 August 2012), available at: http://www.businessinsider.com/map-chinese-investments-in-africa-2012-8#ixzz23U4FwqeP.

13. Suisheng Zhao, 'The China model: can it replace the Western model of modernization?', *Journal of Contemporary China* 19(65), (2010), p. 433.

ability. Diplomatically, China has built the largest number of embassies and consulates, and each year the first overseas trip of the Chinese Foreign Minister is to Africa. Africa's place at the heart of China's foreign policy agenda was highlighted when Chinese President Xi Jinping chose to tour the continent on his inaugural foreign trip in 2013. To help Chinese companies' African ventures, the Chinese government launched the Forum on China–Africa Cooperation (FOCAC) in October 2000 and has built a framework of multilateral institutions. The FOCAC came into the spotlight when more than 50 African heads of states gathered in Beijing at the Third FOCAC summit in November 2006. In addition, the China–Africa Business Conference, the China–Africa Business Council, the China–Africa Development Fund and the Sino–Africa Business and Investment Forum have promoted direct exchange and cooperation between Chinese and African entrepreneurs, encouraged Chinese investment and trade with Africa, and helped Chinese companies learn about business and investment opportunity in Africa.

Culturally, Beijing has rapidly expanded a network of Chinese media outlets covering the entire continent and is encouraging exchange and cooperation between the African and Chinese media to enhance mutual understanding and balanced media coverage of each other. In the meantime, Chinese education and training programs have targeted students from across the continent and welcome many African students to study in China or in Chinese language classes. At the Fifth FOCAC in the summer of 2012, then Chinese President Hu Jintao announced an expansive program that would offer 18,000 government scholarships and train 30,000 Africans 'in various sectors' by 2015.[14] Chinese training programs vary in type and duration, from three-week political tours for government officials to advanced degree programs for students and young professionals. Education is regarded as a 'long-term investment to win the hearts and minds of Africa's future leaders'.[15] African students patronized these education programs because they saw in them a rich value for career enhancement; a lower cost compared to American and British equivalents; and exposure to the inner workings of a rising power.

Financially, while aid from OECD countries stagnated or shrank under the pressure of budgets and an increasingly skeptical public, China played an important role in closing funding gaps of development finance in Africa. Investing billions of dollars in mineral and massive infrastructure projects across Africa, China offered a mix of grants, interest-free and concessional loans as development assistance. Pledging to double assistance to Africa in the next three years and cancel debts owed by 31 African states at the Third FOCAC 2006, Chinese Premier Wen Jiabao at the Fourth FOCAC in 2009 further pledged to provide US$10 billion in concessional loans to African countries and set up a special loan of US$1 billion for small- and medium-sized African businesses.[16] At the Fifth FOCAC in July 2012, Chinese President Hu

---

14. 'President Hu: important progress made in realizing China–Africa new strategic partnership', *Xinhua*, (19 July 2012), available at: http://news.xinhuanet.com/english/china/2012-07/19/c_131725431.htm.

15. Jonathan Kaiman, 'Africa's future leaders benefit from Beijing's desire to win hearts and minds', *Guardian*, (29 April 2013), available at: http://www.guardian.co.uk/global-development/2013/apr/29/africa-future-leaders-china-aid-programme.

16. 'Chinese Premier announces eight new measures to enhance cooperation with Africa', *Xinhua*, (8 November 2009), available at: http://news.xinhuanet.com/english/2009-11/08/content_12411421.htm##.

Jintao pledged US$20 billion in loans to Africa, doubling the size of China's lending pledge made just three years before. While transport, storage and energy initiatives accounted for the largest part, China put hundreds of millions of dollars towards health, education and cultural projects. China also sent thousands of doctors and teachers to work in Africa and rolled out a continent-wide network of sports stadiums and concert halls. Of course, China's assistance in Africa is not mercenary, but motivated by the combination of economic interests and the need to expand its political influence. These concessionary loans must be paid back, frequently with raw materials, and are often used to construct large infrastructure projects tied to Chinese companies. These assistances nevertheless gave the Chinese companies a great advantage over companies from other countries that cannot access such facilities from their governments. They also helped China cement alliances and drum up business.

China's efforts paid huge dividends. China's direct investment in Africa skyrocketed from under US$100 million in 2003 to multi billions of dollars in 2012, covering more than 50 countries.[17] Surpassing the United States in 2009 to becoming Africa's single largest trading partner, China's trade with African countries soared from less than US$10 million in the 1980s to US$198 billion in 2012. China is a major exporter of cheap manufactured goods, such as electronics and clothes, to Africa. During the Global Financial Crisis in 2008–2009, Africa became part of China's effort to diversify its export markets away from their dependence on Western profligates. Among the US$198 billion plus trade, a significant proportion was made up of Africa-bound Chinese consumables. In addition to the state-owned enterprises, many individuals, including plumbers, electricians and small shopkeepers, also made their way to Africa. Today there is barely an African country that does not have a sizeable Chinese presence. Involved in oil, mining, timber and fishing for its resource hungry economy, the Chinese are credited with building crucial infrastructure and providing easy loans and assistance for some of the world's poorest nations as well as internationally condemned dictators.

While America still reigns in terms of cumulative direct investment in a few countries, Chinese involvement provided a welcome infusion of competition that raised prices for African resources and reduced prices for the construction projects on which many Chinese firms bid. Beijing's development projects, from infrastructure to debt relief to providing medical support, are also part of a public diplomacy strategy to build goodwill and international support. As the influence of former colonial powers waned, the China model proved popular particularly among authoritarian leaders in Africa. They spoke openly about China's offer of an alternative to the edicts of Western-dominated institutions like the International Monetary Fund and the World Bank. Senegal's president declared that 'today, it is very clear that Europe is close to losing the battle of competition in Africa', and Nigeria's president said that

---

17. The Chinese government releases very little information on its investments in Africa. Some US researchers launched the largest public database of Chinese development finance in Africa, detailing almost 1,700 projects in 50 countries between 2000 and 2011. According to the database (AidData), the level of Chinese investment was over US $260 billion during 2000–2011. There were 1,422 official projects to 50 African countries over the 2000–2011 period that at least reached commitment stage, totaling US$75.4 billion. See Rob Minto, 'Chart of the week: tracking China's investments in Africa', *Financial Times*, (30 April 2013), available at: http://blogs.ft.com/beyond-brics/2013/04/30/chart-of-the-week-tracking-chinas-investments-in-africa/#ixzz2SB3IX5Vc.

'This is the century for China to lead the world. And when you are leading the world, we want to be very close behind you'.[18] President Robert Mugabe of Zimbabwe also praised that 'China has been able to develop its economy without plundering other countries, and the Chinese economic miracle is indeed a source of pride and inspiration'.[19] President Kagame of Rwanda commented that 'The Chinese bring what Africa needs: investment and money for governments and companies. I would prefer the Western world to invest in Africa rather than hand out development aid'.[20]

The critiques of China's engagement in Africa

China's new engagement in Africa has come not only with praise but also suspicion. As China's positive image is created often 'through political discourses that emphasize difference with Western countries',[21] it has evoked the image of a benign, postcolonial West being outfoxed and marginalized by a ruthless and unscrupulous authoritarian power. Phrases such as 'exploiters', 'harsh employers' and 'neo-colonialists interested only in stripping the continent of its mineral wealth without proper environmental or social precautions' have become typical descriptors of China's role in the African continent.

Human rights

While Beijing's no-strings-attached policy in the name of non-interference is attractive to some African governments, it is criticized for threatening to wipe out the efforts by global institutions to push African governments to improve human rights and government transparency and undermining Africa's forward march towards democracy. This is because the Chinese government worked with any government that could help secure its investments in mining and drilling rights, including those accused of rampant corruption or severe human rights violations. While China played a critical role in supporting African decolonization struggles, its laissez-faire policy raised questions about its moral and ethical commitment to Africa's sustainable socio-economic and political development. In particular, China's deals with countries under Western sanctions, such as Sudan and Zimbabwe, were tackled with vigor in the international media and drew frustration from human rights advocates who would prefer to use investment as carrots in an effort to stop violence and repression in these countries. China's heightened influence in Africa thus raised concerns that China is not only challenging the Western countries' historic dominance but also undermining the efforts to promote democracy, good governance and human rights on the continent.

---

18. Richard Behar, 'China surpasses US as leader in sub-Sahara', *FastCompany*, (9 May 2008), available at: http://www.fastcompany.com/magazine/126/endgame-hypocrisy-blindness-and-the-doomsday-scenario.html.
19. Antoaneta Bezlova, 'China: latest Africa foray: altruism or hegemony?', *Inter Press Service News Agency* (IPS), (9 November 2009), available at: http://www.ipsnews.net/news.asp?idnews=49191.
20. Jonathan Clayton, 'China tightens grip on Africa with $4.4bn lifeline for Guinea junta', *The Times*, (13 October 2009), available at: http://www.timesonline.co.uk/tol/news/world/africa/article6871943.ece#.
21. Sven Grimm, 'China–Africa cooperation: promises, practice and prospects', *Journal of Contemporary China* 23(90), (2014), doi: 10.1080/10670564.2014.898886.

An international pariah, Sudan was out of bounds for years to Western companies because of its policy in Darfur where hundreds of thousands of people were killed, countless numbers raped and tortured, and millions displaced. Chinese state oil companies, however, had no qualms about doing business in Sudan, pumping hundreds of thousands of barrels of oil a day from the Red Sea port into Chinese ships. As Sudan's Energy Minister Awad al-Jaz said, 'With the Chinese, we don't feel any interference in our Sudanese traditions or politics or beliefs or behaviors. Business is business. There is no other business but the business'.[22] In the midst of what many Americans and Europeans dubbed as genocide in Darfur, China continued to be not only the biggest importer of Sudan's oil but also the supplier of weapons used by the government forces and militia against rebellions in Darfur. While it is arguable if China's close relations with Sudan allowed the regime in Khartoum to carry out atrocities, in 2004 China did threaten a veto when Britain and the US pushed for a punitive UN Security Council resolution against Sudan for the mass killing of civilians in Darfur. Working with China, the Sudanese government was untroubled by the sanctions that prevented Western investment.

China also avowed an all-weather friend with the Mugabe government in Zimbabwe. Arguing that China's support for Zimbabwe was based on the close historical ties dating back to the struggle for independence, Chinese companies clinched lucrative deals in mining, aviation, agriculture, defense and other sectors. China's veto, together with Russia, against the UN sanctions on the Mugabe regime for its human rights violations in the midst of an election crisis in June 2008 angered those who worked for substantive political change in Zimbabwe. China was described in international media as 'shrewd, selfish, calculating, greedy and primitive because it prioritizes its economic and political interests over ordinary people's human rights in its dealings with African countries'.[23] Zimbabwe's government would have collapsed during the global financial meltdown, but 89-year-old President Robert Mugabe declared another victory in the 2013 election, then oversaw an extraordinary economic boom thanks largely to Chinese investments and economic aids. It is from this perspective that one commentator suggested:

> It's not that China's money is single-handedly reviving Zimbabwe, but that its willingness to do business (and sell weapons) makes a mockery of attempted Western sanctions. Zimbabwe's options are not simply Western-style freedom or penury. The Beijing model of 'state capitalism' is available as well, and it pays.[24]

China is also blamed for enabling human rights abuse in other African countries. One example is a multibillion dollar deal struck by China International Fund for oil and mineral extraction rights in Guinea under a military junta. The deal came just weeks after the 28 September 2009 massacre, in which soldiers opened fire on protesters after Captain Camara, who seized power in December 2008, announced

---

22. Lindsey Hilsum, 'We love China', *Granta 92: The View from Africa*, (Winter 2005), available at: http://www.granta.com/extracts/2616.
23. Last Moyo, 'China's role in African politics appalling', *Zimbabwe Independent*, (17 July 2008).
24. Fraser Nelson, 'Zimbabwe is booming—but its future lies in Chinese hands', *Telegraph*, (2 August 2013), available at: http://www.telegraph.co.uk/news/worldnews/africaandindianocean/zimbabwe/10216403/Zimbabwe-is-booming-but-its-future-lies-in-Chinese-hands.html.

that he would run for the presidency. Killing 157 people and raping women in the streets, the incident drew international condemnation and prompted international sanctions. While Guinea became a no-go area for reputable companies, China enhanced its position in Guinea, which has the world's biggest deposits of bauxite, as well as gold, diamonds, uranium and iron ore. A Chinese government spokesman insisted that the deal was strictly business, but critics said that China's deal would potentially throw the regime a lifeline, raising questions about the willingness of China to prop up rogue African governments. Indeed, the timing of the deals could not have been worse for China and turned a harsh spotlight on the human rights and geopolitical stakes of the scramble for Africa's natural wealth.[25]

Corruption

China is known for its widespread corruption at home during the reform years. Using bribery to secure government contracts is common in Chinese business practice and China is accused of exporting unconcealed corruption to Africa. In a part of the world prone to corruption, it was described that 'while Europe scrambled for Africa, China which was not part of the fracas is creeping in now like a sweet-talking lover who uses money to seduce, dupe and bribe his way'.[26] This is particularly obvious in the Democratic Republic of the Congo, known as the most corrupt nation on Earth. While Western firms that tried to obey safety laws are constantly targeted by official safety inspectors because they refuse to bribe them, Chinese enterprises could get away with huge breaches of the law because they paid bribes.[27]

The lack of transparency in China's business deals facilitates corruption. Embracing preferential loans from the Chinese government for a host of unmet needs, many African nations have to cope with China's practice of secret government-to-government agreements and the requirement that foreign aid contracts be awarded to Chinese contractors through a closed-door rather than competitive bidding process. In Africa, as in other parts of the world, such secrecy invited corruption and put much of the wealth that China injected into Africa into the pockets of either the Chinese companies or the local elite. The secrecy and elitism that defined China's practice in Africa was poised to usher in toxic intercontinental corruption.

Although for some time in the recent past Western companies conducted similar behavior, US and European companies are increasingly expected to behave in a socially responsible manner. The Extractive Industries Transparency Initiative (EITI) was launched in 2003 and rapidly gained over 30 signatories from governments and companies under the umbrella of the Organization of Economic Cooperation and Development (OECD). The Initiative requires transparent reporting of activities and a concerted effort to clean up their programs and ensures that if they bribe, they commit a criminal offence within their home country. It also demands that foreign

---

25. Tom Burgis and William Wallis, 'China in push for resources in Guinea', *Financial Times*, (11 October 2009).
26. Wonder Guchu, 'China–Africa deal a lose–lose situation', *Informante*, (Thursday, 23 July 2009), available at: http://www.informante.web.na/index.php?option=com_content&task=view&id=4496&Itemid=100&PHPSESSID=8d9e1176ecb2ff808ba0e7504cb8aefe.
27. Peter Hitchens, 'How China has created a new slave empire in Africa', *The Mail on Sunday*, (28 September 2008).

money be awarded and spent transparently, using competitive bidding. Borrowers are given more choice among suppliers and contractors. Donors and lenders can no longer insist that funds be recycled back to their nation's companies.

Not party to the EITI, China defended the murky nature of its practice and claimed that transparency would come only after economic development worked its magic. As Li Ruogu, the head of China's Export–Import Bank, told an audience in Cape Town, 'Transparency and good governance are good terminologies, but achieving them is not a precondition of development; it is rather the result of it'.[28] Although this does not mean that Beijing likes scandals, China's attempts to prevent corrupt practices by its companies overseas are weak to say the least as China does not have a specific law against bribing foreign officials and the government is not eager to investigate or punish companies engaged in shady practices overseas.

Although this position may sometimes give a competitive advantage to Chinese firms, China cannot always get away with these practices. For example, in 2009, Namibia charged Nuctech, a Chinese multinational whose CEO was Hu Haifeng, son of former Chinese President Hu, for funneling US$4.2 million in illegal kickbacks to a front company set up by a Namibian official. The official then split the funds with her business partner and Nuctech's Chinese representative. The official, her business partner and the Chinese representative were subsequently arrested. Following the allegations of bribery involving Nuctech and suspecting a 'pattern of corruption on deals with China',[29] Namibian anti-corruption officials expanded their investigation to include allegations that China National Machinery & Equipment Import and Export Company had agreed to pay the same Namibian company 10% of the final contract price for help in sealing a deal to build a 38-mile-long rail link. As a result of these investigations, Namibia became the first African country to indict Chinese companies for corruption and suspend Chinese loans.

Unsustainable trade pattern and resource exploitation

One reason for the accusation that China is engaged in a form of new colonialism is its importing of resources from Africa and dumping manufactured goods there. While China's main imports from Africa are primary products with little added value for Africa, including crude oil, copper, ores and minerals, Africa is a large buyer of manufactured products such as machinery and textiles from China. This pattern skewed economic ties and undermined linkages between Chinese investment and local economies because resource exports do not necessarily translate into widespread job creation to help improve living standards for ordinary people. Even South Africa, the continent's biggest and most industrialized economy, as well as a major coal and iron ore producer, faced an imbalance. With an official unemployment rate of 25%, South Africa's ruling ANC regularly talked about the need to boost domestic industry. In particular, it would like to see its minerals processed at home. But parts of South Africa's manufacturing sector, nearly 20% of the economy, were in direct competition with China, and in many cases were losing the battle against a

---

28. Behar, 'China surpasses US as leader in sub-Sahara'.
29. Sharon LaFraniere and John Grobler, 'Namibians say inquiry on China will expand', *New York Times*, (1 August 2009).

much cheaper producer. Labor-intensive industries such as textiles, clothing, footwear and furniture, were the hardest hit, with more than 40% of footwear and knitted fabrics purchased in South Africa coming from China.[30]

China is criticized for taking advantage of African countries' need for investments and financing in return for natural resource extraction rights. In addition, rights activists accuse the Chinese of cutting corners, exploiting corrupt local officials and ignoring health, safety and environmental concerns. While China procures the resources it needs to continue its rapid rise, at the end of the process, African nations will stagnate and their people will no longer have the precious and scarce resources they need to rise out of poverty. Indeed, cutting deals and securing supplies of natural resources, China has moved aggressively into the continent. Big state-owned energy companies such as China Petrochemical Corp. and China National Petroleum Corp. have all invested in major oil and gas projects in Africa. Sinohydro Corp has also capitalized on the continent's vast hydropower resources. Chinese companies have pushed deeper into the forests of Mozambique although China itself had introduced widespread logging bans at home in 1999 in order to stop the deforestation that was blamed for soil erosion and severe flooding. China also established a huge operation in Zambia, which has the second-largest reserves of raw copper in Africa. Building a special economic zone in the Copper Belt, China is both the world's biggest user of copper and the eighth largest exporter of refined copper products. Although China is by no means alone in prioritizing its economic relationships with countries of strategic or commercial significance to itself, China's snapping up of the resources in Africa's mountains, forests and offshore waters led some critics to conjure up the vision of a zero-sum competition for finite resources that could trigger a new wave of global conflict and massive environmental destruction in Africa.

Insensible labor practices

One of the most damaging factors to China's image and reputation is the insensitive labor practices linked to Chinese projects in Africa. Complaints range from low wages and labor discrimination to almost despotic treatment. Chinese companies, in contrast with Western companies that relied primarily on local labor, tended to keep local hiring to a minimum and recruited mostly their own professionals and laborers. For example, Chinese companies in Angola brought 70–80% of their labor from home in 2009. While nearly 90% of Chevron's workers were Angolan, including specialized personnel such as engineers and managers, Chinese oil companies employed fewer than 15% Angolan labor and they were usually at the bottom of the pay scale. In 2006 at a Portuguese-run construction site in Maputo, Mozambique, there were only five Portuguese out of 120 workers, while nearby, a Chinese-run site had 78 Chinese workers and only eight locals, three of which were night watchmen.[31]

---

30. Ed Cropley and Michael Martina, 'In Africa's warm heart, a cold welcome for Chinese', *Reuters*, (18 September 2012), available at: http://www.safpi.org/news/article/2012/africas-warm-heart-cold-welcome-chinese.
31. Loro Horta, 'China and Africa', *Asia Sentinel*, (20 November 2009), available at: http://www.asiasentinel.com/index.php?option=com_content&task=view&id=2154&Itemid=422.

In Zambia, where Chinese workers, supervisors and technicians were a common sight at copper mines and on building sites, locals protested that 'the Chinese are not here as investors, they are here as invaders'.[32]

This practice is partly due to the fact that, despite China's convulsive growth and new wealth, it still suffers gravely from the problem of a huge under-employed and unemployed rural population. It sees Africa as a potential for China's unemployed rural workforce to make their fortunes. After a visit of Ethiopia and Guinea to explore possibilities for agricultural cooperation, an agricultural specialist from Hebei province proposed to the annual session of China's National People's Congress that Africa's vast land and underdeveloped agriculture could provide employment for up to 100 million Chinese laborers. He suggested Beijing draft a long-term strategy for dispatching Chinese laborers to Africa in order to solve two of China's greatest challenges—food security and unemployment.[33] Although his proposal is not officially endorsed, hundreds of thousands of Chinese nationals, drawn from the poor rural communities, are already working in Africa, legally and illegally. A dream of wealth in a far-off land was turned on its head for hundreds of Chinese gold miners in Ghana in June 2013, as hundreds of them were rounded up by the government after being accused of sneaking into the country and overstaying visas to illegally mine one of Africa's richest gold fields.

Tensions between the Chinese workers and African locals have occurred often because many Chinese tend to live in isolation with little or no contact with the local population. They don't have incentives or opportunities to learn about local languages and cultures and, therefore, are insensible to local concerns. The official cultural exchange programs of the Chinese government were not helpful in correcting the problem. For example, when President Hu visited Mauritius in 2009, he spoke repeatedly of the benefit of cultural exchanges, but Chinese news reports did not mention anything about his attempts to learn something about Mauritian culture. He did not sample some local cuisine, learn a local aphorism, talk with local artists or even partake in a local festival. Instead, Hu visited a Chinese Culture Center there and all he talked about was what Africans can learn from China and Chinese culture.[34] Nothing was in the news that encouraged the Chinese workers to learn the local cultures. When Chinese companies did hire locals, they tended to flout local laws, had atrocious safety records and no welfare, and paid less than the prescribed minimum wage in these countries. Chinese employers discouraged their workers to join labor unions and did not give workers adequate leave days.

China's policy adjustments and rebalancing actions

China's reaction to the criticism is very delicate. On the one hand, Beijing brushed off accusations by denying any intentions of colonial expansion and dismissed the criticism as being mostly from Westerners, who do not like to see China coming to the continent they once considered to be the preserve of Western powers. The Westerners are charged

---

32. Hitchens, 'How China has created a new slave empire in Africa'.
33. Bezlova, 'China'.
34. 'Culture plays a vital role in promoting Sino–African relations', *Xinhua*, (18 February 2009).

with hypocrisy as their own adventure in Africa was not benevolent and exemplary. On the other hand, as its problematic practices have caused a serious backlash against Beijing's increasing diplomatic and commercial clout in Africa, China has taken cautious actions to adjust its practices and made some positive changes, moving beyond resource grab to address local needs and sensitivities, avoiding politically instable countries and being cautious with repressive regimes. These rebalances are motivated mostly by China's changing economic and strategic interests.

Moving beyond resource exploitation to address local sensitivities

China's heavy-handed pursuit of natural resources and the insensitive practices of Chinese businesses have caused complaints, even anti-Chinese riots, in some African countries. For example, workers in Zambia have protested against mismanagement, lax safety standards and other problems in Chinese-owned mines for many years. In 2004, Zambian authorities had to ask Chinese managers at the Zambia–China Mulungushi Textiles Ltd to stop locking workers in the factory at night and shut down Collum Coal Mining Industries in southern Zambia, saying miners were forced to work underground without safety clothing and boots. The next year, Zambian workers at the Chinese-owned Chambishi Mine rioted, demanding higher wages, observation of safety regulations and better working conditions after an accident that killed 46 people. Coming to Zambia in 2006 for a groundbreaking ceremony for the new 'zone' in the Copper Belt, Chinese President Hu Jintao had to cancel his visit to the area and cut the ribbon from the safety of Lusaka, 200 miles away, because of threats of riots. Underscoring the labor issues, two Chinese managers opened fire at demonstrating Zambian workers and injured 13 of them seriously in 2010. In 2011, a Chinese-owned miner in 2011 fired 1,000 workers from a Zambian copper mine for participating in a strike over wages. These disputes accumulated into another tragedy in August 2012 when Zambian miners killed a Chinese manager during a riot at the Collum Coal Mine. Two other Chinese managers were injured, as were several Zambians. It was reported that the riot started after Chinese management refused to approve workers' demands for salary arrears following the revised minimum wage.[35] Tapping into rising resentment and anger about the deplorable conditions in Zambia's Chinese-run mines, Michael Sata campaigned on a platform heavily laden with anti-Chinese rhetoric. He portrayed the Chinese as competitors rather than investors and was voted in as president of copper-rich Zambia in 2011. Although Sata toned down his fiery language and went some way to repairing the bruised relationship after taking office, he is still known as a champion of the poor, including the workers in Chinese-run mines.[36]

Waking up to the fact that insensitivity to local needs may potentially put Chinese company operations in jeopardy, citizens at risk, and damage China's reputation at

---

35. 'A Chinese miner killed by Zambian employees in a minimum wage riot', *Lusaka Times*, (5 August 2012), available at: http://www.lusakatimes.com/2012/08/05/chinese-miner-killed-zambian-employees-minimum-wage-riot/.

36. One alternative interpretation of the shooting contends that singling out Chinese deflects attention away from neoliberal structural ills and says more about racial and ideological pre-conceptions that politicians and media bring to bear in bashing 'the Chinese' than about the actual Chinese presence in Africa. Barry Sautman and Yan Hairong, 'Bashing "the Chinese": contextualizing Zambia's Collum Coal Mine shooting', *Journal of Contemporary China* 23(90), 2014, doi: 10.1080/10670564.2014.898897.

both the local level and internationally, China began to move beyond natural resources exploitation and focus on addressing local needs. As a result, a growing number of Chinese entrepreneurs have set up manufacturing in the continent. A partial response to the complaint that the flood of cheap Chinese imports undercut Africa's weak manufacturing base, this move is facilitated by new developments in the Chinese economy, including rising Chinese wages in the manufacturing sector, rising purchasing power among the Chinese, pressure to shift from resource-heavy capital spending to a more refined, consumer-driven growth, and the shift of Sino–African relations from being essentially government-to-government to business-to-business. As China's domestic business environment becomes more expensive, Chinese firms have to look overseas for profitable investment opportunities and are shifting an increasing number of manufacturing operations to cheaper locations abroad. As early as 2009, Robert Zoellick, then the president of the World Bank, found that the Chinese government had shown 'strong interest' in 'moving some of the lower-value manufacturing facilities to sub-Saharan Africa', helping the continent develop a manufacturing base and boost its economy.[37]

The shift of Beijing's trade pattern is ultimately determined by market forces.[38] Africa became an attractive destination not only because of the low labor cost and fast-growing consumer base but also because of the trade pacts, such as the African Growth and Opportunity Act, which give African-based firms preferential access to American markets. Many Chinese manufacturers, especially those in clothing and textiles, began to transplant some low-value-added industries to Africa in the hope of escaping labor-cost increases at home and finding easier export routes to America. For example, Huajian Shoes, one of China's leading shoe exporters, launched its first overseas operations in Ethiopia in 2012. While the Chinese group's vice-president who oversaw the Ethiopian move attributed the move to the gradual appreciation of the Renminbi and rising labor costs squeezing margins, the biggest incentives are the preferential tariffs Ethiopia has to access the EU and US markets. In some cases these tariffs give the company a 27.5% advantage over manufacturers in China. As a result, in just over a year, the company employed 1,750 people at its factory outside Addis Ababa, the Ethiopian capital.[39]

In addition to the manufacturing sectors, services such as banking, finance and insurance, were part of the new sectors that China invested in the continent. China has also been involved in setting up special economic zones (SEZs) in several African countries, including Egypt, Ethiopia, Mauritius, Nigeria and Zambia. These are intended to replicate China's own growth experience, whereby a number of export-oriented industrial hubs successfully attracted foreign investment, driving national economic growth. Both China's authorities and African governments hope that the SEZs in Africa will develop into new growth nodes by creating an enabling environment into which Chinese, foreign and African companies can move

---

37. Tania Branigan, 'China wants to set up factories in Africa', *Guardian*, (4 December 2009), available at: http://www.guardian.co.uk/world/2009/dec/04/china-manufacturing-factories-africa.

38. Joshua Eisenman, 'China–Africa trade patterns: causes and consequences', *Journal of Contemporary China* 21(77), (2012), p. 793.

39. William Wallis, 'Chinese shoemaker takes road less travelled to Africa', *Financial Times*, (3 June 2013), available at: http://www.ft.com/intl/cms/s/0/a97492d4-cc5b-11e2-9cf7-00144feab7de.html#ixzz2al6jOgIu.

and gradually form industrial clusters. Each zone focuses on a few key industries, mainly in manufacturing and services, with only one zone concentrating on mineral processing.[40]

Opportunities are thus presented as a move towards a more balanced trade between China and Africa in order to help African countries sustain development and overcome poverty. One example was that Sinopec teamed up with South African counterpart PetroSA in 2012 to explore building a US$11 billion oil refinery on the country's west coast. Since South Africa has no significant oil or proven gas reserves, the proposed plant would depend on imports, and would have to serve the local market to be viable. The plant would therefore serve the South African market and not be used to process exports to China. According to one observer, 'It shows that China's dragon safari is about more than just sourcing commodities for export'.[41] As a result, Africa is now more often seen by Chinese firms as a place to do business rather than just digging stuff out of the ground.

Chinese corporations have also made efforts to localize their operation and adopt sensitive approaches to engage local communities. As Antoine Kernen and Katy Nganting Lam's fieldwork in Ghana discovered, country directors of Chinese state-owned enterprises expressed their intention to hire as many locals as possible because they considered that workforce localization was necessary for practical reasons ranging from reducing labor cost to adapting to local political context.[42] Chinese companies have also started to 'give back' to the local communities. According to one research, CNPC in 2009 provided US$700,000 to set up an Education and Training Fund at Juba University and in 2010 donated a computer laboratory to Juba University in South Sudan. It also donated a conference and cultural center in Malakal and basketball courts in Juba for social community development. In April 2013, a series of seminars, workshops and public lectures was held in Juba that brought together senior government officials, and local non-governmental organizations from both South Sudan and China, along with a variety of Chinese companies working in South Sudan, to discuss operational questions related to managing company impacts on local conflict dynamics: 'While these are initial steps, they reflect new commitment among Chinese and South Sudanese actors to improve the way that commercial projects are delivered in the world's newest state'.[43] In a move to repair its relationship with the new Zambian government that won the 2011 election by tapping into anti-China sentiments, Beijing received President Michael Sata in April 2013.

These developments may eventually make true Edward Friedman's predictions about the possibility of a 'flying geese' effect of development in Africa. The Japanese

---

40. 'China and Africa: a maturing relationship', *The Economist*, (9 April 2013), available at: http://country.eiu.com/article.aspx?articleid=760356060&Country=China&topic=Economy&subtopic=Forecast&mkt_tok=3RkMMJWWfF9wsRoiuqrPZKXonjHpfsX67estUK6g38431UFwdcjKPmjr1YcERMN0dvycMRAVFZl5nQlRD7I=.

41. 'Chinese premier announces eight new measures to enhance cooperation with Africa', available at: http://www.chinaview.cn/index.htm.

42. Antoine Kernen and Katy Nganting Lam, 'Workforce localization among Chinese state-owned enterprises (SOEs) in Ghana', *Journal of Contemporary China* 23(90), (2014), doi: 10.1080/10670564.2014.898894.

43. 'Conflict sensitivity in South Sudan: ensuring economic development supports peace', *SaferWorld*, (30 August 2013), available at: http://www.saferworld.org.uk/news-and-views/comment/103?utm_source=smartmail&utm_medium=email&utm_campaign=South+Sudan+Monitor+August+2013.

started the flying geese pattern and rapidly moved up in technology and value-added production and outsourced much of its lower-end production to other Asian countries. Increasing wages and costs and a higher valued currency in China is now beginning to force the workshop of the world to outsource much of its lower-end production. The flying geese of Japan's Asia could become the flying geese of China's Africa, extending Asian dynamism to Africa. Plugging into Chinese dynamism, African nations can move up a value-added ladder and take advantage of the opportunities inherent in China's rise to join the flock of flying geese. From this perspective, Friedman believed that poor Africans would be lifted out of poverty by Chinese investment and benefit from playing a role in a world economy largely structured by an Asian motor.[44] If it happens, China would become less of a model and more of a catalyst for African development.

Avoiding politically instable countries

As newcomers, Chinese companies often operated in more marginalized countries not only because the markets in more secure countries were already taken by Western companies but also because 'Chinese companies did not regard the political instability, ethnic schisms, or social discords in some African countries as an obstacle'.[45] When Chinese companies first began showing up in large numbers in countries like Angola and Rwanda, their ability to tolerate a high level of risk, not to mention discomfort, made them stand out in contrast to Western competitors. With a no-strings-attached approach, China frayed into the continent's resource-rich but fragile states, a void left by Western companies, and offered Africa an economic and political alternative to the heavily conditioned aid and business operations of Western countries.

As their investment in resource-rich but often politically unstable African countries increases, Chinese companies have come to the realization that there are limits to what they can put up with in the interest of protecting their investment and personnel. For example, when Western Companies such as Royal Dutch Shell were at loggerheads with the Nigerian government, CNOOC bought a 45% stake in Akpo field in 2006 and signed more than US$10 billion of contracts. A change in the government the following year and the Nigerian government's lack of follow-up mechanisms made it impossible to enforce these deals. Chinese companies were also exposed to attacks on personnel, pipelines and other infrastructure by the militants Movement for the Emancipation of the Niger Delta in Nigeria's volatile oil-producing region. The movement arose because most residents lived in intense poverty while oil facilities in the area earned billions of dollars for foreign companies and the Nigerian government. The spokesman of the movement specifically warned

> the Chinese government and its oil companies to steer well clear of the Niger Delta. Chinese citizens found in oil installations will be treated as thieves. The Chinese government by investing in stolen crude places its citizens in our line of fire.

---

44. Edward Friedman, 'China-driven development. As China pours billions into Africa, other countries are trying to keep up', *Beijing Review*, (1 February 2009), available at: http://www.bjreview.com/world/txt/2009-02/01/content_176304.htm.

45. Sigfrido Burgos and Sophal Ear, 'China's oil hunger in Angola: history and perspective', *Journal of Contemporary China* 21(74), 2012, p. 353.

The statement came after Chinese President Hu Jintao's week-long tour of Africa in which he reached a series of deals securing access to oil and other resources, including one deal signed with Nigerian President Olusegun Obasanjo that offered China four oil exploration licenses.[46] Chinese companies were also caught up in the maelstrom of political changes in North Africa during the Arab Springs. More than US$4 billion worth of projects were suspended in Libya and more than 35,000 Chinese workers were evacuated after the fall of Colonel Gadaffi in 2011.[47] The Chinese government also had to cope with the emergency of the kidnapping of 29 Chinese workers in Southern Sudan in early 2012.

As a result, Chinese companies began to avoid some of the most chaotic corners of the continent and refrain from investing in politically instable countries. For example, the Guinean government sought a multibillion-dollar deal with China to build desperately needed infrastructure in exchange for access to its large reserves of bauxite and iron ore. Although Guinea has some of the world's largest reserves of the resources needed for making aluminum, in 2009 China backed away from what Guinean officials portrayed as a done deal to build a much-needed hydroelectric dam because the political situation was not very stable. Another example was China's US $9 billion deal in 2008 with Congo for access to its giant trove of copper, cobalt, tin and gold in exchange for developing roads, schools, dams and railways that were needed to rebuild a country shattered by more than a decade of war. This deal came in to doubt in 2009 in part because Congo's political and ethnic turmoil remained deep, and its economy was near collapse.

Backing away from some of its riskiest and most aggressive plans, Chinese companies began looking for the same guarantees that Western companies have long sought for their investments: infrastructure development, political and economic stability, regional trade and economic integration among different African countries. Chinese investors have realized, like others before them, that good governance norms are not luxuries but necessities in protecting investments. It is no surprise that the bulk of Chinese investment has remained in South Africa, the continent's largest democracy and best-developed economy. Also, while Africa's farming sector appealed to many Chinese investors, they have hesitated to invest in the land because the idea of selling off cherished ancestral lands to foreigners remains a sensitive issue in Africa. Poor legal protection for property rights has added to the risk of putting money into agriculture. This explains in no small part why farming accounted for less than 4% of Chinese investment activity in Africa. Given China's vast appetite for food imports, China's farming investment would likely increase exponentially if legal issues surrounding land ownership were firmed up.[48]

---

46. Craig Timberg, 'Militants warn China over oil in Niger Delta', *Washington Post*, (1 May 2006), p. A15.
47. Shaio H. Zerba, 'China's Libya evacuation operation: a new diplomatic imperative—overseas citizen protection', *Journal of Contemporary China* 23(90), (2014), doi: 10.1080/10670564.2014.898900.
48. Gavin du Venage, 'China as a vital force for Africa', *Asia Times Online*, (8 June 2012), available at: http://www.atimes.com/atimes/China/NF08Ad01.html.

Cautious with repressive regimes

Being accused of supporting a string of despots and oppressive regimes responsible for human suffering, China often faced difficulties in dealing with international human rights organizations as well as an increasing number of democratic countries in Africa. To address the call for taking more international responsibility as a rising global power, China's relations with Western countries can no longer be disentangled from certain difficult issues involving China's relations with African countries. For example, after failing to convince Khartoum to accept the peacekeeping force and organize negotiations to find a political solution to the conflict after Darfur, some US and European activists called for a boycott of the 2008 Olympic Games in Beijing as a means of protesting Beijing's position as a major buyer of Sudanese oil. Film director Steven Spielberg even pulled out as artistic adviser to the opening ceremony. Beijing was also embarrassed in 2008 when a Chinese ship carrying weapons and ammunition manufactured by Poly Technologies, a front for China's military–industrial complex, to violence-riddled Zimbabwe was nicknamed in African newspapers as the 'ship of shame' and blocked by South Africa, Mozambique and Zambia and spent weeks failing to get permission to offload.[49]

As a Chinese scholar discovered, although China did a great deal of work on the ground by building schools, hospitals, stadiums and conference facilities, and by undertaking other public projects in resource exporting countries, 'China needs to learn to improve its international image in "softer" areas, such as winning the support of intellectuals in recipient countries'.[50] As a result, 'Beijing has been quietly overhauling its policies toward pariah states ... China is now willing to condition its diplomatic protection of pariah countries, forcing them to become more acceptable to the international community' because 'China's fears about a backlash and the potential damage to its strategic and economic relationships with the United States and Europe have prompted Beijing to put great effort into demonstrating that it is a responsible power'.[51]

Adopting a more nuanced perspective to deal with unpopular regimes, Beijing's engagement with Khartoum has 'evolved from a rather passive posture, to taking a clear position, and finally, to active persuasion and mediation'.[52] Naming a special Darfur envoy in 2007, the Chinese government has made increasingly public statements to persuade Sudanese leaders to be more cooperative in international efforts to bring an end to the fighting in the embattled Darfur region. Reported on the front page of *People's Daily*, President Hu, in meeting with Sudanese Vice President Ali Taha in June 2008, urged the Sudanese government to do whatever possible for the early deployment of a mixed UN–African Union peacekeeping force. In the same month, China announced that it would deploy 320 military engineers to conflict-

---

49. Celia W. Dugger, 'Angola allows Chinese ship to dock, but not unload arms for Zimbabwe', *New York Times*, (27 April 2008).

50. Yao Yang, 'Chinese investment: a new form of colonialism?', *East Asia Forum Quarterly*, (24 July 2012), available at: http://www.eastasiaforum.org/2012/07/24/a-new-form-of-colonialism/.

51. Stephanie Kleine-Ahlbrandt and Andrew Small, 'China's new dictatorship diplomacy', *Foreign Affairs*, (January/February 2008), pp. 38–39.

52. Jonathan Holslag, 'China's diplomatic manoeuvring on the question of Darfur', *Journal of Contemporary China* 17(54), (2008), p. 71.

ridden Darfur to help with the construction of refugee and resettlement camps and roads and the generation of clean and safe water through the drilling of wells.[53] China's change in position with Sudan is not only because it was under heavy pressure from Western countries to play a positive role, but also because continuation of the conflict could undermine the Comprehensive Peace Agreement that ended decades of war in Southern Sudan and lead to new fighting that could shut off oil production entirely and damage China's oil interests there.

Beijing also put pressure on Mugabe to hold a meeting with Morgan Tsvangirai, the leader of the opposition Movement for Democratic Change, after the signing of a memorandum of understanding mediated by South Africa's president to form a government of national unity. Although the meeting seemed to be a triumph for South African diplomacy,

> the power behind the curtain is China. Mugabe was told in clear terms by his Chinese friends that he has to behave and act in a way that will help dampen international outrage over the recent elections in the run-up to the Olympics.[54]

To make it clear, at a news conference in June 2008, the spokesman of the Chinese Foreign Ministry told reporters that while China respected the integrity of domestic affairs in Zimbabwe,

> we hope the political parties will act in such a way as to put the interests of the people of Zimbabwe first through engaging in further dialogue. We believe that through increased and further dialogue a peaceful solution to the problems in Zimbabwe is possible.[55]

Beijing has many sources of leverage on Mugabe because it is Zimbabwe's most important economic partner and has helped train the military there for more than two decades. By exerting diplomatic pressure on Mugabe, China is protecting its own interests, given the threat to their own economic investments in the country.

Conclusion

As one field study found, 'Beijing has acquired substantial goodwill in Africa yet is developing deep issues and facing uncertain challenges and growing obstacles'.[56] Apparently the top leadership in Beijing has also come to the realization of the necessity of the rebalance to serve its evolving economic and political interests. During his first overseas visit in March 2013 as Chinese President, Xi Jinping said in a conference center built with Chinese loans in Tanzania that 'China will face squarely and sincerely the new developments and new problems confronting China–Africa relations. We should properly handle any problem that may arise in a spirit of mutual respect and win–win cooperation'. In defense of China's economic

---

53. 'Official: US doesn't see "zero-sum" competition with China in Africa', *Xinhua*, (15 October 2008).
54. Ian Evans, 'Robert Mugabe forced into talks with opposition after China told him "to behave"', *Telegraph*, (26 July 2008), available at: http://www.telegraph.co.uk/news/worldnews/africaandindianocean/zimbabwe/2461693/Robert-Mugabe-forced-into-talks-with-opposition-after-China-told-him-to-behave.html.
55. Michael Appel, 'China commits 320 military engineers to Darfur', *BuaNews*, (27 June 2008), available at: http://www.buanews.gov.za/view.php?ID=08062710451003&coll = buanew08.
56. Fei-Ling Wang and Esi A. Elliot, 'China in Africa: presence, perceptions and prospects', *Journal of Contemporary China* 23(90), (2014), doi: 10.1080/10670564.2014.898888.

stake in many African countries, President Xi sought to assure his African audience that China would heed complaints that relentlessly competitive Chinese companies were suffocating African efforts to nurture industry and jobs. He also promised aid, scholarships and technology transfers in an effort to counter those fears.[57] According to one observer, President Xi was more candid than his predecessor, Hu Jintao, in acknowledging that the relationship faced strains.[58] Although this is a welcome development, the Chinese government and Chinese companies are still struggling to balance and rebalance its policies in an effort to find its place and harvest profits on the increasingly crowded continent. China has never had a grand strategy and may never have one in the continent. China's advance on Africa has appeared to be, and will continue to be, ad-hoc at least in the foreseeable future.

---

57. Wu Jiao and Li Xiaokun, 'Xi highlights "shared destiny"', *Xinhua*, (26 March 2013), available at: http://www.cdeclips.com/en/nation/fullstory.html?id=78366.

58. Chris Buckley, 'China's leader tries to calm African fears of his country's economic power', *New York Times*, (26 March 2013), available at: http://www.nytimes.com/2013/03/26/world/asia/chinese-leader-xi-jinping-offers-africa-assurance-and-aid.html?_r=0#h.

Why Do We Need 'Myth-Busting' in the Study of Sino–African Relations?

MIWA HIRONO and SHOGO SUZUKI

The literature on Sino–African relations has debated whether or not China's growing presence is a threat to Western or African interests, and has come to the conclusion that China's behavior is not uniquely immoral. Many countries, including Western liberal democracies, similarly give aid to local autocrats to secure natural resources. Why, then, has so much effort been made to come to this perhaps unsurprising conclusion? We argue that the literature on Chinese foreign policy remains heavily influenced by Western states' policy interests, resulting in an impoverished debate that is primarily concerned with the idea of a China threat. In order to recover the diversity in our research on Chinese foreign policy, we argue for the need to go beyond the confines of Western strategic interests.

Introduction

No topic of China's international relations in recent years has captured the imagination of both popular and academic audiences more than China's relations with the African continent. The People's Republic of China's (PRC's) relations with Africa date back much earlier. Beijing was particularly active there from the late 1950s to the 1970s, when it was attempting to escape from the diplomatic alienation imposed on it by both the Western and Soviet-led Eastern blocs.[1] The Chinese poured

---

* Miwa Hirono is RCUK Research Fellow at the School of Politics and International Relations and Senior Research Fellow at the China Policy Institute, at the University of Nottingham, UK. She is also a visiting fellow at the Australian National University. Her publications include *China's Evolving Approach to Peacekeeping* (Routledge, 2012) and *Civilizing Missions: International Christian Agencies in China* (Palgrave Macmillan, 2008). Shogo Suzuki is Senior Lecturer in the Department of Politics at the University of Manchester, UK. He is the author of *Civilization and Empire: China and Japan's Encounters with European International Society* (Routledge, 2009), as well as several articles on Chinese foreign policy and Sino–Japanese relations. This research was funded by the RCUK Fellowship Recruitment Fund and written in the framework of the CoReach project on 'Europe and China: Addressing New International Security and Development Challenges in Africa', funded by the British Academy and the European Union. The earlier version of this article was presented at the International Studies Association (ISA) Annual Convention, San Diego, 1–4 April 2012. For their help in writing this article, we would like to thank Catherine Gegout (leader of the Co-Reach project), Katherine Morton, Deborah Brautigam, Tadokoro Masayuki, Thomas Wheeler and Kitamura Akihiko. We are also grateful to the anonymous reviewers for their helpful comments on the earlier version of this article.

1. Peter Van Ness, *Revolution and Chinese Foreign Policy: Peking's Support for Wars of National Liberation* (Berkeley, CA: University of California Press, 1973); Alaba Ogunsanwo, *China's Policy in Africa 1958–1971* (Cambridge: Cambridge University Press, 1974); Philip Snow, *The Star Raft: China's Encounter with Africa* (Ithaca, NY: Cornell University Press, 1989).

substantial resources into providing aid (the Tan–Zam Railway being the most famous example), and trumpeted its support for the anti-colonial and counter-hegemonic struggles of the developing world. After Deng Xiaoping initiated the opening up policy in the 1980s, however, Sino–African relations faded from China's diplomatic priorities. Eager to achieve economic growth and bolster the Chinese Communist Party (CCP) regime's legitimacy, Beijing concentrated on deepening economic and diplomatic ties with the industrialized world. Aid to Africa was cut back, and the revolutionary rhetoric of Third World solidarity was muted.[2] For a while—bar the early 1990s, when Beijing again found itself isolated following its suppression of the 1989 demonstrations in Tiananmen Square—the PRC's relations with the African continent appeared to be put on the backburner.

All this has seemingly changed since the beginning of the twenty-first century. The growth of the PRC's influence and presence in Africa has indeed been remarkable. Politically, the PRC has sought to enhance its relations by convening a number of high-level summits between Chinese and African leaders, as seen by the convocation of the Forum on China–Africa Cooperation (FOCAC) since 2000. The Chinese government has also increased its aid to Africa, and cancelled debts.[3] Meanwhile, under the slogan of 'going out' (*zou chu qu*), Chinese firms have sought to expand overseas, including Africa. China's growing need for natural resources has meant that Chinese enterprises are increasingly active in developing mines and oil wells, and today China's oil import from Africa accounts for 30% of the country's entire oil supply.[4]

This growth in Chinese political, security and economic activity has generated a flurry of literature, which can broadly be divided into two genres. The first generally sees the PRC as a threat to the interests of both Africa and the international community, because of their alleged role in supporting corrupt African autocrats and preventing the spread of 'good governance' in the region. These negative views find traction within the highest levels of the Western political elite—British prime minister David Cameron echoed these themes when he was quoted as being 'increasingly alarmed by Beijing's leading role in the new "scramble for Africa"', and 'warned African states over China's "authoritarian capitalism" ... claiming that it is unsustainable in the long term'.[5]

This literature has, in turn, generated the second group of works whose primary purpose is to engage in 'myth-busting' in the study of Sino–African relations.[6] Scholars put Chinese foreign policy in Africa in context, and point out that not only

---

2. Philip Snow, 'China and Africa: consensus and camouflage', in Thomas W. Robinson and David Shambaugh, eds, *Chinese Foreign Policy: Theory and Practice* (Oxford: Clarendon Press, 1994).
3. Hu Jinshan, *Feizhou de zhongguo xingxiang [Africa's Image of China]* (Beijing: Renmin chubanshe, 2010).
4. US Energy Information Administration, *China Energy Data, Statistics and Analysis: Oil, Gas, Electricity, Coal* (2011), available at: http://www.eia.gov/cabs/china/Full.html. Also see Joshua Eisenman, 'China–Africa trade patterns: causes and consequences', *Journal of Contemporary China* 21(77), (2012), pp. 793–810.
5. Jason Groves, 'Cameron warns Africans over the "Chinese invasion" as they pour billions into continent', *Mail Online*, (20 July 2011), available at: http://www.dailymail.co.uk/news/article-2016677/Cameron-warns-Africans-Chinese-invasion-pour-billions-continent.html (accessed 23 February 2013).
6. Chris Alden, *China in Africa* (London: Zed Books, 2007); Erica S. Downs, 'The fact and fiction of Sino–African energy relations', *China Security* 3(3), (2007), pp. 42–68; Shogo Suzuki, 'Chinese soft power, insecurity studies, myopia and fantasy', *Third World Quarterly* 30(4), (2009), pp. 779–793; Ian Taylor, *China's New Role in Africa* (Boulder, CO: Lynne Rienner, 2009); Deborah Brautigam, *The Dragon's Gift: The Real Story of China in Africa* (Oxford: Oxford University Press, 2009).

Chinese but also Western actors are responsible for their lack of political will to solve Africa's security and governance problems. It is indeed not too difficult to search for examples to illustrate the point that China and many Western governments often behave very similarly in the realm of foreign policy. Western governments have provided support for regimes with notorious records for suppressing human rights— British and American support for the Mubarak regime in Egypt is a case in point. Similarly, despite its rather convenient change in tune of late, the British government was also complicit in cultivating closer relations with the Gaddafi regime in Libya. In sum, the recent debates on China in Africa have taught us that China is actually not that 'different' compared to many other Western states—while the effects of the PRC's growing influence in the African continent have been mixed, this does not make China a uniquely pernicious presence.

These studies have certainly enriched our knowledge of Chinese foreign policy behavior in Africa. However—and without wishing to downplay the contribution of these works—should we be surprised by these conclusions? While realism has come under considerable criticism within the discipline of International Relations (IR) of late, one of its most enduring and important insights is that *states are not radically different from one another*, in that they pursue their own strategic self-interests, often with scant regard for the international norms they rhetorically claim to adhere to. Yet, this arguably predictable point is presented almost as a surprise or a discovery in the literature on Sino–African relations.

Why is this so? In this article we suggest that this is because the study of Chinese foreign policy continues to be structured by a powerful discourse which claims that China's rise to power presents a unique and almost unprecedented challenge to the maintenance of the Western-dominated world order. This consequently results in a myopic and dichotomized debate of whether or not Chinese policies in Africa are 'bad' or not. We aim to problematize this starting assumption that has colored the study of Sino–African relations. In doing so, we forward two arguments. First, the study of China has been closely linked to the national interest of Western states, and this structural dynamic means that the field of Chinese foreign policy revolves around one primary axis that is chiefly concerned with whether or not the PRC is becoming a threat to Western power and interests. Second, it argues that the enduring Eurocentrism in IR means that the rise of non-Western powers is under-theorized, resulting in an impoverished vision of a world order where Western hegemony is no longer guaranteed. This, in turn, generates suspicions and fears analogous to the 'Yellow Peril' thesis: the rise of an Asian power is implicitly seen as a 'unique' and 'unknown' development that would somehow threaten the moral fabric of the international order that has historically been constructed and dominated by the West. We will conclude by suggesting an alternative research agenda in Sino–African relations that go beyond national security and Western-centric agenda.

Academia and the national security agenda

For Western scholars, studies of China's international relations are generally still in the service of the national security agenda of the Western policy community.

Historically, the study of the non-West has been closely linked to Western government policies. Long before the term 'area studies' came into existence in the United States (US) in the 1950s, there was a notion that scholarship should serve the political goals of the elite. 'Oriental studies' was recognized as a discipline in the eighteenth and nineteenth centuries, due to Western colonialists' needs to control non-Western people.[7] For example, Sir William Jones, a British legal scholar,

> founded the Asiatic Society of Bengal in 1784, and also worked as an official of the British East India Company; he felt absolutely no conflict of interest in serving imperialism and set the pattern which later Orientalists and area studies experts emulated.[8]

In addition to those who engaged in commercial activities, Christian missionaries laid the foundation for the development of Orientalism, with a belief that the West should help the non-West adopt 'superior' Western civilization.[9] Edward Said rightly states that the purpose of Orientalism was 'to understand, in some cases to control, manipulate, even incorporate, what is a manifestly different world'.[10] Later, social anthropologists joined the Orientalists by bringing more in-depth understanding of the customs and lifestyles of the colonized, in order to aid colonial administrators and missionaries.[11]

However, it was only after World War II (WWII) that area studies really flourished. One of the earliest and best known examples of this genre of scholarship was *The Chrysanthemum and the Sword: Patterns of Japanese Culture*, written by American anthropologist Ruth Benedict in 1946. This was the product of Benedict's involvement with the US Office of War Information during WWII.[12] As the Cold War and the subsequent standoff between the US and the Soviet Union became entrenched in the 1950s, area studies served to fulfill the West's strategic need to understand the enemy or processes by which states hostile to Western interests could be brought into the so-called 'free world'. Research on Asia, particularly China and Japan, was a major beneficiary of this development. As Bruce Cummings states:

> Japan got a favored placement as a success story of development, and China got obsessive attention as a pathological example of abortive development. The key processes were things like modernization, or what was for many years called 'political development' toward the explicit or implicit goal of liberal democracy.[13]

---

7. Asaf Hussain, Robert Olson and Jamil Qureshi, *Orientalism, Islam and Islamists* (Beltsville, MD: Amana Books, 1984), p. 11.
8. Stuart Schaar, 'Orientalism at the service of imperialism', *Race and Class* XXI(1), (1979), p. 68.
9. Miwa Hirono, *Civilizing Missions: International Religious Agencies in China* (New York: Palgrave Macmillan, 2008).
10. Edward W. Said, *Orientalism* (London: Penguin, 1977), p. 12.
11. Hussain et al., *Orientalism, Islam and Islamists*, p. 11.
12. Ruth Benedict's wartime involvement with the US Office of War Information is presented in two US government reports published in 1943 and 1944. Sonia Ryang, *Chrysanthemum's Strange Life: Ruth Benedict in Postwar Japan*, Occasional Paper No. 32 (Japan Policy Research Institute, University of San Francisco Center for the Pacific Rim, July 2004), available at: http://www.jpri.org/publications/occasionalpapers/op32.html (accessed 12 November 2012).
13. Bruce Cummings, 'Boundary displacement: area studies and international studies during and after the Cold War', *Bulletin of Concerned Asian Scholars* 29(1), (January–March 1997), p. 9.

Cummings also documents various area studies departments' close working relations with US government agencies in the early Cold War period, particularly their role in providing a steady source of new recruits and specialist consultants: 'For those scholars studying potential enemy countries, either they consulted with the government or they risked being investigated by the FBI; working for the CIA thus legitimized academics and fended off J. Edgar Hoover'.[14]

While the Cold War has officially ended with the collapse of the Soviet Union, there still remains ample state demand for the study of China, partly because of the 'concern' of Western elites that China is the only strategically competitive peer that could pose a real threat to the West's power and dominance. China is the last remaining communist great power, and its antipathy to liberal democratic governance, coupled with its steady military build-up, has made it a latent 'threat' to Western interests. Therefore, the research agenda of Chinese foreign policy in Africa continues to be influenced by this national interest. For example, the US Congressional Research Service (CRS) has published five reports between 2008 and 2009 on China's activities in Africa, Latin America and Southeast Asia, showing heightened US interests and anxieties in this field.[15] Furthermore, China is the only non-African country to feature in the CRS's reports on Africa. Studies of the US or Europe in Africa are conspicuously absent.[16]

Scholars and analysts have also jumped on this policy bandwagon, and published a series of works that confirm China's 'threat' in Africa. Authors of these works voice their disquiet that the Chinese government is trying to sabotage Western attempts to introduce 'good governance' (such as liberal democratic governance or improved transparency) by propagating its model of 'authoritarian capitalism'. Many also warn darkly that an important part of China's objectives in Africa is to challenge US global hegemony.[17] China's non-conditional aid—denounced as 'rogue aid' by some critics—and trade-oriented relations with 'rogue states' such as Sudan and Zimbabwe are frequently criticized, because they undermine attempts to introduce democracy to the region, provide a lifeline for autocratic rulers, and encourage and exacerbate human rights abuses by them.[18] In the words of Gemot Pehnelt, 'Chinese engagement enables African governments to reject demands made by the IMF, the World Bank and other donors for enhancing transparency, implementing anti-corruption

---

14. *Ibid.*, p. 11.
15. Congressional Research Service, *China's Foreign Policy and 'Soft Power' in South America, Asia and Africa*, (2008), available at: http://fpc.state.gov/documents/organization/104589.pdf; Congressional Research Service, *Comparing Global Influence: China's and US Diplomacy, Foreign Aid, Trade, and Investment in the Developing World*, (15 August 2008), available at: http://fpc.state.gov/documents/organization/109507.pdf; Congressional Research Service, *China's Foreign Policy: What Does it Mean for US Global Interests?*, (18 July 2008), available at: http://www.fas.org/sgp/crs/row/RL34588.pdf; Congressional Research Service, *China's Foreign Aid Activities in Africa, Latin America, and Southeast Asia*, (25 February 2009), available at: http://www.fas.org/sgp/crs/row/R40361.pdf; Congressional Research Service, *China's Assistance and Government-sponsored Investment Activities in Africa, Latin America, and Southeast Asia*, (25 November 2009), available at: http://fpc.state.gov/documents/organization/133511.pdf (all accessed 20 April 2012).
16. US Department of State, *Africa*, (n.d.), available at: http://fpc.state.gov/c20410.htm (accessed 20 April 2012).
17. Horace Campbell, 'China in Africa: challenging US global hegemony', *Third World Quarterly* 29(1), (2008), pp. 89–105.
18. Moisés Naím, 'Rogue aid', *Foreign Policy* 159, (2007), pp. 95–96; see also Joshua Kurlantzick, *Charm Offensive: How China's Soft Power is Transforming the World* (New Haven, CT: Yale University Press, 2007).

strategies, and furthering their democratization efforts'.[19] China's priority is, these scholars argue, simply to secure energy resources rather than to improve human rights conditions in those states.[20] Chinese firms (often bundled together under the somewhat misleading label of 'China' or 'China Inc.') are also accused of neocolonial behavior, such as exploiting African workers, flooding the African market with cheap Chinese consumer goods and ruining the local economy, or stripping African states of their resources.[21]

It is, of course, necessary to acknowledge that not all works portray China as a threat, as evidenced from the 'myth-busting' literature. Brautigam demonstrates that much of China's allegedly pernicious political influence is greatly exaggerated, and that its aid can, at times, actually deliver real benefits to the recipient states.[22] China's aid to authoritarian leaders has not been as vast as is often claimed, and is not as susceptible to being misused. With regard to weapons trade, Western corporations also engage in arms trade with rogue states (at times far more than the Chinese), making Western criticisms of Chinese weapons sales ring somewhat hollow—in fact, a recent study has concluded that the US 'tends to transfer conventional arms to authoritarian regimes to a greater extent than does China, which in turn tends to export more to African democracies and regimes that generally respect human rights'.[23] Studies on China's development activities in Africa have also found that Beijing's role in propping up isolated African autocrats is greatly exaggerated: there is, for instance, no concrete evidence of a systematic attempt to export China's development model of authoritarian capitalism, whose existence is highly debatable.[24] African scholars such as Adekeye Adebajo have also pointed out that the US has provided 'support for a cantankerous warlord's gallery',[25] which again reminds us that many governments support undemocratic regimes, provided that it is in accordance with their national interests. With regard to economic exploitation of

---

19. Gernot Pehnelt, *The Political Economy of China's Aid Policy in Africa*, Jena Economic Research Papers No. 051 (Friedrich-Schiller-University and The Max Planck Institute of Economics, Jena, Germany, 2007), p. 8.

20. Ngaire Woods, 'Whose aid? Whose influence?: China, emerging donors and the silent revolution in development assistance', *International Affairs* 84(6), (2008), pp. 1205–1221; Clemens Six, 'The rise of postcolonial states as donors: a challenge to the development paradigm?', *Third World Quarterly* 30(6), (2009), pp. 1103–1121; Stefan Halper, *The Beijing Consensus: How China's Authoritarian Model will Dominate the Twenty-First Century* (New York: Basic Books, 2010); Jonathan Holslag, 'China's diplomatic manoeuvring on the question of Darfur', *Journal of Contemporary China* 17(54), (2008), pp. 71–84; Jonathan Holslag and Sara Van Hoeymissen, eds, *The Limits of Socialization: The Search for EU–China Cooperation towards Security Challenges in Africa* (Brussels: Brussels Institute of Contemporary China Studies, 2010).

21. 'The Chinese in Africa: trying to pull together', *The Economist*, (20 April 2011), available at: http://www.economist.com/node/18586448; 'Africa and China: rumble in the jungle', *The Economist*, (20 April 2011), available at: http://www.economist.com/node/18586678 (both accessed 4 December 2012). On 'China Inc.', see Bates Gill and James Reilly, 'The tenuous hold of China Inc. in Africa', *The Washington Quarterly* 30(3), (2007), pp. 37–52.

22. See Brautigam, *The Dragon's Gift*.

23. Indra de Soysa and Paul Midford, 'Enter the dragon! An empirical analysis of Chinese versus US arms transfers to autocrats and violators of human rights, 1989–2006', *International Studies Quarterly* 56, (2012), p. 844. Also see Taylor, *China's New Role in Africa*.

24. Suzuki, 'Chinese soft power, insecurity studies, myopia and fantasy'. Also see Suisheng Zhao, 'The China model: can it replace the Western model of modernization?', *Journal of Contemporary China* 16(65), (2010), pp. 419–436; and Scott Kennedy, 'The myth of the Beijing Consensus', *Journal of Contemporary China* 19(65), (2010), pp. 461–477.

25. Adekeye Adebajo, 'An axis of evil? China, the United States and France in Africa', in Kweku Ampiah and Sanusha Naidu, eds, *Crouching Tiger, Hidden Dragon? Africa and China* (Scottsville: University of KwaZulu-Natal Press, 2008), p. 233.

African labor, Chinese enterprises are found to be neither better nor worse than many of their Western counterparts, and their buying up of African natural resources often pales into insignificance compared to Western purchases. Yet, it is important to note that even this type of literature is essentially an extension of the same question that dominates the Western policy community: does China's rise present a threat to Western interests in Africa? The starting point of their enquiry is the same as that of their respective governments' national security concerns.

This close link between the academic and national security agendas in the West suggests that there is still a key governmental interest in understanding the Chinese 'enemy'. In the latter half of the twentieth century, the core objective of US foreign policy was to combat communist regimes and advance liberalism and capitalism. After the end of the Cold War, US foreign policy has strived for the maintenance of US hegemony or, at least, US strategic superiority vis-à-vis China. Therefore, the US government has encouraged social scientists to study subjects that assist such policy purposes. This is demonstrated, for example, by the fact that state funding for Asia-related topics has been linked to government programs such as the National Security Education Act.[26] In addition, since 2008, the US Department of Defense has provided a number of million-dollar-level Minerva Research Initiative funds to university-based social science research programs, focusing on 'areas of strategic importance to US national security policy'.[27] It has seven priority research topics, including science, technology and military transformation in China and developing states. China is the only country specifically mentioned in all seven priority research topics.

This tendency of government funding policies fostering close links between academic and policymaking communities is replicated in other states as well. In the United Kingdom (UK), the Research Excellence Framework (REF) and Research Councils UK pay particular attention to the importance of 'non-academic impact',[28] such as that on political decision making. This includes 'fostering global economic performance, and specifically the economic competitiveness of the United Kingdom; increasing the effectiveness of public services and policy; enhancing quality of life, health and creative output'.[29] According to this academic strategy, one of the most effective ways to make non-academic impact in area studies (as well as other academic disciplines, for that matter) would be to demonstrate their utility in achieving these goals set by the policy community. The prospect of a positive evaluation of research impact, which leads to increased state funding to universities, coupled with greater opportunities for applying for external funding, could serve to encourage research that addresses the interests of the policy elites.

---

26. Cummings, 'Boundary displacement', pp. 19–22.
27. US Department of Defense, *Minerva Research Initiative: Program History & Overview*, (2012), available at: http://minerva.dtic.mil/overview.html (accessed 12 December 2012).
28. The REF is a grading exercise carried out throughout the UK to assess the quality of research undertaken by UK universities. Government funding is allocated partly on the basis of the level of 'excellence' universities achieve in the REF.
29. Economic and Social Research Council, *What is Impact?*, (2012), available at: http://www.esrc.ac.uk/funding-and-guidance/tools-and-resources/impact-toolkit/what-how-and-why/what-is-research-impact.aspx (accessed 23 February 2013); Economic and Social Research Council, *UK Strategic Forum—Areas for Further Consideration*, (2012), available at: http://www.esrc.ac.uk/about-esrc/information/five-areas.aspx (accessed 23 February 2013).

Nevertheless, Stephen Walt, Joseph Nye and Alexander George argue that the gap between scholars and policy community is ever growing.[30] They claim that policymakers tend to ignore academic research because of its irrelevance to their day-to-day work of policymaking.[31] While this gap may be the case for the disciplines of Political Science and International Relations since WWII in general, China-specific discussions have remained, as discussed above, closely related to the strategic interests of the state. When area studies falls into this trap of policy research, academic research tends to end up focusing on a predictable and stereotyped agendum that fit with national interests. This results in what Amitav Acharya calls a state of 'entrapment', which

> ... occurs when scholars, after having offered consequential intellectual input at an early stage of policymaking ..., remain beholden to the choices made by officials and thereby [become] unwilling or incapable of challenging officially sanctioned pathways and approaches for the fear of losing their access and influence.[32]

It is not our intention to claim that the close link between academia and policy community is necessarily problematic. However, what we need to be vigilant about is the tendency for policy needs to influence the academic research agenda in the study of Sino–African relations, rather than the other way around (i.e. academic research agenda influencing policy direction). Other options for academia include deliberately maintaining intellectual distance from the policy community, so that scholars can freely advance their research without being constrained by structural and political obstacles that the policy community faces. It is problematic that much of the scholarship on Sino–African relations remains focused on whether or not China is a threat: such a debate is influenced by and remains confined to Western governments' strategic interests, and could crowd out the intellectual space for alternative research topics.

This is not to say that all literature that revolves around the China threat and 'myth-busting', necessarily seeks to inform Western policymakers and their interests.[33] Particularly, scholars outside Western academic circles often do not have ties with Western policymakers and may regard the latter as antagonistic to African and/or Chinese interests. However, research on Sino–African relations undertaken by Chinese analysts has frequently been reactive to *Western* debates of Sino–African relations, which results in defensive essays refuting Western criticisms of the PRC's role in Africa.[34] Ironically, this only serves to further entrench the 'myth-busting'

---

30. Stephen M. Walt, 'The relationship between theory and policy in international relations', *Annual Review of Political Science* 8, (2005), pp. 23–48; Joseph S. Nye Jr, 'Bridging the gap between theory and policy', *Political Psychology* 29(4), (2008), pp. 593–603; Alexander L. George, *Bridging the Gap: Theory and Practice in Foreign Policy* (Washington, DC: United States Institute of Peace Press, 1993).

31. For example see Walt, 'The relationship between theory and policy in international relations', pp. 23–24.

32. Amitav Acharya, 'Engagement or entrapment? Scholarship and policymaking on Asian regionalism', *International Studies Review* 13(1), (2011), p. 12.

33. We are grateful to the anonymous referee for this point. One good example of critical work on China–Africa relations led by African scholars includes Ampiah and Naidu, eds, *Crouching Tiger, Hidden Dragon?*

34. He Wenping, 'Zhongfei guanxi fazhan chudongle shei de shenjing?' ['Who's nerves have the development of Sino–African relations touched?'], *Shijie zhishi* [*World Knowledge*] 19, (2006), pp. 30–32; Li Anshan, 'Wei zhongguo zhengming: zhongguo de feizhou zhanlüe yu guojia xingxiang' ['Establishing a name for China: China's Africa strategy and national image'], *Shijie jingji yu zhengzhi* [*World Economy and Politics*] 4, (April 2008),

narrative, irrespective of their intention to inform or not to inform Western policymakers.

Western exceptionalism and international relations

The close link between the national security agenda and the academic literature in the West can be also seen as a by-product of Western exceptionalism that remains prevalent in the discipline of IR. As a field of study which emerged in the West as a self-conscious academic discipline, Acharya and Barry Buzan argue, it is almost a truism to say that 'the main ideas in this discipline are deeply rooted in the particularities and peculiarities of European history, the rise of the West to world power, and the imposition of its own political structure onto the rest of the world'.[35] This cultural/geographical bias has often resulted in a somewhat one-sided interpretation of global order, in that Western dominance is seen as progressive and thus the *only* form of hegemony that matters historically and normatively. The rise of a non-Western state or non-Western hegemony is both poorly theorized and almost axiomatically seen as a threat. The in-built Eurocentric biases of IR theory have resulted in the European regional order being conceptualized as a product of something rational and liberal, as it ensured the survival of individual sovereign states and prevented the emergence of a (universal) empire.[36]

Thus, the global expansion of this order in the nineteenth century is celebrated as an event where 'Europe expands outwards and graciously bequeaths sovereignty and Europe's panoply of civilised and rational institutions to the inferior Eastern societies'.[37] Consequently, IR scholars working in the constructivist and English School approaches have a tendency to interpret international normative change as a two-step process, where 'international norms' are essentially seen as emanating from the West and are transmitted via socialization to the 'non-West'. This narrative 'allows for the continued imagination and invention of Europe's intellectual and political superiority, treating the West as a perennial source of political and religious tolerance'.[38] This line of thinking can also be discerned in Buzan and Richard Little's work on international history when they note that there is a 'story of unevenness ... on the difference between the West and the rest', where 'the Western states began to

---

Footnote 34 continued
pp. 6–15; Li Ruogu, 'Xifang gui zhongfei hezuo de waiqu ji qi zhengwei' ['The West's distortions and falsifications of Sino–African cooperation'], *Shijie jingji yu zhengzhi* [*World Economy and Politics*] 4, (2009), pp. 16–25; Luo Jianbo and Zhang Xiaomin, 'Multilateral cooperation in Africa between China and Western countries: from differences to consensus', *Review of International Studies* 37(4), (2011), pp. 1793–1813.

35. Amitav Acharya and Barry Buzan, 'Why is there no non-Western international relations theory? An introduction', *International Relations of the Asia Pacific* 7(3), (2007), p. 293.

36. Edward Keene, *Beyond the Anarchical Society: Grotius, Colonialism and Order in World Politics* (Cambridge: Cambridge University Press, 2002).

37. Benjamin de Carvalho, Halvard Leira and John M. Hobson, 'The big bangs of IR: the myths that your teachers still tell you about 1648 and 1919', *Millennium* 39(3), (2011), p. 22; see also Hedley Bull and Adam Watson, eds, *The Expansion of International Society* (Oxford: Clarendon Press, 1984).

38. Turan Kayaoglu, 'Westphalian Eurocentrism in international relations theory', *International Studies Review* 12(2), (2010), pp. 195–196.

develop a much more intense set of shared rules, norms and institutions amongst themselves'.[39] The Western world is depicted as a 'more highly developed core', and again serves to demarcate the West from 'the rest'.[40]

In the context of international politics today, this intellectual tradition frequently manifests itself in a sense of Western exceptionalism.[41] Proponents of American exceptionalism paint 'the world as a hostile place, an environment in which America must constantly strive to control and eliminate evildoers before their malevolent acts hit the American homeland'.[42] This 'permanent aura of exaggerated insecurity may also help create and sustain the role and efforts' of the US 'to liberate others' and guide them to the trappings of liberal democracy and capitalist development.[43] Given this ideological belief, it is perhaps not surprising that the US is seen by some IR scholars as an almost uniquely benevolent state in the contemporary international system, as 'no other country can make comparable contributions to international order and stability'.[44] The US is seen as 'the only major power whose national identity is defined by a set of universal political and economic values',[45] and is frequently praised for its almost selfless distribution of public goods, such as its provision of aid for the restructuring of Germany or Japan after WWII, or the opening up of its markets that would allow other states' economies to flourish.[46] Even more altruistically, the US is said to have chosen to surrender part of its power to a series of international institutions, making its actions predictable and allaying other states' fears. Thus it is claimed that

> ... the creation of rule-based agreements and political–security partnerships were both good for the United States and for a huge part of the rest of the world. The result by the end of the 1990s was a global political formation of unprecedented size and success—a transoceanic coalition of democratic states tied together through markets, institutions, and security partnerships.[47]

The case of Europe is somewhat different, as European states individually do not have the same degree of political, economic and military clout that the US enjoys. Nevertheless, Europe does count itself as part of the West, which still remains the dominant force in the international order. Furthermore, the same sense of moral superiority can also be found in European visions of itself as a 'civilian power' that is

---

39. Barry Buzan and Richard Little, *International Systems in World History: Remaking the Study of International Relations* (Oxford: Oxford University Press, 2000), p. 338.

40. *Ibid.*

41. This is not to suggest that exceptionalism is something unique to the West: for an interesting discussion of *Chinese* exceptionalism vis-à-vis Africa, see Chris Alden and Daniel Large, 'China's exceptionalism and the challenges of delivering difference in Africa', *Journal of Contemporary China* 20(68), (2011), pp. 21–38.

42. K. J. Holsti, 'Exceptionalism in American foreign policy: is it exceptional?', *European Journal of International Relations* 17(3), (2011), p. 392.

43. *Ibid.*, p. 394.

44. Samuel P. Huntington, 'The clash of civilizations?', *Foreign Affairs* 72(3), (1993), p. 82.

45. *Ibid.*

46. G. John Ikenberry, *After Victory: Institutions, Strategic Restraint, and the Rebuilding of Order after Major Wars* (Princeton, NJ: Princeton University Press, 2000); G. John Ikenberry, *Liberal Leviathan: The Origins, Crisis, and Transformation of the American World Order* (Princeton, NJ: Princeton University Press, 2011).

47. G. John Ikenberry, 'Power and liberal order: America's postwar world order in transition', *International Relations of the Asia Pacific* 5(2), (2005), p. 139.

characterized by its ability to lead through ethical example.[48] The European Union's (EU) existence as a post-Westphalian political entity has spawned an implicit belief that it is an inherently progressive, cosmopolitan entity that has overcome and transcended conflict arising from nationalism. In this sense, it is a pioneer that 'changes the norms, standards and prescriptions of world politics away from the bounded expectations of state-centricity'.[49] Furthermore, in an age where the US is either seen to be in decline or has squandered its moral authority through its unilateral policies during the Bush presidency, the EU is seen to be uniquely positioned to 'share in the responsibility for global security and in building a better world',[50] primarily through the promotion of 'sustainable peace, freedom, democracy, human rights, rule of law, equality, social solidarity, sustainable development and good governance'.[51] Such policies are a strong reflection of the EU's own self-identity as an 'ethical' actor whose historical trajectory of development is universal and highly worthy of emulation across the world.

Of course, not all theories of IR would necessarily see Western states as somehow uniquely 'ethical' than the rest: as noted above, structural realism assumes that all states are equally *amoral*, regardless of their cultural or ethnic makeup.[52] Yet, even in these theories, a curious sense of Eurocentrism lingers, albeit implicitly. As Isabelle Grunberg has noted with reference to realist theories of hegemonic stability, while most theorists concede that 'great powers ... pursue hegemony and open trade policies for self-seeking motives', it is assumed that in the cases of *Western* hegemony 'the effects of their policies are beneficial to other states as well .... This is because they help create a structure that profits everybody by promoting growth'.[53] The end of American hegemony thus becomes something not only bad for the US—it is to the detriment of the entire world.

The big unknown of IR: the rise of a non-Western power

IR's Eurocentrism, which originally derives from its almost exclusive focus on historical periods of Western dominance, produces other problems. Crucial for our argument here is the fact that the significantly long period of international history where Europeans interacted with non-European polities and people *from a position of military inferiority* is all but ignored.[54] Consequently, the rise of a non-Western power somehow becomes an 'unknown', an unprecedented and potentially dangerous development. When Asian states are perceived to be on the ascendant, Western leaders have claimed that the state in question 'is an adversary who does not respect

---

48. Hartmut Mayer, 'Is it still called Chinese whispers? The EU's rhetoric and action as a responsible global institution', *International Affairs* 84(1), (2008), p. 62.
49. Ian Manners, 'The normative ethics of the European Union', *International Affairs* 84(1), (2008), p. 45.
50. European Council, *A Secure Europe in a Better World*, (12 December 2003), available at: http://www.consilium.europa.eu/uedocs/cmsUpload/78367.pdf (accessed 23 February 2013).
51. Manners, 'The normative ethics of the European Union', p. 66.
52. John J. Mearsheimer, *The Tragedy of Great Power Politics* (New York: W. W. Norton, 2002); Kenneth Waltz, *Theory of International Politics* (New York: McGraw-Hill, 1979).
53. Isabelle Grunberg, 'Exploring the "myth" of hegemonic stability', *International Organization* 44(4), (1990), p. 440.
54. But see John M. Hobson, *The Eastern Origins of Western Civilisation* (Cambridge: Cambridge University Press, 2004).

the rules of the game and whose overwhelming desire is to conquer the world'.[55] Worse still, its peoples have been depicted as 'yellow dwarfs' who 'sit up all night thinking of ways to screw us'.[56]

Crucially, these words were not uttered in the late-nineteenth century, but in 1991 by then French prime minister Édith Cresson. Furthermore, they were not referring to China, but *Japan*, which is not an authoritarian state, but a liberal democracy. These statements have interesting similarities with the 'Yellow Peril' discourse, which can be defined as a discourse that

> ... orders the peoples and phenomena of the Far Eastern 'Orient' into a praxiologically constituted modal moral 'logic' .... In the matter of the 'Yellow Peril', the Asian aggregate, or some subsegment of it ... are feared because the dominant group ... believes that it, i.e. the particular element of the Asian aggregate under discussion, 'is not keeping to its [appropriately subordinated] place but threatens to claim opportunities and privileges from which it has been excluded'; even more fearsome is the belief 'felt [that the Asian aggregate or its subset is] ... a threat to the status, security, and welfare of the dominant ethnic group'.[57]

In the context of contemporary international politics, this translates into an assumption that relations between 'Asia' and the West are of a highly competitive, zero-sum nature. As can be discerned from Cresson's remarks above, the rise of an Asian power is a grave challenge to the security of the West. Regardless of the type of regime, the Asian state in question is (somewhat mysteriously) assumed to be opposed to Western values and interests, and will inevitably seek to overturn the international status quo of Western dominance. This will also rob the West of its privileged (but apparently deserved) position as *the* 'norm maker' in the international community. Cresson was not alone in trumpeting the Japan threat theory in the early 1990s. Such views were gaining some traction in the US, with books such as *The Coming War with Japan* declaring that the Japanese would overthrow American hegemony in the Asia–Pacific.[58] A report drafted for the CIA also 'described the Japanese as "creatures of an ageless, amoral, manipulative and controlling culture"',[59] and darkly warned of 'the potential of a Japanese–Soviet alliance that would give Japan a hedge against an "American backlash"'.[60]

While these views may seem sensationalist and idiosyncratic in the context of Japan's subsequent economic stagnation and China's rise today, it is important to remind ourselves that the potential threat of Japan was taken seriously in academic circles as well. Political scientists such as Christopher Layne claimed that Japan was beginning 'to develop the capability to gather and analyze politico-military and

---

55. 'Outspoken Edith Cresson appointed as prime minister', *The Economist*, (18 May 1991), retrieved from Factiva database.
56. Robert Cottrell, 'Profile: France's Nicholas Ridley, Edith Cresson', *Independent on Sunday*, (21 July 1991), p. 26.
57. Stanford M. Lyman, 'The "Yellow Peril" mystique: origins and vicissitudes of a racist discourse', *International Journal of Politics, Culture and Society* 13(4), (2000), p. 687.
58. George Friedman and Meredith LeBard, *The Coming War with Japan* (New York: St. Martin's Press, 1991).
59. Laurie Goodstein, 'Cap and gown, cloak and dagger at RIT; university's CIA links generate stormy debate', *Washington Post*, (20 June 1991), p. A3.
60. 'Japan lacks global responsibility—CIA-funded study', *Jiji Press*, (10 June 1991), retrieved from Factiva database.

economic intelligence independently of the United States',[61] and concluded that this was preliminary evidence of Japan's desire to 'acquire the full spectrum of great power capabilities and its desire to seek international recognition of its great power status'.[62] Similarly, Samuel P. Huntington also warned:

> In the 1930s Chamberlain and Daladier did not take seriously what Hitler said in *Mein Kampf*. Truman and his successors did take it seriously when Stalin and Khrushchev said, We will bury you. Americans would do well to take equally seriously both Japanese declarations of their goal of achieving economic dominance and the strategy they are pursuing to achieve that goal.[63]

The assumptions behind the 'Yellow Peril' discourse against Japan in the 1990s are shared by more alarmist analyses of Sino–African relations today. The first of these assumptions is that China and Japan are/were both perceived by the West as a threat to the Western traditional dominance. Recent debates surrounding Sino–African relations may be more than just a simple concern over China's threat to the welfare of Africa and African peoples. Rather, they implicitly reflect deeply-rooted Western anxieties that their traditional dominance in Africa is about to be overthrown by a non-Western power (a trend perhaps accelerated by the recent global financial crisis), and that this is part of a broader trend of American and European decline.

This is why academic and political debates surrounding Sino–African relations are obsessed by the question of whether or not China's growing role on the continent is another manifestation of the 'China threat'. The underlying assumption of zero-sum competition between Asia and the West, which is a hallmark of the 'Yellow Peril' thesis, means that the PRC is axiomatically seen as a threat to the West's privileged position in Africa. In this context, many critics of China's Africa policy frequently fail to contextualize their arguments by considering the possibility that the EU could also be seen as a threat to the region. Europe's dominance in Africa has often been criticized for neocolonialism by African analysts,[64] and it is also worth noting that its economic presence still overshadows that of China's.[65]

It is, of course, possible to argue for the existence of a 'China exception' rule, in the sense that it is not Asian powers that get treated as a threat: rather, it is *China* that tends to be—perhaps unfairly—at the receiving end of the bulk of Western criticisms and fears, while other Asian states are hardly criticized. There is certainly a grain of truth to this point. For instance, Japan has hardly been criticized for its trade ties with Sudan, even though it was one of Sudan's key export partners along

---

61. Christopher Layne, 'The unipolar illusion: why new great powers will rise', *International Security* 17(4), (1993), p. 38.
62. *Ibid.*, p. 37.
63. Samuel P. Huntington, 'Why international primacy matters', *International Security* 17(4), (1993), p. 76.
64. Chukwuma Charles Soludo, 'From Berlin to Brussels: will Europe underdevelop Africa again?', *New African* 516, (2012), pp. 10–17; see also Adebajo, 'An axis of evil?'.
65. In 2009, the EU's share of trade in Africa was 63.5%, as opposed to China's 13.9%. In the case of FDI flows to African countries, the EU's share between the years 2005 and 2010 was 43.7%, as opposed to China's 0.9%. See African Development Bank, Organisation for Economic Co-operation and Development, United Nations Development Programme and The United Nations Economic Commission for Africa, *African Economic Outlook 2011: Africa and its Emerging Partners* (Paris: OECD Publishing, 2011), pp. 97, 101.

with China.[66] Yet, such 'China-bashing' has interesting parallels with 'Japan bashing' that was prevalent in the early 1990s, and this suggests a thread of continuity in these discourses that cannot simply be reduced to factors unique to China. It is telling that some of the literature—consciously or not—uses the term 'China Inc.', which originates from 'Japan Inc.' and was used in the 'Japan threat' theses. It was seen as 'synonymous with a broader Japanese undertaking to pursue economic success regardless of the social and environmental costs at home or the detriment to relations with other nations, particularly the United States',[67] just like the PRC is accused of doing today in Africa.

The second of the assumptions shared by the Western study of Sino–Africa relations and the 'Yellow Peril' discourse is a curious sense of *cultural* bias which casts Asian states—be they democratic or authoritarian, as seen from the examples of the China and Japan threat theses—as entities that are somehow fundamentally different from the West, and therefore a threat. The result is an almost knee-jerk reaction that sees any influences in Africa other than Western ones as immoral and undesirable.

One famous example of this line of thinking comes from Samuel P. Huntington's Clash of Civilizations thesis, which posited that future conflicts would take place between different civilizations, rather than states. Here, states were deemed potential threats not only because of their potential to be strategic peer rivals to the US, but primarily because they were *culturally fundamentally different*. It is interesting to note that Germany—which like Japan was at times touted as a potential challenger to US hegemony in the early 1990s by some structural realists[68]—somehow gets taken off the list of potential enemies of the US, by the simple virtue of it apparently belonging to the zone of Western Christianity. Japan and China remain in different civilizational entities. Huntington assumes that different civilizations will almost inevitably clash, and the fact that Japan and China are different means that they will necessarily be a threat to Western civilization.[69] Although Huntington's controversial argument has been subjected to a wide range of criticisms,[70] other analysts seem to share Huntington's assumption that cultural difference somehow produces a threat to global stability. Emma Mawdsley's research into British broadsheet newspapers' representations of China and Africa shows that many pundits seem to reproduce implicitly the dichotomy between an inherently 'unethical' non-West and an 'ethical' West whose dominance is ultimately good for the rest of the world. Thus:

---

66. The 2005 CIA World Fact Book, for instance, places Japan second behind China as Sudan's export partner, with 10.7% of the share of Sudanese exports. China had 66.9%, and Saudi Arabia, at 4.4%, came third. These can be accessed from http://permanent.access.gpo.gov/lps35389/ (accessed 9 February 2013).

67. Narrelle Morris, *Japan Bashing: Anti-Japanism since the 1980s* (Abingdon: Routledge, 2011), pp. 23–24. In the China–Africa case, see for example: Gill and Reilly, 'The tenuous hold of China Inc. in Africa'. We are grateful to the anonymous reviewer's comments in helping us to develop this point further.

68. John J. Mearsheimer, 'Back to the future: instability in Europe after the Cold War', *International Security* 15 (1), (1990), pp. 5–56; Layne, 'The unipolar illusion', pp. 5–51.

69. Huntington, 'The clash of civilizations?', pp. 22–49.

70. For example Robert W. Cox, *The Political Economy of a Plural World: Critical Reflections on Power, Morals and Civilization* (London: Routledge, 2002); Gerard Delanty, 'The making of a post-Western Europe: a civilizational analysis', *Thesis Eleven* 72(1), (2003), pp. 8–25; John Mandalios, 'Civilizational complexes and processes: Elias, Nelson and Eisenstadt', in Gerard Delanty and Engin F. Isin, eds, *Handbook of Historical Sociology* (London: Sage, 2003).

Western actors—businesses, governments, national and international development NGOs—are typically portrayed as benign ... the mistakes of the past have been addressed, and the West is now the architect and energizer of a new drive towards good governance and development, with aid now accompanied by ethical conditionalities, while reformed commercial practices promise investment and trade that will enhance development rather than line the pockets of kleptocratic elites.[71]

The growing stature of the PRC in the African continent therefore not only becomes a strategic threat to the West: it also challenges deeply-rooted notions of the Western 'self'. As Mawdsley notes, 'Western political imaginaries of itself in relation to Africa remain dominated by an enduring notion of trusteeship, despite a long and ongoing history of exploitation, and lack of sufficient action to address systemic inequalities and injustice'.[72] US Secretary of State Hilary Clinton seemed to echo this theme when she stated that China must join the US efforts to support democracy and increase transparency in economic activities, otherwise we will see a 'new colonialism' on the continent:

> We are ... concerned that China's foreign assistance and investment practices in Africa have not always been consistent with generally accepted international norms of transparency and good governance, and that it has not always utilised the talents of the African people in pursuing its business interests.[73]

The 'Yellow Peril' discourse is also related to the issue of a perceived lack of China's 'responsibility' in its Africa policy. The PRC is criticized for its 'security free riding' and unwillingness 'to work with the international and African regional community',[74] which leads to the limited degree of Sino–Western cooperation in providing stability and security to Africa.[75] While it is true that the Chinese side does display reluctance to coordinate their Africa policies with a third party such as the US and EU, Western analysts frequently fail to question whether or not the uncritical acceptance of Western moral superiority, assumed benevolence towards Africa, or intellectual influences of the 'Yellow Peril' discourse result in a problematic *Western* refusal to coordinate its policies with the Chinese. This argument again shares a similarity with so-called 'Japan bashing', where a decidedly 'selfish' Japanese identity was constructed in order to warn against the threat a rising Japan would pose to Western dominance. Japan was, in similar fashion to China in Africa, accused of free-riding on American security provision, and 'reneging on its moral obligations to the United States, which it allegedly owed as a result of the United States' post-war

---

71. Emma Mawdsley, 'Fu Manchu versus Dr Livingstone in the dark continent? Representing China, Africa and the West in British broadsheet newspapers', *Political Geography* 27(5), (2008), pp. 519–520.

72. *Ibid.*, p. 512; see also William Pfaff, 'A new colonialism? Europe must go back into Africa', *Foreign Affairs* 74(1), (1995), pp. 2–6.

73. Cited in Andrew Quinn, 'Clinton warns Africa of China's economic embrace', *Reuters*, (10 June 2011), available at: http://uk.reuters.com/article/2011/06/10/us-clinton-africa-idUSTRE75962920110610 (accessed 23 February 2013).

74. Holslag and Van Hoeymissen, eds, *The Limits of Socialization*, p. 11.

75. Exceptions include UK–China–Africa trilateral cooperation. See UK Department for International Development website, available at: http://www.dfid.gov.uk/Where-we-work/Asia-East-Pacific/China/, and UK National Archives, *Minutes for the Africa–Britain–China Conference on Agriculture and Fisheries*, available at: http://webarchive.nationalarchives.gov.uk/+/http://www.dfid.gov.uk/documents/china-africa/mins-afr-brit-ch-conf-agr-fish.pdf (accessed 18 February 2013).

contributions to and support of Japan'.[76] It was also blamed for using its growing power irresponsibly, as epitomized by then *Newsweek* editor Robert J. Samuelson, who 'argued ... that Japan had acquired global responsibilities "before being capable, psychologically and politically, of discharging them"'.[77]

What is to be done? Burning questions of Sino–African relations

This article has analyzed why so much ink has been spilt to assess whether or not China is a unique threat to Africa, even though the conclusions have often indicated that the Chinese have acted neither better nor worse than many Western actors. We have argued that the study of Sino–African relations is closely linked with the Western governments' geostrategic interests, and that both sides of the debate remain deeply influenced by the Eurocentrism and the 'Yellow Peril' discourse that continue to linger in IR. These studies bring with them a myopic agenda that revolves almost exclusively around the question of China's 'challenge' to the West.

What, then, needs to be done to find new questions in Sino–African relations that go beyond national security and Western-centric agendas? We suggest a number of possible paths. First, the research agenda needs to be defined with attention paid to historical continuity, rather than simply following policy/academic 'fashions'. As mentioned at the beginning of this article, China's Africa policy has transformed every decade or two: from its support for proletarian revolutions in the 1960s, counter-hegemony to the US and to the Soviet Union in the 1970s, promotion of African self-reliance (a rhetorical term for indicating China's waning of interest in Africa) in the 1980s and 1990s, and to its return to Africa.[78] China's increasing interest in Africa may be 'hot' in scholarship today, but who can be sure that China will not lose interest in Africa again in the next decade?

Second, Chinese perspectives need to be explored further. Despite its official stance, China's primary foreign policy goals and interests are still based on its relations with the developed powers, particularly the US and Japan, not in the developing world.[79] It therefore does not have many Africanists compared to specialists on the developed powers. Yet, there is a growing number of institutions specializing in the study of Africa (such as Zhejiang Normal University, Xiangtan University, Beijing University, Chinese Foreign Affairs University, Yunnan University and the Chinese Academy of Social Sciences). Furthermore, the nature of Chinese studies of Sino–African relations is undergoing a change. As mentioned earlier, previous research was frequently reactive to *Western* debates of Sino–African relations, and often resulted in defensive essays refuting Western criticisms

---

76. Morris, *Japan Bashing*, p. 25.
77. Ibid.
78. Joshua Eisenman, 'China's post-Cold War strategy in Africa: examining Beijing's methods and objectives', in Joshua Eisenman, Eric Heginbotham and Derek Mitchell, eds, *China and the Developing World: Beijing's Strategy for the Twenty-First Century* (Armonk, NY: M.E. Sharpe, 2007).
79. Peter Van Ness, 'China as a third world state: foreign policy and official national identity', in Lowell Dittmer and Samuel S. Kim, eds, *China's Quest for National Identity* (Ithaca, NY: Cornell University Press, 1993); Shogo Suzuki, 'Journey to the west: China debates its "great power" identity', paper presented at *Sino–Australian Security Relations: Regional Cooperation in an Interdependent World conference*, Australian National University, Canberra (October 2007).

of the PRC's role in Africa. However, there is emerging Chinese literature that goes beyond defending and promoting Beijing's Africa policies. Some have been critical of Chinese reliance on (outdated) Western stereotypes to understand Africa,[80] while others have critiqued Chinese officials for their lack of initiative in realizing cooperation agreements.[81] There is also increasing appreciation of exploitation of Africans by rapacious Chinese entrepreneurs, and that this is harmful to Chinese foreign policy goals.[82] Greater engagement is needed with these critical voices that exist in China. Furthermore, a number of Chinese scholars are now conducting field research in Africa for a lengthy period, and they bring an empirically rich and critical research to the study of Sino–African relations that go beyond the confined nature of the debate that serves the strategic interests of governments.[83]

Relevant to the second point, the third path is to encourage *Chinese and African scholars* to set the research agendas of China–Africa relations. The majority of the literature on Sino–African relations in English is written by US and European scholars who work on Africa, security and development, and international relations, with a heavy reliance on English-language materials. This, as Daniel Large correctly concludes, has 'the danger of playing out a self-referential logic',[84] as well as reproducing the national security agendas of Western states. The study of Sino–African relations therefore needs to be diversified further and reflect more African perspectives, rather than Western strategic anxieties. Sino–African relations are indeed the result of generalization of the relationship between China and individual African states. Yet, among 54 countries in Africa, there are only about a dozen countries that often attract attention in the study of Sino–African relations, such as Sudan, Angola and Zimbabwe. In recent years, however, there has been a welcome

---

80. Li Xiaoyun, 'Zhongguo yao gaibian ziji de feizhou guan' ['China needs to change its own views of Africa'], *Fenghuang zhoukan*, (15 June 2011).

81. Zhou Haijin, 'Lun xin shiqi zhongke wenhua yu hezuo' ['Sino–Cameroon cultural exchange and cooperation in the new era'], *Zhejiang shifan daxue xuebao* [*Journal of Zhejiang Normal University*] (Social Sciences) 36(4), (2011), pp. 16–20.

82. Zhao Minghao, 'Zhongfei minjian jiaowang: jinzhan ji mianlin de tiaozhan' ['Sino–African civilian interactions: developments and challenges faced'], *Guoji zhanwang* [*International Perspective*] 6, (2010), pp. 49–62.

83. Examples include Chen Fenglan, 'Wenhua chongtu yu kuaguo qianyi qunti de shiying celüe: yi nanfei zhongguo xinyimin qunti weili' ['Cultural differences and strategies for adaptation by transnational immigration group: a case study of new Chinese immigrants in South Africa'], *Huaqiao Huaren Lishi Yanjiu* [*Overseas Chinese History Studies*] 3, (2011), pp. 41–49; and Guo Zhanfeng, Li Xiaoyun and Qi Gubo, 'Zhongfei xiaonong jiating nongye shengchan zuzhi guocheng duibi fenxi: jiyu feizhou sancun de tianye diaocha' ['Comparative analysis of the processes of small-scale household agricultural production organizations in China and Africa: based on fieldwork in three villages in Africa'], *Guangxi Minzu Daxue Xuebao* [*Journal of Guangxi University for Nationalities*] (Philosophy and Social Science Edition) 34(2), (2012), pp. 82–89. A number of Chinese scholars also advocate the conducting of field research as one of the key methods for 'Chinese African studies'. See, Ma Yankun, 'Renleixue feizhou yanjiu ji zhongguo xueke jiangou de xianshi suqiu' ['Anthropolitical African studies and the realistic seeking for its course building in China'], *Xiya Feizhou* [*West Asia and Africa*] 7, (2010), pp. 26–31; Liu Hongwu, 'Zai guoji xueshu pingtai yu sixiang gaodishang jiangou guojia huayuquan: Zailun jiangou you tese zhi "zhongguo feizhouxue" de teshu shidai yiyi' ['To construct Chinese national discourse rights on international academic platform and ideological high ground: a further study on the special significance of the era of constructing a unique "Chinese African studies"'], *Xiya Feizhou* [*West Asia and Africa*] 5, (2010), pp. 17–23; and Liu Hongwu, 'Guoji sixiang jingzheng yu feizhou yanjiu de zhongguo xuepai' ['International competition of thoughts and the Chinese school of African studies'], *Guoji Zhengzhi Yanjiu* [*International Politics Quarterly*] 4, (2011), pp. 89–97.

84. Daniel Large, 'Beyond "dragon in the bush": the study of China–Africa relations', *African Affairs* 107(426), (2008), p. 58.

trend of global research collaboration among scholars from China, Africa, the West and other parts of the world. For example, 'The Chinese in Africa/Africans in China Research Network'—a global independent network which has been hosted by the University of Johannesburg and Rhodes University—aims to facilitate discussions and collaboration on social, economic and political research on issues relating to people-to-people encounters.[85] More of these kinds of efforts will be necessary to diversify and deepen the study of Sino–African relations.

Research by African scholars may find both Western and Chinese policies on the continent contrary to African interests,[86] or could even demonstrate that support for Chinese policies in Africa is not limited to 'isolated autocrats' alone.[87] Whichever way, these works have the potential to serve as a powerful critique to the dichotomous and myopic debates of the 'China threat' that has impoverished the debate on Sino–African relations. It will also help overcome complacent Western views that Western ideological influences are axiomatically 'universal' and will be welcomed by the peoples throughout the world.

Fourth, it is important to take an anthropological perspective to what is often regarded by the West as universal concepts, such as democracy, human rights, security, poverty and development. While these are the major themes of the study of Sino–African relations, the meaning of these terms varies depending on the region. When using terms that originated in the West in a different context, researchers could fall into the trap of using a square peg in a round hole. The study of Sino–African relations is even more complicated, given that it requires examination of such terms in at least three different contexts—one in Africa, one in the West and one in China. It is essential to scrutinize how each actor conceptualizes these basic terms.

But more fundamentally, we need to overcome the deep sense of Western exceptionalism that has continued to color the lenses by which we view international politics. As has been noted with reference to Orientalism in strategic studies, it

> ... is not enough to show that myths about the enemy are empirically flawed. The exotic eastern warrior will not stop being a silhouette in the Western imagination. We therefore need to ask why, to understand what motivates our fascination in the first place, and recognise that these myths are powerful codes through which Westerners debate about themselves .... One-dimensional caricatures of Oriental warfare reflect the anxieties, fears, ambitions, confidence or self-doubt of Western observers ...[88]

In many respects, the same critique could be applied to the study of Sino–African relations today. Much of the literature that voices fears that China is systematically undermining Western influence and power on the African continent (as well as the

---

85. Yoon Jung Park, e-mail messages to the authors, 26 February and 10 August 2013.
86. See for example Adebajo, 'An axis of evil?'; Lloyd Sachikonye, 'Crouching tiger, hidden agenda? Zimbabwe–China relations', in Ampiah and Naidu, eds, *Crouching Tiger, Hidden Dragon*.
87. See Barry Sautman and Yan Hairong, 'African perspectives on China–Africa links', *The China Quarterly* 199, (2009), pp. 728–759.
88. Patrick Porter, *Military Orientalism: Eastern War through Western Eyes* (London: Hurst & Co, 2009).

stream of literature that refutes it) is more a debate about the West's own deep-seated anxieties, rather than what the PRC is actually doing or not doing in Africa. Such thinking, however, not only serves to impoverish our scholarship and understanding of China and Africa's interactions, but also has the danger of unnecessarily creating a new 'China threat' edifice which is nothing but a by-product of Western fears that its influence and thinking are no longer regarded as 'universal' and 'authoritative'.

China in Africa: presence, perceptions and prospects

FEI-LING WANG and ESI A. ELLIOT

This article reports and analyzes China's presence in Africa with an emphasis on how that has been perceived by the Africans. Based on the findings from surveys and field research conducted in eight sub-Saharan African countries and interviews with scholars and practitioners from other African countries as well as Chinese and Americans in Africa, we outline the diverse, complicated and evolving African perceptions about China's explosive presence in general and the booming Chinese business activities in particular that now range from love to suspicion. Our findings about how China is perceived in Africa suggest that Beijing has acquired substantial goodwill in Africa yet is developing deep issues and facing uncertain challenges and growing obstacles.

Introduction: China in Africa

China's explosive presence in Africa, especially its phenomenal expansion of business activities since the 1990s, has begged for answers to profound questions. How is China perceived and received by the local hosts? Is Beijing acquiring sustained power and influence in Africa as a result? Is China's presence in Africa a grand and comprehensive competition with the West? And, if so, what are the prospects for that competition?

This article seeks to address those questions through examining how China is now perceived in Africa. Based on the findings from surveys and field research conducted in eight African countries (Botswana, Ghana, Kenya, Madagascar, Namibia, Tanzania, South Africa and Zimbabwe) and interviews with African scholars and practitioners from other African countries as well as Chinese and Americans in Africa, we outline the diverse, complicated and evolving African perceptions about China's presence that now range from love to suspicion and worse. The general patterns and key features of how China is perceived in Africa seem to suggest that China is poised to further extend its presence in Africa and Beijing has acquired substantial goodwill among Africans

---

* Fei-Ling Wang is a professor at Georgia Institute of Technology. Esi A. Elliott is an assistant professor at Suffolk University. The authors wish to thank many colleagues and friends (and the participants of the Denver and Macau workshops on 'China in Africa' in 2013) for research assistance and comments. We are especially grateful for the generous support of a Minerva Grant from the Office of the Secretary of Defense of the United States through the US Air Force Academy. The views presented here are the authors' alone and do not represent that of the US military or the US government.

yet is developing deep issues and facing uncertain challenges and growing obstacles. China has encountered uphill battles that are often quite cost-ineffective in its much-speculated competition with the West in Africa, if Beijing indeed had a coherent grand strategy for such a rivalry. The opportunistic and extractive nature of Chinese business activities and Beijing's inability to offer a distinctively different yet attractive value-norm system appear to be some of the key hurdles limiting China's winning of hearts and minds in Africa.

An explosive presence

The People's Republic of China (PRC since 1949) is not a newcomer in Africa. Because of its ambitious agenda of promoting world revolution, and its pragmatic and often desperate need for international recognition deemed crucial to regime survival, Beijing went to Africa during the Mao Era (1949–1976) with massive financial and military aid.[1] The Chinese efforts yielded considerable political payoff as the PRC made advances culminating in its membership of the United Nations in 1972 with significant support from African countries.[2] China's ventures in Africa were also a major component of Beijing's effort of isolating Taiwan, its political opponent. After decades of a diplomatic bidding war, Beijing succeeded in persuading most African countries to switch their diplomatic recognition from Taipei to Beijing, although today Taiwan still maintains an unofficial but strong presence throughout Africa.[3]

However, China's going to Africa in the 1950s–1970s was almost entirely favor procuring and thus was financially costly, expedient and unsustainable. When Beijing started to trade-away its world-revolution goals for economic ties with the West, Chinese activities in Africa were scaled back and even gutted. The Maoist presence in Africa, politically motivated and narrowly focused, nonetheless left behind a tangible legacy like Tazara (Tanzania–Zambia Railroad) and the lauded service of Chinese medical teams, generating credit and good feelings.[4]

By 1980, Sino–African trade was only about US$1 billion.[5] Only in the 1990s did China's own capitalistic economic reform lead to the rising need for new markets for its exports, especially when the trade relationship with its main market, the United States, remained constrained and unstable due to the then annual US Congressional review of the Sino–US trade status after 1989. Furthermore, China's booming manufacturing sector has developed insatiable demands for raw materials and energy that Africa could provide in abundance. Therefore, chiefly for profit and resources, China went back to Africa with massive orders, investment and exports.[6]

---

1. Philip Snow, *The Star Raft: China's Encounter with Africa* (New York: Weidenfeld and Nicholson, 1988).
2. Chang Ailing, 'From "brothers" to "partners": China, Africa build strategic ties', *China Daily*, (9 September 2007).
3. Peter Brookes and Ji Hye Shin, 'China's influence in Africa: implications for the United States', *Heritage Foundation Report*, Washington, DC, (22 February 2006).
4. Since the early 1960s, China has sent medical teams to 46 African countries with over 18,000 medical workers, and treated 200 million patients. PRC State Council, *China–Africa Economic and Trade Cooperation* (Beijing, 2010).
5. Peter Wonacott, 'In Africa, US watches China's rise', *The Wall Street Journal*, (2 September 2011).
6. Chris Alden, Daniel Large and Ricardo Soares De Oliveira, *China Returns to Africa: A Rising Power and a Continent Embrace* (New York: Columbia University Press, 2008).

Bilateral trade has grown rapidly ever since, by double digits annually, reaching US$55 billion in 2006, to make China Africa's second largest trade partner (surpassing France).[7] With a trade explosion of 160 times over 30 years and surpassing the US in 2011, China became Africa's largest trading partner with a total volume of US$199 billion and a trade deficit of US$28 billion in 2013.[8]

Chinese investment in Africa has boomed too. Growing at an annual rate of 114%, foreign direct investment (FDI) from China has covered most of Africa, active in just about every industry, especially mining, energy exploration and construction.[9] Chinese IT companies have been pursuing 'near monopoly' positions in half a dozen African nations.[10] Huawei, for example, has scored a 20–25% annual growth in West Africa and over 10% in the whole of Africa, beating Erickson for the top position.[11] By 2013, China had a total of US$193.8 billion contracted FDI and over 2,000 state-owned and private companies operating in Africa, with US$40 billion new FDIs in 2011 alone, including some mega projects.[12]

China has also become a major provider of aid in education, medical services and civic projects in 50 African nations, funding 31 Confucius institutes in 23 African countries.[13] China also plans to build 1,000 'Hope Schools' in Africa as a gift, with nearly 100 already funded by 2012. A large Chinese media center based in Nairobi produces news and entertainment programming for African markets. Tens of thousands of African students are now studying in China, many funded by the Chinese government.[14] Through a Beijing-funded 'Ten with Ten' program, ten Chinese think tanks are pairing up with ten African think tanks to enable structured exchanges of ideas.

Beyond economic and sociocultural arenas, China has sent thousands of soldiers to participate in peacekeeping missions in Liberia and DR Congo and sustained naval assets to patrol off the Horn of Africa against maritime pirates.[15] Making major policy shifts away from 'non-interference', Beijing has acted to protect its interests and image in Africa with direct moves such as the massive evacuation of Chinese in Libya in 2011,[16] the 'quiet' demands made on behalf of Chinese state enterprises in

---

7. Wonacott, 'In Africa, US watches China's rise'.

8. Mike King, 'China–Africa trade booms', *Journal of Commerce*, (18 July 2012); US Census Bureau, http://www.census.gov/foreign-trade/balance/c0013.html; 'China and Africa: a maturing relationship', *The Economist Intelligence Unit*, (9 April 2013).

9. Xiaoyang Tang, 'Bulldozer or locomotive? The impact of Chinese enterprises on the local employment in Angola and the DRC', *Journal of Asian and African Studies* 45(3), (2010), pp. 350–368; Sigfrido Burgos and Sophal Ear, 'China's oil hunger in Angola: history and perspective', *Journal of Contemporary China* 21(74), (2012), pp. 351–367.

10. Email interviews with Cisco employees in Africa, November 2012.

11. Huawei's employment in Ghana jumped from 60 in 2008 to 1,000 in 2013 (half are Chinese). Interview with Huawei regional managers, Accra, Ghana, February 2013.

12. 'Mozambique–China trade continues to grow', available at: http://allafrica.com/stories/201212090143.html (accessed 12 December 2013); Samuel Mungadze, 'PetroSA, Sinopec sign deal for $10bn refinery at Coega', *Business Day*, Johannesburg, (26 March 2013), p. 18.

13. PRC State Hanban figures, available at: www.hanban.edu.cn. Beijing (accessed 15 April 2013).

14. During 2013–2015, Beijing was set to provide 1,800 full scholarships to African students and bring over 30,000 Africans for short-training programs. Kenneth King, *China's Aid and Soft Power in Africa: The Case of Education and Training* (Suffolk: Boydell & Brewer, 2013).

15. 'Seychelles invites China to set up anti-piracy base', *Agence France-Presse*, Victoria, (2 December 2011).

16. Shaio H. Zerba, 'China's Libya evacuation operation: a new diplomatic imperative—overseas citizen protection', *Journal of Contemporary China* 23(90), (2014), doi: 10.1080/10670564.2014.898900.

Tanzania and Zambia, and the pressure backed by official aid to hush criticisms against illegal activities by a princeling-run Chinese company in Namibia.[17] Clearly disapproving of two of its neighbors' (India and Japan) aspirations for permanent seats at the UN Security Council, China has publicly supported similar bids by all three of the possible African candidates: South Africa, Egypt or Nigeria.[18] Beijing has largely funded the China–Africa Forum of Cooperation since 2000.

In short, China's presence is expanding rapidly in Africa as an active donor, financier, contractor and builder, marketer, buyer, and investor.[19] At least a million merchants, investors, farmers and workers from the PRC now live in Africa, up from negligible numbers 30 years ago (this number increased ten times in 2002–2011).[20] Over 40,000 Chinese state-employees now work in Luanda, Angola alone.[21] At least 10,000 Chinese now live in Tanzania, up from barely 1,000 a decade ago.[22] Chinese in Zambia are estimated in late 2013 to number from 11,000 to the hard-to-believe 100,000.[23]

Features and patterns

At a very rapid rate of growth, China is now Africa's largest trade partner, exporting to basically all 54 African countries with no clear reliance on any single market, yet four countries (Angola, Libya, South Africa and Sudan) provide over 71% of China's imports from Africa. Taken together with the market of four other exporters (Congo, DR Congo, Equatorial Guinea and Zambia), 84% of China's imports from Africa is oil (one-third of China's total oil import), mineral ores and timber.[24] This is a rather typical Africa–outside country trade pattern with manufactured goods flowing in and raw materials and energy going out—a *bona fide* colonial style economic relationship as termed by the Governor of the Nigerian Central Bank.[25]

In addition to the rapid growth and the rather typical and traditional trade patterns, the third characteristic of China's booming economic relationship with Africa is that much of China's business activities are state actions even though most of the

---

17. Beijing offered huge loans to quietly make the Namibian government drop its anti-bribery case against the Chinese company run by then PRC President Hu Jintao's son. Author's interviews in Windhoek, Namibia, March 2013.
18. Jean-Christophe Servant, 'China's trade safari in Africa', *Le Monde Diplomatique*, Paris, (11 May 2005).
19. Jenifer L. Parenti, *China–Africa Relations in the 21st Century* (Washington, DC: National Defense University, 2009). For more analysis of China's presence in Africa, see Suisheng Zhao, 'A neo-colonialist predator or development partner? China's engagement and rebalance in Africa', *Journal of Contemporary China* 23(90), (2014), doi: 10.1080/10670564.2014.898893; and Sven Grimm, 'China–Africa cooperation: promises, practice and prospects', *Journal of Contemporary China* 23(90), (2014), doi: 10.1080/10670564.2014.898886.
20. 'Zhongguo maibu feizhou' ['China strides in Africa'], *Guoji caijing shibao* [*International Financial Times*], Beijing, (22 August 2011).
21. BBC, *The Chinese are Coming*, documentary (London: BBC, 2012).
22. Wu Ming, 'Feizhou guniang yanzhong de zhongguo nanren' ['Chinese men in the eyes of African girls'], *Renmin wang* [*People's net*], Beijing, (5 October 2012); Marc Francis and Nick Francis, *When China Met Africa*, documentary (Speakit Films: 2010).
23. Author's correspondence with field-study scholars, November 2013.
24. 'Zhongguo maibu feizhou', *Guoji caijing shibao*; *The Economist*, (20 April 2011); 'China and Africa', *The Economist Intelligence Unit*.
25. Lamido Sanusi, 'African must get real about Chinese ties', *Financial Times*, (11 March 2013). For more in-depth analysis about this issue, see Joshua Eisenman, 'China–Africa trade patterns: causes and consequences', *Journal of Contemporary China* 21(77), (2012), pp. 793–810.

2,000 plus Chinese firms and business entities in operation in Africa are not exactly state-owned.[26] Unlike Western states, the Chinese government actively finances, encourages and organizes Chinese business ventures into Africa. Interest-free or low interest government loans are common such as the US$2 billion credit line at 1.5% for 17 years to Angola in 2005 for oil exploration and the US$2 billion subsidized loans to Ghana for oil and gas projects in 2010. Also, Beijing has shown significant generosity in dealing with African states with, for example, the cancelation of over US$10 billion in government debt in 2000–2005.[27] In 2012, PRC President Hu Jintao pledged US$20 billion new credit to Africa. In 2013, the new PRC President Xi Jinping pledged billions more.

The active and aggressive Chinese government-backed business dealings, financial aid and charitable activities in Africa are often free of political and ideological strings. In fact, China has made lucrative deals with African states such as Zimbabwe, Sudan and Angola that are chastised by Western sanctions on the grounds of human rights violations. Beijing has used its veto power in the UN Security Council to provide defense and protection to the leaders of some of its African partners who are challenged by the West. This type of seeing-no-evil and hearing-no-evil business attitude has clearly contributed to China's explosive growth in Africa and earned significant goodwill from many of the African leaders and political elites who have grown tired of the political 'nanny' from the West.[28]

Not only ignoring local human rights and other political issues in Africa, Chinese officials and businessmen practice personal connections (*guanxi*) that often involve bribery. This Chinese way of business effectively matches with some traditional social norms in many African countries and greatly oils the wheels of bureaucracies in host countries to facilitate deals. The Chinese way of business also helps to finance local government officials' affluent life style and contributes to the coffers of the rulers to sustain their current, democratic or not and popular or not, regimes. Well known examples include the relationship between Beijing and Harare and the controversial case of China's secretive, eye-popping 2007 investment contract for minerals with DR Congo, reportedly worth US$40–120 billion.[29]

Finally, due to China's own peculiar position in the global chain of production, the expansion of its business activities in Africa is not necessarily a displacement of the West, as Chinese goods and technology have yet to compete directly with, let alone begin replacing, goods and technology from the West. Although US officials are concerned that Chinese government loans and counterfeits may disadvantage US firms, direct competition between Chinese and US firms has been limited as they 'general[ly] operate in different sectors'.[30] As some Africa-based business leaders have asserted, the infrastructure and mining projects that the Chinese are building in

---

26. We learned in the field that there is a complicated and opaque ownership and cooperative relationship between Chinese government and many non-state business ventures in Africa.
27. Servant, 'China's trade safari in Africa'.
28. Suisheng Zhao, 'A neo-colonialist predator or development partner?'.
29. 'China's oil trade in Africa', *The Economist*, (13 August 2011); 'China international fund: the Queensway Syndicate and the Africa trade', *The Economist*, (13 August 2011); Global Witness on DR Congo, *China and Congo: Friends in Need* (London: 2011).
30. US GAO, *Sub-Sahara Africa: Trends in US and Chinese Economic Engagement* (Washington, DC: USGAO, February 2013), pp. 49, 54–55.

Africa, competing against some Western firms for sure, also create new opportunities for more Western exports to Africa such as automobiles and heavy machinery.[31] We noticed in the field that many Chinese construction sites were using non-Chinese equipment and the booming traffic on Chinese-built roads was mostly vehicles made in Germany, Japan, Korea and the United States.

Reflections in the literature

Not surprisingly, China's presence in Africa has drawn great attention.[32] Scholars ponder the strategic and global implications of China's going to Africa.[33] Some explore the Chinese incentives and motives behind its active ventures in Africa as part of China's new global strategy for power and influence that may actually provide an opportunity for Africans to develop their economy.[34] Others view China's ventures in Africa as potentially new challenges to China itself as well as to the West.[35] Dispelling some 'fuzzy facts' about China's oil ventures in Africa, one study argues that the Chinese presence in Africa is quite similar to that by other multinational corporations,[36] the colonial or neo-colonial predators acting aggressively for influence and wealth. Some wonder if the Sino–Western (specially European) competition in Africa may already be in China's favor.[37]

An international team of experts has offered a comprehensive report on the state and prospects of China's relationship with Africa.[38] Another edited volume presents the Chinese, Western and African perspectives on China's presence in Africa in the light of the post-Cold War economic globalization.[39] A European study investigates the 'three most important instruments China has at its disposal in Africa'.[40] 'Returning' or not, argued another work, China has probably entered a new 'scramble for Africa' and begun a 'new Chinese imperialism', with profound implications for the continent and beyond.[41] Concerning China's rapidly expanding ties with Africa, especially its mining and timber interests, there are several scholarly works that offer rich statistical and anecdotal evidence and analyses.[42] Noted experts have tried

---

31. Jaco Maritz, 'Six thoughts on China's involvement in Africa's infrastructure sector', *Foreign Investment*, (7 March 2013).
32. 'The Chinese in Africa: trying to pull together', *The Economist*, (20 April 2011).
33. 'Contrasting rhetoric and converging security interests of the European Union and China in Africa', a special issue of *Journal of Current Chinese Affairs* 40(4), (2011).
34. Marcel Kitissou, ed., *Africa in China's Global Strategy* (London: Adonis & Abbey, 2007).
35. Sarah Raine, *China's African Challenges* (London: Routledge, 2009).
36. Erica Down, 'The fact and fiction of Sino–African energy relations', *China Security* 3(3), (Summer 2007).
37. Thierry Bangui, *China: A New Partner for Africa's Development—Are We Heading for the End of European Privileges on the Black Continent?* (Hauppauge, NY: Nova Publication, 2011).
38. Robert I. Rotberg, ed., *China into Africa: Trade, Aid, and Influence* (Washington, DC: Brookings, 2008).
39. Julia C. Strauss and Martha Saavedra, eds, *China and Africa: Volume 9: Emerging Patterns in Globalization and Development* (The China Quarterly Special Issues) (New York: Cambridge University Press, 2010).
40. Meine Pieter van Dijk, *The New Presence of China in Africa* (Amsterdam: Amsterdam University Press, 2011).
41. Ricardo Soares de Oliveira, Christopher Alden and Daniel Large, *China Returns to Africa: A Rising Power and a Continent Embrace* (New York: Columbia University Press, 2008).
42. Serge Michel and Michel Beuret, *China Safari: On the Trail of Beijing's Expansion in Africa* (New York: Nation Books, 2009); Ian Taylor, *China's New Role in Africa* (Boulder, CO: Lynne Rienner Publication, 2010).

to answer the question of whether Beijing is helping or hurting Africa through its peculiar, extensive, often apolitical aid.[43]

The now politically contentious issue of Chinese migrating to work and live in Africa has been studied, albeit still mostly on national or sub-regional levels. There has been a history of racist discrimination against Chinese in Africa especially under the apartheid regimes since the nineteenth century.[44] Studies have found that contemporary Chinese migrants in South Africa, the largest host of Chinese in Africa, appear to be a 'differentiated grouping of migrants' and have a highly interesting 'transitional, unstable, and fluid nature'.[45]

Many African scholars and journalists have reflected on China's booming presence in Africa. We see here the general pattern of positively assessing and appreciating China as rewarding economic opportunities for Africa and as a non-confrontational political and ideological partner, in sharp contrast to the old Western relations with the continent. Growing doubts, criticism and resentment among Africans against the Chinese presence are also reported.[46] Some Africans have pondered if China has become a development partner, an economic competitor or simply a new hegemony to Africa.[47] Another group of African scholars and practitioners analyze specifically China's role in Southern Africa's extractive industries with thoughtful policy suggestions calling for innovative actions aiming at a win–win–win partnership for the Chinese, local rulers and elites, and local populace.[48] African researchers have also published impressive works to examine China in Africa with a highly commendable goal of understanding China more.[49]

Gauging the African perception: our field research

Armed with the lessons from the existing literature, in 2012 and 2013 we conducted field research in sub-Saharan Africa to gauge and examine how China's explosive presence has been perceived by the African hosts. We commissioned an opinion survey in Ghana and utilized another opinion survey done in Madagascar. One of us visited seven countries (Botswana, Ghana, Kenya, Namibia, Tanzania, South Africa and Zimbabwe) from February to April 2013, met many dozens of African, Chinese and American politicians, officials, businesspersons, workers, scholars, students and

---

43. Deborah Brautigam, *The Dragon's Gift: The Real Story of China in Africa* (New York: Oxford University Press, 2011).

44. Karen L. Harris, 'Not a Chinaman's chance: Chinese labour in South Africa and the United States of America', *Historia* 52(2), (2006), pp. 177–197.

45. Philip Harrison, Khangelani Moyo and Yan Yang, 'Strategy and tactics: Chinese immigrants and diasporic spaces in Johannesburg', *Journal of Southern African Studies* 38(4), (December 2012), pp. 899–925; Yoon Jung Park, 'Boundaries, borders and borderland constructions: Chinese in contemporary South Africa and the region', *African Studies* 69(3), (2010), pp. 457–479.

46. Ali Askouri et al., *African Perspectives on China in Africa* (Cape Town: Pambazuka Press, 2007); Axel Harneit-Sievers, Stephen Marks and Sanusha Naidu, eds, *Chinese and African Perspectives on China in Africa* (Cape Town: Pambazuka Press, 2010).

47. Chris Alden, *China in Africa: Partner, Competitor or Hegemon?* (London: Zed Books, 2007).

48. Garth Shelton and Claude Kabemba, eds, *China, Southern Africa and Extractive Industries* (Johannesburg: SARW, 2012), pp. 228–237.

49. A good outlet of such works has been the South African Journal *African East-Asian Affairs—The China Monitor*.

journalists there, and visited many sites and institutions related to Chinese activities in Africa. We also carried out interviews or discussions with scholars from several other African nations (Burkina Faso, DR Congo, Togo and Zambia).

The findings of our field research tend to support much of the existing literature and, we believe, have also generated new understandings about African perception of China's presence in Africa. Our research has offered clues to help answer questions about the implications and long-term sustainability of China's presence in Africa, and about the prospects for the much speculated China–West competition for power and influence in Africa.

Major findings from the field

There is clear evidence that Beijing is making extensive efforts to influence African views and improve its position, to win favors, in Africa. In 2013, PRC President Xi Jinping outlined in Africa his new initiatives for exchanging ideas of governance with Africans, enhancing personnel exchanges and movements, and educating young Africans and cultivating future African leaders to protect Sino–African friendship.[50] The Chinese media network, including Sida International TV (based in Nairobi and headquartered in Beijing), is covering many African communities. Beijing has funded numerous educational and cultural institutions in Africa. Chinese diplomats have actively forged largely cooperative partnerships with the Chinese diaspora throughout Africa. The real size of China's aid to Africa during 2000–2011 was revealed in 2013 to be US$75 billion, 'significantly larger than previously estimated'.[51]

Yet, our research suggests that either China does not really have a coherent grand strategy (beyond basically opportunistic and economy-driven efforts of 'going global') to compete with and replace the West in Africa, or that strategy has been poorly designed and even more poorly implemented. Beijing's massive and expensive efforts to influence Africans are often self-defeating and largely cost-ineffective so far. The Ghanaian and Zimbabwean elites and their offspring educated in China (for graduate degrees for free) we met were eager to get 'useful' further training in the United States, for instance. Few university students we met in Africa made China their top choice for further schooling (many wished that the US could offer the same kind of scholarships as the PRC). A senior African politician dryly commented about his colleague (a Chinese-trained Ph.D. who now assists the country's vice president) 'people educated in China tend to run away from China fast when they return home'.

The explosive growth of Chinese business activities and personnel inflow easily cause people to suspect that China is aiming towards some kind of neo-colonialism scheme in Africa. But, beyond significant yet largely still opportunistic gains like profits, markets and raw materials, China's effort at winning hearts and minds in Africa seems to be a tough uphill battle with an uncertain prospect. Chinese citizens

---

50. Xi Jinping, 'Remaining reliable friends and faithful partners forever', speech at the Nyerere Center, Da es Salaam, Tanzania, 24 March 2013.
51. The US aid to Africa during the same period was larger (US$90 billion). Claire Provost and Rich Harris, 'Soft power, hard cash, how China has spent billions on aid and development in Africa', *The Guardian*, (29 April 2013).

in Africa, now estimated to be between 1 and 2 million,[52] mostly segregated from the local populace,[53] are often treated as second-class foreigners by the local hosts, hardly colonial masters.[54] Sometimes, like in Ghana, local police used lethal force to control the activities of Chinese migrant workers.[55] Beijing's political influence in Africa is at best limited and obscure as leaders and officials happily take Chinese money but do little special for the Chinese beyond window-dressing and lip-services, such as the eye-pleasing Chinese signs at the Harare Airport.

It may be just an issue of time—the clumsiness of Beijing as a newcomer. But the low-yield of China's great efforts and massive transfer of funds in Africa seems to be more a structural problem: some key structural and ideational features of China's own internal sociopolitical economy have seemingly contributed to the great expansion of the Chinese presence in Africa but appear to be fundamentally hindering the growth of Chinese power and influence on the African peoples. Almost everywhere in Africa we found eager appreciation of the inflow of China's 'easy' money and praise for the infrastructure projects the Chinese have built, often as gifts, and Beijing's official policy of non-intervention in local sociopolitical affairs,[56] but much less so about the coming of the Chinese people who indeed in many ways seem to be effectively neutralizing Beijing's massive and expensive charm diplomacy.[57]

---

52. Due to the many illegal immigrants from China, the size of the Chinese diaspora living and working in Africa is a guesstimate at best and even unknowable. In South Africa, for example, depending on which government agency or research institution you ask, there were 150,000 or 350,000 Chinese in the country in early 2013.

53. We observed in the field that Chinese predominantly lived by themselves, often in heavily guarded compounds, with limited sociocultural contacts with the local community. In Windhoek, the compound named China Town was like a fortress or jail. We heard very few cases of Chinese–African dating or marriage, and virtually none of the Chinese we interviewed in Africa planned to make Africa their permanent home. We heard of only a handful cases of Chinese being naturalized and most of them were from Taiwan or Southeast Asia. Many Chinese we met in Africa basically spoke no English, let along the local native languages, even after many years.

54. PRC citizens often could not get the convenient landing visa, let along visa-waiver. In Zimbabwe and Tanzania (dubbed by one West African scholar as mini-China), both heavily financed by Beijing, we saw that Chinese pay more in advance for the same visa granted to Americans on landing and they were routinely harassed at the customs for petty bribes.

55. Ghanaian police killed one and arrested over 100 PRC citizens in 2010, for unlawfully working in small gold mines. Chu Xinyan, 'Jiana junjing qiansha zhongguo gongmin' ['Ghanaian military and police shot and killed Chinese citizens'], *Xin Jingbao* [*New Capital News*], Beijing, (15 October 2010). The Ghanaians we met in Ghana in 2013 unanimously blamed the Chinese workers, as did the Chinese officials there. Another major crackdown happened in June 2013 with 169 Chinese arrested [Adam Nossitter, 'Ghana arrests Chinese in gold mines', *The New York Times*, (6 June 2013)], their properties looted and burned, and 'several killed' according to Chinese media that were reporting based on the desperate call for help via social media. Peng Yinru *et al.*, 'Zhongguo taojinzhe de jiana weiji' ['Crisis for Chinese gold diggers in Ghana'], *Diyi caijing shibao* [*First Financial Times*], Shanghai, (6 June 2013).

56. Unconditional aid and trade/investment packages have been a key for Beijing to open doors and gain access in Africa. The political elites in African nations we met seemed to unanimously appreciate and praise this as a welcoming contrast to the West's annoying 'nanny' attitude. However, some African politicians and more intellectuals mentioned the negative impact of Beijing's nice-guy policies, suspecting that 'Beijing actually has other selfish plans' such as to displace the West, retard the much needed sociopolitical reforms in Africa, and extract resources through opaque deals with corrupted local officials.

57. The bulk of negative opinions about China we heard in Africa involves complaints against the activities, business or otherwise, of the many Chinese merchants, workers and opportunists: from counterfeiting, smuggling, tax-evasion, corruption, harsh labor practices, predatory extraction of minerals, to driving local merchants and traders out of business and poaching wild life. See Erin Conway-Smith, 'Chinese eat up Zimbabwe's endangered wildlife', *globalpost.com*, (12 April 2012); Aisia Rweyemamu, 'Kagasheki unveils Chinese ivory haul in Dar es Salaam', *IPP Media*, (5 November 2013); David Brown, *Hidden Dragon, Crouching Lion: How China's Advance in Africa is Underestimated and Africa's Potential Under Underappreciated* (SSI, US Army War College, 2012), pp. 64–71.

While Beijing has worked hard to create and maintain its image of being an 'all-weather' genuine friend and faithful partner to Africans forever,[58] more than a few Chinese shop owners we met simply refused to hire locals for clearly racist reasons. One Chinese grocer in Nairobi told the author that he never even allowed any black customers in his shop because 'they always shoplift'. The lack of good coordination between the cautious and friendly Beijing and the increasingly numerous cowboy-like Chinese fortune seekers in the wild west of Africa is so obvious that Chinese officials in charge of the China–Africa Development Fund were calling it 'a major problem for us'. This reality contrasts interestingly with a rather popular African view that 'the Chinese are so well-disciplined and organized by Beijing to come here to do XYZ at our expenses'.[59] The many Chinese businessmen and workers we met in Africa not only ignored and disobeyed Beijing's official policies, they openly thought that the Chinese officials were in corrupt collision with African officials and 'never care about us' or 'cannot do anything to protect us'.[60] Other research has also confirmed that the massive number of Chinese with very diverse backgrounds coming to Africa are mostly 'independently motivated by their desires to improve their lives' rather than organized agents of the PRC state.[61] It makes us wonder about the nature and future of China's presence in Africa and what that may mean for China's own domestic politics.

Somewhat as expected, we found very little genuine enthusiasm in Africa for Chinese culture and values or the Chinese way of governance and politics, although there were quite a few Africans we met who thought that an emulation of Chinese style state-capitalism and mercantilism would be good for economic growth in Africa. Several African intellectuals we met believed that the so-called China Model could be a good alternative to the Western style governance to enable real economic growth in Africa, yet they also subsequently doubted the sustainability of that model in China itself.[62] There are now Beijing-facilitated experiments of special economic zones (SEZs) in Egypt, Ethiopia, Mauritius, Nigeria and Zambia to hopefully replicate the success story of the Chinese SEZs in the 1980s–1990s.[63] China has also offered to train civil servants for African countries. Beyond that, Beijing's political and policy influence in Africa is largely reduced to gestures of goodwill and *ad hoc* deals, even though there seems to be a widespread speculation in many African countries that Beijing has already unduly bought many officials and officers of the local governments and militaries.

---

58. Xi Jinping, 'Remaining reliable friends and faithful partners forever', speech. He reiterated the same ideas six days later in Brazzaville, Republic of Congo.

59. Interviews with bankers and academicians in Accra, February 2013.

60. When asked 'why come here', Chinese non-state businesspeople and workers in Africa usually list the following: it's freer here, officials are easier to deal with here, easy to make money here, air is cleaner here, or just want to get out of China citing corruption, pollution, over-crowded space, declining returns of investment and the low job-prospects at home.

61. Tu T. Huynh, Yoon Jung Park and Anna Ying Chen, 'Faces of China: new Chinese migrants in South Africa, 1980s to present', *African and Asian Studies* 9, (2010), pp. 286–306.

62. Some interviewees asserted that the Chinese model was 'unsuitable in Africa as Africa has a very different state–society relationship'. Interviews with African economists in Accra, Ghana, Nairobi, Kenya, and Cape Town, South Africa, February–March 2013. We heard similar views from politicians in Namibia, Tanzania and Zimbabwe.

63. 'China and Africa', *The Economist Intelligence Unit*.

We have seen few Africans could go beyond saying *nihao* (hello) in Chinese or show any possession of Chinese artifacts. The audience for Chinese TV programming was basically only the Chinese diaspora. Perhaps due to the relatively high price, Chinese restaurants in Africa, unlike in most other places of the world, tend to serve very few locals. Contrary to the fairly common use of the Star-Spangled Banner as decoration by many Africans, we saw no instance of the public displaying the PRC flag in Africa outside of official functions and Chinese-owned buildings. Compared to the ubiquitous presence of Western cultural or merchandise icons such as Coca Cola, KFC or Apple, even the infamously massive flood of Chinese goods seems to be rather stealthy in Africa: in the University of Nairobi, when asked about the maker of their cellphones, all six students initially insisted that their Chinese-made devices were from Korea.

A great dichotomy

There appears to be a great dichotomy in African perceptions about China's presence: China is widely viewed in Africa as a beneficial business partner, investor and donor, contributing significantly to the economic development in many host nations. The so-called Chinese model of authoritarian state-capitalism has been gaining an audience and followings in Africa, as has Chinese culture. Many, including some white Africans, hope that China is representing a new outside influence in Africa that will be decisively different from the role of 'resource-extraction plus political-nanny' of the West in the past.[64] On the other hand, there are growing African complaints against the opportunistic even predatory behaviors by the Chinese in Africa—acts that resemble very much the old Western resource extractions that damage the local business environment, ruin social fabric, and undermine labor rights and employment opportunities. Certain peculiar business practices by the Chinese have created resentments over the depletion of non-renewable resources and bio-diversity.[65] Moreover, there are creeping suspicions about China's geopolitical objectives and the rapidly growing influx of Chinese migrants, colored with nationalistic and populist reasons, racial and ethno-culturally xenophobic or Sino-phobic feelings.[66] An African leader from Kenya commented in private that Beijing acted nicely and generously to really promote its 'long-term strategic and political objectives in Africa' and the Chinese people who came to Africa as a result of that were only there for quick-profits and resource-extraction, not treating Africans as their equal business partners, so there is now a 'fundamental mismatch between Chinese approaches and African perspectives'.

---

64. Interviews with scholars in South Africa and Botswana, March 2013. An economist from Kenya in Accra in February 2013 strongly echoed this view; three other economists (from Burkina Faso, DR Congo and Togo) present, however, disagreed.

65. Many Africans frequently commented to the author about how 'China is doing just the same as the old Western colonialists'. Also see Sanusi, 'African must get real about Chinese ties'.

66. In Zimbabwe, where a visiting South African NGO activist labeled it 'a part of the new Chinese empire in Africa', a senior leader bluntly told the author that China is treating Africans as peoples lesser than Europeans or Americans and 'we know that and we have lot to be suspicious about China's motives and objectives in Africa'. One DR Congo scholar asserted that many of his countrymen now simply 'hate' the Chinese, especially Chinese retailers. In early 2013, a South African journal *Noseweek* published repeatedly racist essays against Chinese immigrants.

Table 1. Visa fees in Zimbabwe (2013)

Landing visa	
US citizen	US$30
Russian citizen	US$30
Japanese citizen	US$30
UK citizen	US$55*
Other EU citizen	US$30
Canadian citizen	US$75*

Advance visa (14 days before arrival)	
PRC citizen	US$65–161 (400–1,000 RMB)

Note: * The reason why the British are treated less favorably is obvious. Harare dislikes Canadians because Ottawa, since 2008, has frozen the assets of President Robert Mugabe and banned arms transfer to Zimbabwe. (Author's field notes, Harare, Zimbabwe, March 2013.

As a result, Chinese people are often treated less favorably than Europeans and Americans in Africa even in some of China's long-time friends like Zimbabwe where PRC citizens are treated lesser than the openly hostile British and Canadians concerning visa privileges. Zimbabwean politicians and officials agreed that China had quiet but considerable political clout in Harare, even in the military, procured through opaque and massive financial aid. Yet, both Zimbabwean officials and Chinese miners have confirmed that Harare has now quietly adopted polices to restrict Chinese immigration and presence through raising annual license fees and rents to squeeze the Chinese out of the lucrative gold and diamond mines (see Table 1).[67]

China's peculiar sociopolitical values and norms have visibly led to cultural and value frictions and conflicts with local populations, suggesting an outside influence that is somewhat challenging the existing Western-influenced institutions and values regarding governance and social rights. But the Chinese challenge has yet to be meaningfully strong, much less attractive. As evidenced throughout Africa, China seems to offer little lasting alternatives superior or even distinctive to colonialism or neo-colonialism from the West. The Chinese way of real estate development in Angola, for example, has proven that Beijing wastes African money in Africa just like it does with its own at home.[68] Being increasingly Westernized itself and changing constantly at home, China seems to have a hard time developing its own brand of soft power in Africa, despite its explosive presence throughout the continent.

In short, China is well liked and welcomed by African host states as a new source of economic opportunities and an alternative political support. Yet Beijing also faces

---

67. Anecdotally, such fees and rents have been raised '10–25 times' in the past ten years so some Chinese simply abandoned their 99-year land leases and mining licenses. Harare declared that it would start to arrest Chinese business owners in 2014 for working in 'sectors reserved for local people', mostly services including wholesale and retail activities. 'Zimbabwe: Nigerians, Chinese business owners face arrest in January', available at: www.zimeye.org/?p=95055 (accessed 8 January 2014).

68. China built Nova Cidade de Kilamba in Luanda, for half a million residents, in three years. But it is now an empty ghost town, priced way above the local income level and has cost US$3.39 billion (to be paid with Angolan oil). 'Why has China built a ghost town in Africa?', *Daily Mirror*, London, (9 July 2012).

rising suspicion, grievances and even resentments that have led to quiet resistance and policy changes at many levels and sometimes outbursts of open demonstrations and protests labeling China as an economic and cultural imperialist power. While not many in Africa believe the assertion that Beijing is pursuing a neo-colonialist conspiracy to take over and colonize the continent, China is rapidly being viewed and treated by ever more Africans as just another outside power coming to Africa pursuing self-interests, remarkably similar to the West (or to the 'disliked' Indians in East Africa). Some Africans hope for a 'positive competition' between China and the West in Africa so 'we Africans may have more bargaining power in negotiations with the West' but they also wonder if that is possible at all.[69] Beijing's signature, often refreshingly different, approaches, such as building infrastructure for resources, refraining from criticizing or commenting on hosts' internal affairs, and sprinkling generous and unconditional official aid, have earned nearly unanimous warm receptions from the rulers and leaders in Africa. The picture becomes more complicated, however, if the general public and the actual government policies of the local hosts are examined. The activities of the million-strong Chinese who now live and work in Africa have worked to significantly neutralize Beijing's charm diplomacy and cause growing backlashes, stirring up local concerns about Chinese colonialism. Many governments openly (such as Zambia and Algeria) or quietly (such as Zimbabwe and Tanzania) have adopted measures to 'slow down and restrict Chinese immigration' and reduce and even confiscate Chinese properties such as land and mining leasing rights.[70] One African think tank published tellingly on 'how to fight Chinese imports' to save African jobs.[71] Chinese influence in Africa is likely to grow further but this is unlikely to be at the same pace as the growth in trade and investment. In the long run, Beijing faces a steep uphill battle to win hearts and minds, or to replace the reputation and influence of the West in Africa.

Tales from two nations

To further illustrate African perceptions of China's presence, we report below on opinion surveys and field investigations conducted in two African countries that are literally a continent apart: Ghana in West Africa and Madagascar off the coast of East Africa, representing the English-speaking and French-speaking African countries, respectively.

China perceived in Ghana

Ghana was the first British colony in Africa to achieve independence in 1957 and is now one of the fastest growing economies in the world. The China–Ghana relationship dates back to the 1950s and China has been a significant economic partner

---

69. Interviews with African economists in Accra and Nairobi, February–March 2013.
70. Zambia re-nationalized a Chinese coalmine after high-profile labor disputes in February 2013. See Barry Sautman and Yan Hairong, 'Bashing "the Chinese": contextualizing Zambia's Collum Coal Mine shooting', *Journal of Contemporary China* 23(90), (2014), doi: 10.1080/10670564.2014.898897.
71. Zwelinzima Vavi, 'How to fight Chinese imports', available at: www.safpi.org/news/article/2013/vavi-how-fight-chinese-imports (accessed 7 January 2014).

Table 2. Perceptions about China in Ghana

	Std error	P value	Significance
General attitudes towards Chinese	0.5574	0.1194544	Not significant
Views about Chinese employment	0.5217	0.3354689	Not significant
Views about Chinese social behavior	0.5138	0.3723661	Not significant
Views about Chinese infrastructure	0.6813	2.539596e-06	Significant
Views about Chinese contribution to development	0.7609	5.273849e-12	Significant
Views about Chinese business in Ghana	0.5611	0.1009874	Not significant
	t	Test value = 3 Sig. (2-tailed)	Significance
General differences			
High confidence—Chinese business opportunities	0.4279	0.3346	Not significant
High confidence—Chinese new products	2.55	0.005808	Significant
Confidence—Chinese contributing to economic growth	0.4211	0.3371	Not significant
Low confidence—Chinese businesses contributing to social development	−4.427	8.315e-06	Significant
Low confidence—Chinese businesses contributing to good governance	−6.2652	1.347e-09	Significant
Higher expectation—Chinese investment in Ghana	2.1922	0.01483	Significant
Higher expectation—Chinese development projects	2.8817	0.002223	Significant
Lower expectations—Chinese people in Ghana	−0.1287	0.4489	Not significant
Study preference in China to Japan, EU, US, etc.	−1.5065	0.06687	Not significant
Perceptions of China as very different business partner compared to EU and US	3.4868	0.0003083	Significant
Differences in views by profession			
Positive vs. negative views about Chinese employment	12.1079	0.007022	Significant
Negative views—bankers versus businessmen	3.1849	0.07432	Significant
Negative views—bankers versus students	8.5972	0.003367	Significant
Negative views—bankers versus traders	5.363	0.02057	Significant
Positive versus negative views about Chinese social behavior	13.6998	0.003344	Significant
Negative views—bankers versus businessmen	3.8991	0.04831	Significant
Negative views—bankers versus students	11.6596	0.0006387	Significant
Negative views—bankers versus traders	11.6596	0.0006387	Significant
Positive versus negative views about Chinese infrastructure	0.7604	0.3832	Not significant
Positive versus negative views about Chinese contribution to development	0.1471	0.7013	Not significant
Positive versus negative views about Chinese business in Ghana	4.6984	0.1953	Not significant

to Ghana.[72] Beijing set up its West African office of the China–Africa Development Fund in Accra and provided massive credit for Chinese companies in Ghana.

In the fall of 2012, we commissioned a survey study in Ghana about the Ghanaian perception of the Chinese presence, in comparison to the Ghanaian perception of the Western presence there. This study involved 184 Ghanaian stakeholders of Chinese business in Ghana, all based in Accra. The statistical results of the survey are summarized in Table 2.

---

72. Samuel Kwadwo Frimpong, 'Research on relationship between China and Ghana: trade and foreign direct investment (FDI)', *Journal of Economics and Sustainable Development* 3(7), (2012), pp. 51–61.

Nicely supporting the general findings of our field research, the survey indicates that there is not a direct translation of trade and investment/aid into goodwill. Ghanaian views of China vary widely, from an enthusiastic welcoming to apprehension. Many Ghanaians think that the Chinese are more committed to developments in Africa than the Europeans or Americans. However, there are nuanced reservations in Ghana about Chinese practices there. The bankers (the most savvy group of businessmen) tend to be especially skeptical. During our field interviews, Ghanaian bankers showed concerns about China's 'dumping' style of exports, 'rule-violations' and 'taking over' Ghanaian real estate and land, and harsh labor practices—in 2013, a major Chinese construction company paid its local employees only US$70 per month while the legal minimum wage, albeit poorly enforced usually, was US$300.

As we discovered elsewhere in Africa, the interests of Chinese corporations and that of the Chinese government are interestingly often in conflict. It is the Chinese state that pushed and lured Chinese businesses to Africa in the first place, backing Chinese firms in local biddings and providing cheap credit and skilled low-cost labor from China.[73] However, the multiple oversight bureaucracies and competing companies often undermine the objectives of the Chinese government. Chinese companies were often viewed as acting in a fiscally irresponsible manner because they could rely on China's state-controlled banks for financial support regardless of performance. Some Chinese firms in Ghana have been instructed by their provincial governments to make unprofitable bids just to get a foot in the door. Such practices have led to negative perceptions and even resentment among Ghanaians. Chinese construction firms tended to prioritize completing projects quickly and cheaply, resulting in lax safety, poor quality and bad labor practices.

A major concern has been that Chinese firms are not building much local capacity.[74] The Chinese government does encourage Chinese firms to hire local people to create a good social image and some state-owned firms are working on it,[75] yet many Chinese companies behave differently. Unlike other outside investors, Chinese companies were uninterested in spending time and money training local workers for those usually one-time and short-term (less than three years) projects. The public projects sponsored by government agreements make the Chinese firms mostly concerned with keeping to schedules in order to please the politicians who award the contracts. There has been a displacement of African workers by the tightly managed workers brought from China.[76] A self-claimed 'pro-China' senior politician we interviewed in Accra in 2013 voiced strong complaints that the Chinese investors often insist on a majority (60%) control of joint-ventures. He asserted that Ghanaian

---

73. Gill Bates and James Reilly, 'The tenuous hold of China Inc. in Africa', *Washington Quarterly* 30(3), (2007), pp. 37–52; Hannah Edinger and Christopher Burke, *AERC Scoping Studies on China–Africa Relations: A Research Report on Zimbabwe* (Centre for Chinese Studies, University of Stellenbosch, 2008).

74. Margaret Pearson, 'The business of governing business in China', *World Politics* 57(2), (2005), pp. 296–322.

75. Zhiren Zhou, 'Study of government performance management in China: a historical review and critical assessment', *Journal of Public Administration* 1, (2009), p. 4; Antoine Kernen and Katy Nganting Lam, 'Workforce localization among Chinese state-owned enterprises (SOEs) in Ghana', *Journal of Contemporary China* 23(90), (2014), doi: 10.1080/10670564.2014.898894.

76. Chris Alden and Martyn Davies, 'A profile of the operations of Chinese multinationals in Africa', *South African Journal of International Affairs* 13(1), (2006), pp. 83–96.

workers and rural people simply 'dislike even hate' Chinese for the jobs lost to the 'rule-violating' retailers, miners and construction workers from China.

The survey confirms our field observation that different socioeconomic segments of the hosting societies tend to perceive China differently, in clear correlation to the benefits they receive (or lose) from China's presence. Much like elsewhere, government and political elites appreciate the resources and opportunities China has brought to Ghana and thus view the Chinese presence more positively than the lower social classes who have been negatively affected by the increasingly numerous Chinese migrants. Chinese building of infrastructural projects was widely acknowledged and appreciated as such projects create new business opportunities for the locals.[77] Yet Chinese in Ghana were similarly segregated from local communities and fueled suspicions and resentment, justified or unfounded, among locals especially the less informed working class people. The common practice of violating intellectual property rights has created resentment among Ghanaian manufacturers inundated by counterfeit 'African handcraft' goods from China.

China perceived in Madagascar

On the east side of sub-Saharan Africa, Madagascar has had a long and stable relationship with China. Madagascar established diplomatic relations with the PRC in 1972, relatively late in Africa. Representing the French-speaking Africa, Madagascar showcases diverse and increasingly complicated opinions about China and the Chinese presence.[78]

The Chinese came to Madagascar early, at the latest in the mid-nineteenth century, as construction and plantation workers under French rule. The Chinese population grew steadily to reach 60,000 in the 1990s and jumped to over 150,000 by 2010. The Chinese presence has a visible impact even on the Madagascar diet—'ubiquitous shops dedicated to varieties of noodle soup and stir-fried noodles are a "fast food" standard in every city in the country'.[79] China became the biggest trade partner of the country in 2003, surpassing the old colonial master of France, and is now twice as big a trade partner as the United States.

Among Malagasy people, including the old (came before 1990s) Chinese-Malagasy citizens, there is now a complicated feeling that the newly arriving Chinese from the PRC are both a blessing and a curse to the local society: they have brought with them business opportunities but also created rising tensions. As in other African nations, the 'new' Chinese are generally isolated from the local community (hiring workers from China and living by themselves), speaking limited local languages (French and Malagasy) and behaving like typical outsiders making money in competition with the locals. There have been 'occasional anti-Chinese public

---

77. Giles Mohan and May Tan-Mullins, 'Chinese migrants in Africa as new agents of development: an analytical framework', *European Journal of Development Research* 21(4), (2009), pp. 588–605.

78. Our report on Malagasy perception of China relies primarily on the work by Gregory Veeck and Sokhna H. A. Diop, 'Chinese engagement with Africa: the case of Madagascar', *Eurasian Geography and Economics* 53(3), (2012), pp. 400–418.

79. *Ibid.*, pp. 407–408.

Table 3. Grievances against the Chinese in Madagascar (2011–2012)

	Issues and/or themes	Number of articles	Number of related blog comments
1	Undue political influence by Chinese government and Chinese companies	7	137
2	Mineral resource 'grabs' and excessive access to Madagascar natural resources by China and Chinese	5	42
3	Illegal logging and traffic in hardwoods and endangered species	3	54
4	Social conflicts and cultural insensitivity by Chinese towards Malagasy citizens	6	31
5	Monopolization of trade and investments in various sectors by Chinese firms	8	10
6	Counterfeit and substandard Chinese goods sold in Madagascar	2	14
7	Real estate/agricultural 'land grabs' by Chinese firms	1	1
8	Cultural imperialism by China	2	2

Source: Adopted from Veeck and Diop, 'Chinese engagement with Africa', pp. 414–415.

protests that have sometimes led to looting and threats of physical violence' against both the new Chinese and the old Chinese-Malagasy.[80]

In Antananarivo (the capital), the Chinese businesspeople all live together in a compound of Chinatown, and all plan to return to China after three–five years or after making the fortune they planned to make there, reflecting a rather typical outsiders' attitude, even though they all reported to have enjoyed living in Madagascar, citing clean air, good food and the 'friendliness' of local people,[81] similar to what we saw in English-speaking Africa. In Accra, Ghana and Brazzaville, Congo, one report shows the Chinese businessmen all plan to leave for 'richer' places like Canada after making their fortunes in Africa.[82]

Similar to places like South Africa, Chinese in Madagascar have developed worries about anti-Chinese violence and riots. They tend to hire very few locals (in a Chinese-run shopping mall, all of the 60 plus stores hired only Chinese workers). Local, non-Chinese Malagasy and French online media reflect 'clear and growing frictions' between Chinese and their local hosts. The vast majority of the blogs seemed to show anger and frustration directed either toward the Chinese government and or the 'new' Chinese from the PRC, with harsh words and sharp criticisms lamenting 'Chinese economic imperialism'. Those grievances, seen in eight categories in Table 3, are still mostly words only but have led to a couple of college-student-led open protests in 2011.

Conflicts between the locals and the Chinese are hardly that momentous but some have become 'very inflammatory', showing poorly functioning intercultural relations 'on the ground'. Even the life of the old Chinese-Malagasy citizens has now been affected by the anti-Chinese sentiment that has boiled over into street demonstrations and violence. As expected, the local government consistently views China as a source

---

80. Catherine Fournet-Guérin, 'New Chinese immigration in Antananarivo', *China Perspectives* 67, (September–October 2006), pp. 45–57.
81. Veeck and Diop, 'Chinese engagement with Africa', pp. 410–413.
82. 'The Chinese in Africa', *The Economist*.

of economic and political benefits. Consequently, there is now a growing local suspicion that the Madagascar government has given too much to China and lost control of the influx of Chinese migrants and investment.[83] This kind of mismatch and cost-ineffectiveness in China's cultivation of local goodwill is very much the same as what we have found elsewhere in Africa.

Conclusion

China has had an explosive success in expanding its presence, primarily business activities, in Africa over the past two decades. It is now Africa's top trade partner, major investor and big donor, well perceived by the rulers, leaders and elites throughout the continent. Chinese cultural, even political, influence is also clearly rising. This is driven by the creative efforts of the Chinese state and the hard work of millions of industrious Chinese. Two key ingredients of China's success are seen from the field research: there has been a great push by the Chinese state with its diplomatic power, condition-free trade and investment packages, generous (and secretive) financial aid, and non-interference in local politics. There is also the peculiar Chinese ways of conducting business, such as cultivating personal connections and freely using financial means to get around local bureaucracies, that fit well many traditional social and cultural norms and practices in Africa. There appears to be room for a greater Chinese presence in Africa, given the development stage that China and Africa are at and the rise of China's global power position and ambition.

So far, the impact of and the reaction to China's presence and activities in Africa vary significantly across the different sectors of African society. There is clearly a co-relationship between Chinese orders, investment and aid and the level of positive feelings about the Chinese presence. There is also a co-relationship between the sociopolitical status of the Africans and their attitude towards the Chinese presence—the political and business elites seem to be more welcoming and appreciative whereas the nationalistic and liberal-leaning elites and displaced local traders and business owners tend to be much less positive. An appreciation for the Chinese style of authoritarian state capitalism is now detectable in Africa. Nationalistic African criticisms of China are growing on the grounds of environmental protection, labor and other human rights, anti-resource extraction, and anti-corruption.[84] As our report from the field shows, there is also a growing concern, even suspicion, among local elites about Chinese infringement of African intellectual property rights, extracting local resources, colonizing land, retarding local sociopolitical reforms, displacing the West's development of good governance in Africa,[85] and using Africa in Beijing's global competition with the West, particularly the United States. Some even started to have suspicions about the military ties Beijing is funding and cultivating there.

---

83. Veeck and Diop, 'Chinese engagement with Africa', p. 416.
84. 'Illicit activities and high level of corruption' were deemed common among Chinese businessmen in Mozambique. Mafalda Picarra, 'Revisiting Sino–Mozambican cooperation', *African–East Asian Affairs—China Monitor*, South Africa, 72, (June 2012). Algeria banned both Huawei and ZTE 'due to corruption cases'. See *CCS Weekly China Briefing*, South Africa, (15 June 2012).
85. Brown, *Hidden Dragon, Crouching Lion*, p. 4.

The general public in Africa also seems to have complicated feelings about the Chinese presence. There is a clear appreciation for the orders, goods, investments, tourists and humanitarian assistance from China. There is approval and admiration for Chinese culture and Chinese entrepreneurship and business success. However, due to the significant competition for jobs created by the massive influx of Chinese workers and the human-rights-deprived Chinese labor management on the factory floors, in the mines and in the farming fields, there is a rise of resentment among the masses about Chinese people in Africa.[86] Complaints about the sub-standard quality of Chinese goods and construction works as well as other 'grey' even unlawful practices, just like elsewhere in the world, are very common in Africa.

Furthermore, fueled by the sensational cases of land-grabbing, the environmentally unfriendly resources extraction projects, the typical colonial-style ways of living and working in Africa, and the noticeable collusion between Chinese businessmen and local officials (sometimes at the expense of the national interests of the host countries), suspicion and grievances against the Chinese presence in Africa are growing.[87] These negative attitudes are increasingly mingled with nationalistic, xenophobic and even racist feelings on both sides to create a seemingly new wave of opposition, sometimes in the radical name of anti-Chinese cultural and economic imperialism.[88] Although some Africans with Chinese blood have assumed government positions in Africa, such as Manuel Chang who was appointed Finance Minister of Mozambique in 2009,[89] it was only in the fall of 2013 that the first PRC-born Chinese migrant, Yu Hong Wei (Astan Coulibaly) who was naturalized through marriage 30 years ago, ran for public office (member of parliament in Mali) but lost with just 6.5% of the votes in her home district of Segou.[90]

On balance, as long as the rulers and leaders in Africa are sufficiently happy with the Chinese infusion of funds and goods, and as long as the Chinese government continues to be careful, considerate and generous in its conduct in Africa, African perception of China is likely to remain generally positive and friendly. However, the positive perceptions are often literally purchased with huge sums of Chinese money and hence are increasingly costly to Beijing. They have also to race against the growth of societally negative feelings and the exponential growth of various frictions and conflicts between local hosts and the Chinese traders, investors and workers.

It is noteworthy that the negative attitudes towards the Chinese presence in Africa are now increasingly entangled with the local hosts' domestic politics and tribe rivalries. In more democratic African nations, we now see political leaders and

---

86. In Zambia, local farmers accused the Chinese chicken farmers of taking away '90 percent of business' by flooding the market with sub-standard but cheaper poultry products. BBC, *The Chinese are Coming*, documentary.

87. Such suspicion, as a Cape Town-based African scholar put it, is often aggravated by Beijing's politics such as its stance against the Dala Lama.

88. Chinese media reported tense resentments in Malawi where the huge influx of Chinese merchants 'are not welcome at all'. A local merchant was quoted saying that 'a violent battle will happen sooner or later'. Lu Zhengqing, 'Feizhou minzhong ruhe kandai zhongguo shangren?' ['How African people view Chinese merchants?'], *Ershiyi shiji wang* [*21st Century Network*], Beijing, (8 October 2012).

89. 'Manuel Chang, Finance Minister, Mozambique', available at: www.changes-challenges.org/MChang (accessed 4 December 2013).

90. See www.abamako.com/elections/legislatives/2013/election/cercle.asp?R=4&C = 22#gsc.tab = 0 (accessed 4 December 2013); 'Chinese candidate a Shanghai surprise in Mali polls', *New Vision*, AFP, Uganda, (17 November 2013).

factions seizing the grievances against China and running with them for political gains, as the recent developments in Tanzania and Zambia have demonstrated. In that way, China in Africa may turn out to be eventually not that special or unique after all. Beijing is to be deemed and treated in predictably the same way as others before: outsiders who come for their own self-interests. Furthermore, when well-oiled and well-connected friends in power are somehow toppled or replaced, China stands to lose its investments spectacularly, as has happened in Libya recently.

Interestingly, the most vocally anti-China political leaders, such as the President of Zambia, Michael Sata, who famously stated that 'we want the Chinese to leave and the old colonial rulers to return … at least Western capitalism has a human face' before winning his presidency, seem savvy enough to continue welcoming Chinese investment and workers despite their politically incinerating but scoring rhetoric.[91] It is telling that most of the African intellectuals we interviewed tend to suggest that African states ought to 'bargain harder and better' with Beijing. Equally interesting is the finding that the professed friends, like Zimbabwe and Ghana, have actually already started to curtail and restrict the Chinese presence.

The Chinese presence in Africa, complicatedly perceived by the locals, also develops interesting and profound demands and pressures for Beijing. Many Chinese we interviewed in Africa think Beijing is 'soft' and 'neglecting us' with 'no real help' when they were in dire need. Some Chinese in Beijing have already openly called for more concrete government and even military actions to ensure the security of Chinese people and Chinese interests in Africa, especially those 'struggling' non-state enterprises and millions of workers there: 'when (our citizens) are kidnapped, there is no option other than (military) rescuing' like those done by the American SEAL teams in places like Somalia—as the officially bragged 'traditional friendship' (procured with aid) between China and Africa is 'far from enough'.[92] If Beijing reacts or is forced to react, a more assertive Chinese political even military presence may greatly alter the overall situation.

Finally, China's soft power is growing perceptibly in Africa. But the complicated local reception struggling between wanting financial gain and resenting Chinese conduct, seems to suggest that there are clear limits to how much power and influence China can actually acquire in Africa in the long run. In all the African countries we observed, the locals seem to have shown no significant deference and little difference in their general attitude towards China's explosive presence when compared to the relatively stable and even declining presence of the West, indicating an interesting absence of alternative ideas, norm, institutions, values or even distinctive characteristics from the PRC.[93] This may be an issue of time lag but may also be

---

91. 'Chinese–African attitudes: not as bad as they say', *The Economist*, (1 October 2011). Sata later also visited Beijing in April 2013.

92. Nuan He, Gao Hongyan and Jin Xiaonan, 'Zhongguo touzi feizhou daijia yuelai yuegao' ['The cost of Chinese investment in Africa is getting ever higher'], *Zhongguo maoyi xinwen* [*China Trade News*], Beijing, (15 February 2012). Hong Kong media reported that hundreds in Guangxi demonstrated to protest against the PRC embassy's 'inaction' and lack of sympathy when hundreds of Guangxi miners were arrested (some killed) by Ghanaian police; see *Apple Daily*, (7 June 2013).

93. Despite the fact of Beijing's self-assumed 'exceptionalism'. Chris Alden and Daniel Large, 'China's exceptionalism and the challenges of delivering difference in Africa', *Journal of Contemporary China* 20(68), (2011), pp. 26–38.

more telling about the nature and future of China's presence in Africa. One African leader has already warned that, bringing in only financial benefits but little good governance, human rights and political democracy, China could soon be 'rejected' by Africans.[94] Just like the overall endeavor of the rise of China,[95] the prospects for the much-speculated competition with the West for African hearts and minds, if it were to be the case, appears to be still greatly uncertain, increasingly expensive and seriously challenging to Beijing.

---

94. Rene N'Guettia Koussi (director of economic affairs of the African Union Commission), 'Long-term ways China should underpin its aid to Africa', *Europe's World* no. 19, Brussels, (Autumn 2011), pp. 92–96.

95. David Shambaugh, *China Goes Global: The Partial Power* (New York: Oxford University Press, 2013).

China–Africa Cooperation: promises, practice and prospects

SVEN GRIMM

Chinese engagement in African states has increased tremendously over the last decade, much in line with Chinese globalisation strategies and supported by state encouragement and financial support. The size and potential of China as a world power leads to the level of expectations the country faces from the developing world. However, some elements of these expectations are also created through political discourses which emphasise differences with Western countries. The types of promises that the Chinese leadership makes to create such enthusiastic welcome amongst African political leaders are linked closely to the discourse on South–South cooperation. Albeit different from Western development assistance promises and parallel attempts to produce moderate expectations, the current discourse is thus partly sowing the seeds for future disappointment. This article takes a closer look at the discussions around South–South cooperation in China–Africa relations and at key rhetorical features ('mutual benefit'; 'non-interference') and at the practice of this cooperation. It concludes that the Chinese discourse is creating large public expectations in African countries and while China delivers on many projects, its impact on development is less certain. The overall development success of this strategy builds on longer-term success and is implicitly linked to the occurrence of more reforms in Africa. Chinese policy thus 'bets on the future' in their foreign relations with Africa; the success of this strategy is dependent on political circumstances among the partners that are largely beyond Chinese control. In a number of cases, it can thus be expected that currently up-beat political rhetoric is going to meet obstacles that will require adjustments in a discourse that, in its current form, might undermine Chinese credibility if not the core elements of South–South cooperation altogether.

Introduction

The intensified relationship between China and African states since the late 1990s has been subject to much debate and, in some quarters, also much controversy. While some accounts of the interactions are more balanced and comprehensive than others,[1] the opportunities and challenges that the renewed relationship poses for African economies and for the development of states and societies are still discussed.

---

* Sven Grimm is director of the Centre for Chinese Studies at Stellenbosch University in South Africa.
  1. See Deborah Bräutigam, *The Dagon's Gift. The Real Story of China in Africa* (Oxford: Oxford University Press, 2009); David Shinn and Joshua Eisenman, *China and Africa—A Century of Engagement* (Philadelphia, PA: Pennsylvania University Press, 2012); Kwesi Ampiah and Sanusha Naidu, *Crouching Tiger, Hidden Dragon?—Africa and China* (Scottsville: University of KwaZulu-Natal Press, 2008).

The discussion on China–Africa relations includes the Chinese motivation, official justification and politico-economic drivers for interactions as well as the practice and developmental impact of cooperation in African states and societies. All of these factors shape the expectations of African actors, fuelled and fanned—or cautiously managed—by Chinese rhetoric and its perception in Africa. The relationship between Chinese rhetoric and African expectations is an interesting aspect in itself that merits a closer look.

This article explores the practice of cooperation as well as the promises entailed in the public discourse on China–Africa cooperation. It discusses the implications of the reoccurring rhetorical elements in the discourse, and its perception among African state elites. This article is not meant to provide a mere discourse analysis. Rather, it aspires to cast a light on expectations and perceptions in public policies, including in the international arena. These rather 'soft' elements are understood to have an important effect on the relationship and its assessment on the African continent. Are Chinese state actors successfully managing expectations across the African continent? In a diverse setting of 54 states, it is likely that Chinese actions are not perceived homogenously as only positive or negative. Yet, there is potentially something to learn for the Chinese leadership from instances where promises and practice are seen as two different things. These instances, hardly ever discussed in a Chinese setting with limited academic scope for discussion, are likely to influence the prospects of China–Africa relations.

African dependency and African agency

Developing countries usually are short of funding with regard to the tasks they face.[2] Large infrastructure projects—and sometimes even basic state functions—require additional funding, be they public private partnerships, or borrowing. The latter can be based on commercial credits, or soft loans or grants via development cooperation.

Recent discussions acknowledge the growth potential and the current economic growth in a number of African countries. The growth of an average of 6% between 2001 and 2008 is indeed 'the strongest consistent performance since the early 1970s'.[3] African states have experienced a growth in income, as the commodities upon which much of the continent's formal economic activities are based have experienced a spike in global prices. The price boom for mineral resources is driven by high demand in emerging economies that need raw materials (specifically: energy) to keep their industrial growth on track. The demand, however, is also affecting food prices, as emerging economies are seeking agricultural resources to feed their increasingly better-off population that can afford more (and higher quality) food than ever before. While food prices negatively affect poor urban parts of the population, an increase in overall consumption also means opportunities for farmers. The positive trends, however, are not enough to sustain development and to provide sufficient income to

---

2. China is a notable exception with foreign currency reserves of more than US$3 trillion. For China, macro-economic balancing of state interventions is most crucial, avoiding inflationary tendencies, for instance—or, more importantly, preventing a credit and real estate bubble from bursting.

3. Mills Soko and Jean-Pierre Lehmann, 'The state of development in Africa: concepts, challenges and opportunities', *Journal of International Relations and Development* 14, (2011), p. 101.

meet the large infrastructure needs in Africa. The massive needs are partly due to a backlog in investments, and partly come as post-war reconstruction efforts, as was the case in Angola after 2002 and in most parts of the DRC after 2004, for instance.

Public debt is one way of securing funding. Credits are based on the assumption that the debtor will be able to repay the loan; 'softer' credit lines like in development cooperation make the repayment conditions easier by extending the grace period before repayment and/or reducing the interest rate for the acquired debt when compared to the market rate. The assessment of repayment capabilities is a major element in the decision to provide funding or not. As most of Africa is considered a high-risk lending destination, accessibility of African states to international credit is limited. Chinese policy banks are clearly following a different risk assessment and thus are an option. Their overall finance volumes, however, are nowhere near Western aid, and come with their own limitations, as discussed below. Aid monies, both as grants and as concessional loans, are thus often the only option.[4]

Almost half of Africa (21 African states out of then 53) was highly aid dependent in 2005, if we take a minimum of 10% of a state budget being paid by external donor funding as the benchmark for aid dependency.[5] A high dependency of aid is usually understood to limit the agency of governments, as the realisation of national governments' policies hinge on external actors' funding priorities and will have to heed conditionality imposed by various financing agencies. Aid dependency, however, should not be understood as an absence of agency. Albeit limited in their financial range of manoeuvre, African states are no mere objects in international politics.[6] Aid funds potentially loosen the link to domestic tax collection and thus ease the pressure on government to justify expenditure to its constituencies. Even though these funds often come with conditionalities and thus some form of external accountability, disbursement pressure on the donor's side, to some extent, softens rhetorically harsh stances. Last, but not least, debates about reforms on how aid is given in Paris (2005), Accra (2008) and Busan (2011), provide some levers for donor coordination. Keywords such as 'ownership' can be used for agency. When donors 'talk the talk' of pushing recipient governments for more structured programming in policies, recipient governments, in turn, can demand donors to also 'walk the walk', as, for instance, Rwanda does.[7]

One challenge for African states in the emergence of non-Western cooperation partners is thus to explore Chinese interests in African states and to link own interests to a different 'cooperation discourse' in China and elsewhere in order to use it as leverage.[8]

---

4. The limited access to private capital markets is an element that prevents states from turning away from development aid altogether. Aid critics such as Dambisa Moyo in her book *Dead Aid* (London: Penguin Books, 2009) often overlook the limited access to capital markets.

5. World Bank, *World Development Indicators* (Washington, DC: World Bank, 2007), pp. 348–351.

6. Giles Mohan and Ben Lampert, 'Negotiating China: reinserting African agency into China–Africa relations', *African Affairs* [London] 112(446), (2013), pp. 92–110.

7. Sven Grimm, 'Aid dependency as a limitation to national development policy?—The case of Rwanda', in William Brown and Sophie Harman, eds, *African Agency in International Politics* (Oxon: Routledge, 2013).

8. Sven Grimm, Heike Höß, Katharina Knappe, Marion Siebold, Johannes Sperrfechter and Isabel Vogler, *Coordinating China and DAC Development Partners: Challenges to the Aid Architecture in Rwanda* (Bonn: Deutsches Institut für Entwicklungspolitik/German Development Institute, 2010).

Chinese engagement—economic pragmatism as a key driver

The growing scope of Chinese economic engagements in Africa is a relatively recent trend. They link to older political relations, but have substantially transformed the rationale for interactions: the economic motivation in cooperation is more substantial than ever before. Angola is a good illustration of foreign policy shifts in Beijing: after hurtful experiences with a policy driven by ideology, pragmatism has prevailed in Chinese foreign policy since the end of the Cold War.[9] The political dimension is clearly still present in Sino–African relations, particularly after Beijing's insecurity due to the collapse of Communism in Eastern Europe and in the face of the student protests on Tiananmen Square in 1989. However, hard economic interests have come to the fore since the Chinese economic reforms.

The economic nature of engagement has led to some degree of interdependence between China and some African states, even if some of the current debates on China's Africa engagement have historic predecessors.[10] One of the watershed dates in China–Africa relations is thought to be 1993, when China became a net importer of oil; the quest for resources has since become much stronger in Beijing's foreign policy. Since 2007, China has also become a net importer of coal. African oil sources were the easier way to access resources, and one of the goals is to produce the oil rather than having to import from (volatile) global markets. Consequently, trips by the Chinese prime minister to African states were being dubbed 'energy trips' by the Chinese media.[11] Chinese policies to promote international engagement of Chinese big corporations—often referred to as the 'go out' policy of 1998—have incentivised a market-seeking behaviour of strategically important Chinese enterprises, both state-owned and private. The formulation of policy papers has taken quite a pace in China in recent years, including an 'Africa Policy' (2006), as well as 'white papers' on China–Africa economic and trade cooperation (2010) and on China's Aid Policy (2011), amongst others. All these papers emphasise that China is 'the world's largest developing country', and all of them are cautious to avoid association with Western policy-making towards Africa.

Besides structural drivers, such as the need for energy and other resources as well as the wish to foster political relations, agent drivers have to be added to the discussion of Sino–African relations.[12] Actors on the Chinese side nowadays include state agencies, state-owned enterprises (both central and provincial government

---

9. Sofia Fernandes, 'China and Angola: a strategic partnership?', in Marcus Power and Ana Alves, eds, *China and Angola—A Marriage of Convenience?* (Cape Town: Pambazuka Press, 2012), p. 69; Shinn and Eisenman, *China and Africa*, p. 343.

10. In an article published in 1975 in *African Affairs*, Martin Bailey was already discussing the 'Chinese development model', at a time when that model was very different from the growth successes since the late 1980s. At the time, Bailey noted: 'The relationship between Tanzania and China appears to be very one-sided since it is China which is exporting goods, capital, technicians, and ideas to Tanzania while the traffic in the opposite direction is small' [see Martin Bailey, 'Tanzania and China', *African Affairs* 74(294), (1975), pp. 39–50]. The *unidirectional* characteristic of the traffic between China and Africa as a whole has certainly changed since.

11. Zhang Baohui, 'Chinese foreign policy in transition: trends and implications', *Journal of Current Chinese Affairs* 39(2), (2010), p. 56.

12. Johanna Jansson, 'Views from the "periphery": the manifold reflections of China's rise in the DR Congo', in Xing Li and Steen Christensen, eds, *The Rise of China—The Impact on Semi-Periphery and Periphery Countries* (Aalborg: Aalborg University Press, 2012), p. 180.

owned), as well as private economic actors (corporations as well as individuals). Within the state administration, various branches are actors in the relations, which—under the overall domination of the Communist party—are mainly managed by the Ministry of Commerce (MOFCOM) as the main provider of funding.[13] Chinese enterprises are not only passively reacting to state policy; projects can, indeed, be initiated and pushed for by Chinese corporations that seek to diversify their activities rather than being purely state initiated and driven.[14] These enterprises can and do also react to local pressure.[15] Implementation, even in the case of development cooperation projects, is mostly by private corporations that operate as contractors to Chinese projects. Some projects, such as medical teams or Confucius Institutes, are attributed to institutions from certain provinces.[16] Financial supervision and project assessments are conducted in cooperation by a department in the Ministry of Commerce and by EXIM Bank. There is, however, no central state agency to implement projects abroad.

The rhetorical positioning of China–Africa relations from Beijing is clearly within South–South cooperation, which allows the current Chinese leadership to present a continuity in Third World commitment and thinking in China and beyond.[17] This positioning opens policy space over quite strongly standardised Western development cooperation, but also creates expectations amongst developing countries about the mode of cooperation.

The rhetoric of cooperation: South–South cooperation's roots and core elements

Under conditions of substantial global interactions and changing patterns of economic power, the debate on South–South cooperation was revived in the early 2000s. The substantial economic rise of the non-OECD economies has given new impetus to the discussion. The rise of China and India as 'Asian drivers' can, indeed, be regarded as a global 'tectonic shift'.[18] In the current debate, China serves as a projection surface in many instances: as a country that has managed a steep increase in economic development, it is considered an example to follow. This is an expectation formulated by African thinkers—often not much substantiated by the specific reforms that were undertaken in China; Chinese officials are rather cautious and play down the 'model' character and rather emphasise the need for African states to find their own paths to development.

---

13. Suisheng Zhao, 'The making of Chinese foreign policy—actors and institutions', in Ampiah and Naidu, eds, *Crouching Tiger, Hidden Dragon?*, pp. 39–52.

14. Jansson, 'Views from the "periphery"', p. 180.

15. Antoine Kernen and Katy N. Lam, 'Workforce localization among Chinese state-owned enterprises (SOE) in Ghana', *Journal of Contemporary China* 23(90), (2014), doi: 10.1080/10670564.2014.898894.

16. For the medical teams, see Li Anshan, *Chinese Medical Cooperation in Africa*, Discussion Paper 52 (Uppsala: The Nordic Africa Institute, 2011).

17. Gregory T. Chin, 'China as a "net donor": tracking dollars and sense', *Cambridge Review of International Affairs* 25(4), (2012), p. 589.

18. Raphael Kaplinsky and Dirk Messner, 'Introduction: the impact of Asian drivers on the developing world', *World Development* 36(2), (2007), pp. 127–209.

The debate on South–South cooperation (SSC) has its roots strongly in the mid-twentieth century, specifically the movement of non-aligned states and an early anti-colonial conference in Bandung in 1955. In the case of China, the often-evoked Five Principles of Peaceful Co-existence were first formulated in Sino–Indian negotiations over Tibet (1954), and are still considered to be guiding principles for Chinese external relations.[19] The principles are:

- mutual respect for each other's territorial integrity and sovereignty;
- mutual non-aggression;
- mutual non-interference in each other's internal affairs;
- equality and mutual benefit; and
- peaceful co-existence.

The cooperation amongst sovereign states is based on a vision of states peacefully co-existing. The wording of the principles is a reminder of the Westphalian peace accord and reflects the assumption of living in a 'state world', which is strongly reflected in the realist school of foreign policy theory. Despite being fundamentally 'old fashioned' in its state-centric view, the socialist rhetoric of 'people-to-people' cooperation somewhat allows for some interpretations of Hill's observations of the 'changing politics of foreign policy'.[20] The reference to SSC is politically opportune also for China's current leadership, as it is making reference to the anti-colonial roots of the discourse. The 'Eight Principles for China's Aid to Third World Countries'—still referred to as the 'basic principle for China's foreign aid' in the 2011 policy paper—were formulated by then Premier Zhou Enlai during a visit to Mali. They built on the principles of peaceful co-existence and define standards for development cooperation.[21]

The anti-colonial background—and the utilitarian reference to it nowadays—often enough leads to SSC being defined by what it is NOT: it is not similar to Western aid, it is not a substitute for Western aid, and it does claim not to come with conditionalities.[22] A stringent positive definition of South–South cooperation is hard to find; the concept remains broad. Central elements in South–South cooperation (SSC) are 'self-reliance' and 'self-help', as Chaturvedi relates from documents of the

---

19. Zhiqun Zhu, *China's New Diplomacy—Rationale, Strategies and Significance* (Farnham: Ashgate Publishers, 2010).

20. Christopher Hill, *The Changing Politics of Foreign Policy* (New York: Palgrave, 2002).

21. Chin, 'China as a "net donor"', p. 590. The principles for cooperation were codified in 1964. The eight principles (as cited by Chin on p. 590) are:

(1) Emphasize equality and mutual benefit.
(2) Respect sovereignty and never attach conditions.
(3) Provide interest-free or low-interest loans.
(4) Help recipient countries development independence and self-reliance.
(5) Build projects that require little investment and can be accomplished quickly.
(6) Provide quality equipment and material at market prices.
(7) Ensure effective technical assistance.
(8) Pay experts according to local standards.

22. Particularly the latter notion can be questioned, based on the understanding of what constitutes a condition; a key condition to engage with the People's Republic of China is to accept the One China policy, i.e. accepting claims of sovereignty over Taiwan by Beijing.

G77 [group of 77 developing countries] dating back to 1964.[23] Debates of SSC were subsequently heavily influenced by the theory of dependency as developed by Raúl Prebitsch and others, and by structuralist thinking that called for a New International Economic Order in the 1970s. These debates are closely associated with the United Nations Conference on Trade and Development (UNCTAD), originating from similar intellectual sources.

Overall, cooperation among states of 'the global South' has a strong preference for bilateral government-to-government cooperation. Three key principles are identified in SSC, namely 'mutual respect, equality and "win–win" situation'.[24] In the Chinese official rhetoric, these principles often come in the wording of non-interference (as a consequence of 'mutual respect'), and mutual benefits.

Non-interference

The principle most in debate—and lately most facing challenges—is that of 'non-interference'. It is a consequence of assumed equality of partners in South–South cooperation. Non-interference is presented as a key principle for the official discourse on China–Africa relations and can be regarded as one of the strong selling points to African elites, as it is a distinguishing feature from 'Western' aid. Interestingly, Chinese policies are now (to their puzzlement) criticised for a change from previous revolutionary strategy of Maoist times: even if strongly criticised by Western NGOs and political actors, 'this principle means a benevolent gesture to not waging [sic] ideological competition with the western countries in Africa'.[25] While China's leadership clearly does not want the country to be regarded as yet another donor or superpower, it clearly also wants to avoid being accused of 'passivity' or 'aloofness' in its African engagement. Therefore, orthodoxy with regard to 'non-interference' can also become 'an albatross that impinges on the conscience of China's peaceful rise',[26] if Chinese policy is seen as selfish and careless and becomes identical to indifference.

China has resisted international discussions on changed interpretations of and limits to state sovereignty in the 1990s. The debate around a responsibility to protect (R2P) might also have taken place much beyond the African continent. However, it is somewhat reflected in African documents, after the genocide in Rwanda in 1994. Besides the principle of non-interference enshrined in its Constitutive Act[27] (Article 4 g) and other principles that are closely following wording from the Bandung conference, such as 'peaceful co-existence' and the 'promotion of self-reliance' (Articles 4i and 4k), the African Union (AU), however, goes further. The successor to the Organisation of African Unity clarifies that 'non-interference' must not be an

---

23. Sachin Chaturvedi, 'Development cooperation: contours, evolution and scope', in Sachin Chaturvedi, Thomas Fues and Elizabeth Sidiropoulos, eds, *Development Cooperation and Emerging Powers—New Partners or Old Patterns?* (London and New York: Zed Books, 2013), pp. 13–36.
24. *Ibid.*, p. 18.
25. Zhang Chun, 'China engaging into Africa—looking for the future', in Liu Hongwu and Yang Jiemian, eds, *Fifty Years of Sino–African Cooperation: Background, Progress & Significance* (Kunming: Yunnan University Press, 2009), p. 135.
26. Richard Aidoo, 'China's "image" problem in Africa', *The Diplomat*, (25 October 2012).
27. Constitutive Act of the African Union, 2002.

excuse for indifference. The principle of 'non-interference' is coupled with provisions for a 'right of the Union to intervene in a Member State pursuant to a decision of the Assembly in respect of grave circumstances, namely war crimes, genocide and crimes against humanity' (Article 4 h). Sovereignty is thus, in principle, no longer absolute on the African continent, but demands respect for higher values, and some argue that the AU is thus in the process of 'norm localization'.[28]

In a changing global context, for its own domestic reasons, Beijing is still advocating an orthodox understanding of sovereignty and resists this norm localisation with regard to any change in its 'non-interference' norm. In Beijing's realpolitik, however, the need to act as a 'responsible' member of the international community has changed Chinese understanding of non-interference and has somewhat softened the orthodox stance of not getting involved. Mediation between the two Sudans[29] and the exertion of some political pressure in the conflict on both sides to seek a peaceful solution[30] is an illustration of this change, as well as an illustration that the stakes are increasing with increasing Chinese investment in Africa. Libya already offered signs of some 'elasticity of non-interference',[31] as Beijing positioned itself for a post-Gaddafi era by meeting with opposition forces in Qatar even while the self-declared 'King of Kings' was still in power. In practice, China has also changed the engagement with the United Nation's operations in Africa 'from an absolute refusal to support peace operations under any circumstances to a permanent commitment to do so',[32] even if a condition *sine qua non* for Chinese participation remains a UN mandate and the acceptance of the mission by the host government.

Furthermore, a challenge to the orthodox doctrine comes from domestic pressure in China. Chinese trade interests and investments are increasing substantially and the stakes that Chinese actors risk in cases of conflict are thus increasing. A strict 'hands-off' attitude becomes more and more difficult with increased global activity and a population that is directed towards nationalism as a 'surrogate ideology'. Additionally, an increasingly wealthy middle class in China demands accountability for international military action (or the lack thereof), as testified by Chinese micro-blogs. This is not necessarily giving in to proactive 'temptations to help shape and sustain the requisite business environment needed for these investments to flourish',[33] but might simply be reactive at the level of safeguarding assets. The evacuation of 35,860 Chinese employees from Libya during the 2011 Civil War[34] was predominantly geared toward domestic politics and was meant to send a signal from Beijing that it will not tolerate Chinese citizens (and assets) being harmed

---

28. Paul D. Williams, 'From non-intervention to non-indifference: the origins and development of the African Union's security culture', *African Affairs* 106(423), (2007), p. 256.
29. See various issues of the Weekly China Briefing 2011/2012, Centre for Chinese Studies, Stellenbosch University, South Africa (http://www.sun.ac.za/ccs).
30. Saferworld, *China and Conflict-Affected States—Risks and Opportunities for Building Peace*, Saferworld Briefing (London: Saferworld, February 2012), p. 5.
31. Aidoo, 'China's "image" problem in Africa'.
32. Ian Taylor, *China's New Role in Africa* (Boulder, CO: Lynne Rienner, 2009), p. 151.
33. Aidoo, 'China's "image" problem in Africa'.
34. Shaio H. Zerba, 'China's Libya evacuation operation: a new diplomatic imperative—overseas citizen protection', *Journal of Contemporary China* 23(90), (2014), doi: 10.1080/10670564.2014.898900.

abroad. President at the time, Hu Jintao issued the unusual statement that he had ordered government workers 'to spare no efforts to ensure the safety of life and properties of Chinese citizens in Libya'.

With an explicit hint to the development in Libya, a Chinese scholar was arguing for change in the official positions: 'Protecting overseas interests is another big challenge for Chinese investment in Africa. The risks derive not only from market uncertainties but also from political and social instability and even competition from other traditional and emerging powers'.[35] Overall, however, there is no appetite for the shaping of new norms.[36] The 'state-oriented, top-down model of stability'[37] is still the mode of operation for Chinese engagement in Africa, despite changes in context and setting.

Mutual benefit

In its cooperation with other states, Chinese leadership emphasises the 'mutual benefit' in the relationship. The understanding of this principle might be one of the gravest misunderstandings of African (and Western) observers who are used to engaging in and assessing *development assistance* projects. It is unduly simplifying to say that African officials rather put the emphasis on 'mutual', while Chinese officials rather stress the 'benefits'; yet, it illustrates differences in expectations. The Chinese side does not claim to be altruistic ('we earn money by building a road; you get a road'); benefits for the African side are expected to be ensured by the African partner in negotiations. In the context of an Eastern African country, a Chinese official particularly emphasised that equality of partners also includes the aspect of 'we don't owe you'.[38] Chinese assessments of the financial sustainability and other impact assessments are certainly conducted—and increasingly sophisticated, given some negative experiences—when funding decisions are prepared. They are, however, not aiming at informing (or even substituting for) host governments assessments.

The rhetoric of mutual benefit, or 'win–win', does not mean that African actors buy the Chinese interpretations wholesale. Those African officials reporting on negotiations do often enough emphasise that the negotiations were tough and demands by their Chinese counterparts were high.[39] Parts of African governments or civil society might be highly sceptical of Chinese engagement, despite official 'friendship' between China and the respective state, and officials are not naïve. *Africa–Asia Confidential* quoted an internal report by ANC Chairperson Baleka Mbete to the South African governing party's National Executive Committee as saying: 'China conducts international relations on a very pragmatic basis and we must keep this in mind in all our interactions. [The Chinese officials] are very clear that the basis for such relationship is the creation of favourable conditions for the advancement of China'.[40]

---

35. Zhang Chun as quoted in *China Daily*, (7 January 2013).
36. Zhengyu Wu and Ian Taylor, 'From refusal to engagement: Chinese contributions to peacekeeping in Africa', *Journal of Contemporary African Studies* 29(2), (2011), pp. 137–154.
37. Saferworld, *China and Conflict-Affected States*, p. 6.
38. Interview in Beijing, July 2013.
39. For the case of Mauritian investments of Tianli, see Bräutigam, *The Dagon's Gift*.
40. *Africa–Asia Confidential*, (November 2012).

In the debate about mutual benefit, however, another layer of difficulty is added for African governments by the fact that discussions about project implementation details are usually held with enterprises, not with state agents. The enterprises engaged in implementation are likely to seek immediate gains and might thus (re)interpret the official framework as set by the state actors.

The practice of cooperation with Africa: leniency, loans, but not necessarily love

In South–South cooperation terms, partners claim to engage in an eye-level partnership, which excludes prescriptions to partners and interference in domestic affairs. The mere idea that, say, Malawi or Mali or Mozambique and China are equals seems ridiculous; the differences in power potential are obvious. Yet, Beijing's symbolic display of 'mutual respect'[41] makes visits to Beijing a treat for Africa's political elite, and for many a stark contrast to trips to Western capitals. Irrespective of human rights record or levels of corruption: when it comes to political decision makers, first-class treatment is guaranteed in Beijing. With police escorts, VIP treatment and national flags in the street, China buys off any criticism of its practice of cooperation policy in Africa, as a member of the African diplomatic corps in Beijing put it.[42]

Also for lower level officials, the tone and treatment is different from Western lectures: training sessions in capacity development activities are reported to consist of Chinese lecturers presenting *their* experiences with projects and state development. Few if any prescriptions are made, and conclusions are often left with African participants. This is often not an analysis of current strengths and weaknesses, and clearly, the failure and human costs of, say, the 'Great Leap Forward' in the 1950s and the Cultural Revolution in the 1970s are not discussed. Obviously, presentations during these trips have a propaganda function and are skewed towards a purely positive image of the development of modern China and its African interests.

South–South rhetoric lumps a range of activities together as 'win–win cooperation' and thereby deliberately blurs the line between aid and investment. Agreements on Chinese cooperation are often made in government-to-government negotiations that result in package deals, including some aid measures, some commercial loans and some support for strategic investments by key Chinese enterprises.[43] The selling point to African partners is mostly on aid as a prominent feature, or on infrastructure deals, irrespective of their modes of finance. These points help in public discussions and present a benevolent picture of Chinese engagement where it, more often than not, follows a business rationale.

---

41. This is irrespective of what the sentiment might be; see Stephen Chan, 'The Middle Kingdom and the Dark Continent: an essay on China, Africa and many fault lines', in Stephen Chan, ed., *The Morality of China in Africa* (London and New York: Zed Books, 2013), pp. 3–46.
42. Interview in Beijing, July 2013.
43. See Bräutigam, *The Dagon's Gift*; Sven Grimm, *Transparency of Chinese Aid—An Analysis of the Published Information on Chinese External Financial Flows,* CCS Research Report (Stellenbosch: Centre for Chinese Studies, August 2011).

The substance of Sino–Africa relations consists of three elements: trade, provision of development finance and investments. They are discussed separately in the following as much as possible.

Trade

Since the late 1990s, trade volumes have gone from height to height year over year. Trade has soared to more than US$220 billion in 2012.[44] An overview of exports to and imports from China for the top six African trade partners—which together account for more than 60% of all trade—is provided by Cissé.[45]

Positive effects on economic growth are likely, not least due to increasing market value of some sought-after commodities due to demand by emerging economies, and particularly China.[46] The effect, however, will be uneven across Africa and varying according to the economic structures and tradable goods per African country. In any case, growth in trade will not automatically lead to development. The structure of trade is still asymmetric—and, irrespective of South–South rhetoric, strongly resembles the trade patterns with developed countries in the North. While raw materials and unprocessed commodities (e.g. oil, coal, iron ore, copper, cotton) are exported to China, African countries import manufactured goods, machinery and electrical equipment, construction material, clothing, vehicle parts, etc.[47] It is specifically the South African manufacturing sector that has lost market shares in Southern Africa between 1997 and 2010 due to Chinese competition.[48] The structural unsustainability of the trade has, in fact, been criticised by South African President Zuma during the Forum on China–Africa Cooperation meeting in July 2012. Zuma was not the first to denounce this imbalance; his predecessor, Thabo Mbeki, had already made similar statements in December 2006.[49]

Development finance

With regard to the provision of finance for development, China has become a major actor, even if often overrated. For African states, indeed, new and additional sources of finance have opened up. This, however, says little about the terms of engagement and their effect on development. China has lent more money to developing countries in the years 2010–2012 than has the World Bank: loans worth US$110 billion were provided by China Exim Bank and the China Development Bank to enterprises and developing countries, while the World Bank made loan commitments of US$100 billion over that period.[50] This comparison with the World Bank was therefore

---

44. *Africa–Asia Confidential*, (November 2012).
45. Daouda Cissé, *FOCAC: Trade, Investment and Aid in China–Africa Relations*, Policy Briefing (Stellenbosch: Centre for Chinese Studies, May 2012).
46. John Humphrey, *European Development Cooperation in a Changing World: Rising Powers and Global Challenges after the Financial Crisis*, EDC2020 Working Paper 8 (Bonn: European Association of Development Research and Training Institutes, 2010), p. 20.
47. Cissé, *FOCAC: Trade, Investment and Aid in China–Africa Relations*.
48. Rhys Jenkins, Chinese Competition and the Restructuring of South African Manufacturing, DEV research briefing 4 (Norwich: University of East Anglia, August 2012).
49. Chin, 'China as a "net donor"', p. 578.
50. Chaturvedi, 'Development cooperation', p. 14.

intended to cause some ripples in the discussion and stuck with quite a few academics in the developing world, given the doubtful reputation of the World Bank in many quarters there. However, it was not comparing items that necessarily were at the same logical level, as the Chinese loans were provided to countries *and (Chinese) enterprises*, and it remains unclear how much of the loans would have qualified as aid.[51]

In terms of aid as per the Western definition, China was estimated to have provided between US$1.4 billion[52] and US$2.2 billion[53] for the year 2007. This figure might have increased to around US$2–3 billion for China by 2012. The figure clearly illustrates that China provides substantial finance in its own right and has become a player to reckon with. It, however, needs to be seen in the context of more than US$100 billion by OECD countries per annum. Additionally, it is counterproductive to only be staring at China in this context; the figures also have to be placed in the context of increasing global assistance from non-OECD sources that are estimated to have been around US$14.5–17 billion.[54] Despite its rapid growth in volume, the value of donations and grants is substantially lower than other forms of engagement; Beijing has good reasons not to be too transparent on how much aid it provides and, for instance, to treat country-specific data as a state secret.[55]

Investments

When it comes to Chinese investments, the often repeated rhetoric on the distinctive nature of South–South cooperation—also popular amongst African governments, including the South African[56]—is not to be confused with negotiation positions. While the need for external finance and capacity limitation might put African governments in a weaker position towards Chinese (and other external) finance providers, some governments are clear about competing agendas between African development endeavours and Chinese commercial activities.[57] Only few, however, go as far as upfront rejecting deals that are not seen as beneficial enough.

One example for the prevalence of commercial considerations is the Chinese engagement in Angola and how Angola—at least in some instances—has managed the relationship. The country is China's biggest trade partner on the continent and an important destination for Chinese investment. The rhetoric of partnership and friendship can obviously be trumped by commercial considerations on the Angolan side, too; negotiations are based on national goals, and negotiations can be tough, as

---

51. Loans need to fulfil certain criteria within the OECD to qualify as aid, and the World Bank follows that OECD definition whereas Chinese institutions do not. Differences thus exist in what is counted; China does *not* include scholarships in its statistics, which make up a sizeable share of Western aid, for instance. Beijing does, however, include some military cooperation (cf. Grimm, *Transparency of Chinese Aid*; Bräutigam, *The Dagon's Gift*).
52. Bräutigam, *The Dagon's Gift*.
53. Carol Lancaster, *The Chinese Aid System. Essay* (Washington, DC: Center for Global Development, 2007).
54. With reference to a text by Homi Kharas: see, Chaturvedi, 'Development cooperation', p. 14.
55. Grimm, *Transparency of Chinese Aid*.
56. Katharina Hofmann, Jürgen Kretz, Martin Roll and Sebastian Sperling, 'Contrasting perceptions: Chinese, African, and European perspectives on the China–Africa summit', *International Politic and Society* 2, (2007), pp. 75–90; Sven Grimm, *South Africa as a Development Partner in Africa*, EDC 2020 Policy Brief No. 11 (Bonn: EADI, March 2011).
57. Interviews in Eastern Africa, July 2013.

the content of the deal—and often, its attraction—is business, not aid delivery.[58] Angola was in the limelight of interest in the early 2000s for having signed deals with Chinese enterprises to refurbish or construct infrastructure in exchange for oil as repayment for the debt. Discussion was about an 'Angola model' of Chinese investment, providing infrastructure in exchange for resources, with China stepping in at a time when Western institutions were hesitant. Angola turned to China with an urgent need for finance to rehabilitate its infrastructure after the long civil war in 2002 after Western donors did not respond to the call for a donor conference for Angola. However, Angola has not fully replaced Western loans with Chinese credit lines, but rather aspires to balance the two.[59] Angola's Chinese debt is being repaid from an account into which the Angolan government receives payment for 10,000 barrels per day for the first two years of the repayment, and 15,000 barrels per day afterwards at fixed prices agreed as part of the deal. While this type of 'barter trade' was the case to some extent, the debt service touched only 1.5–2.5% of the oil sold by Angola in 2008.[60] In this case, the win–win situation for Luanda meant a quick provision of finance based on the potential of the country.

As Deborah Bräutigam pointed out repeatedly, media coverage is often about the signing of memorandums of understanding, i.e. declarations of intent.[61] When no agreement can be reached (even at later stages), deals fall through, despite wordy announcements. Fernandes relates, for instance, disagreement over the target market of a proposed refinery. While Angolan authorities wanted it to also service the national market, Chinese SINOPEC wanted to export production entirely to China. The Angolan state-owned oil corporation prevented these plans by making use of its pre-emptive rights; this was one project that ultimately thus fell through.[62]

Critical points in the current practice of China–Africa relations

With growing African experiences in cooperation with China particularly over the last decade, rhetoric from Beijing is increasingly unlikely to convince large parts of the population in the face of practical problems attributed—rightly or wrongly—to the Chinese engagement in African states. 'Practice is the sole criterion for testing truth' was the title of a newspaper article by the philosopher Hu Fuming of Nanjing University in 1978 which triggered the reform debates in China and assisted in bringing the Cultural Revolution to an end.[63] A debate around practice and results will also increasingly have to replace the broad South–South rhetoric for a number of actors, including for Chinese external relations.

In recent times, the quality of delivered facilities has increasingly become an issue in China–Africa relations, made particularly public by statements by Botswana's President, Ian Khama, in 2013. Chinese supervision of quality delivery by local

---

58. Bräutigam, *The Dagon's Gift*, pp. 4–5.
59. Lucy Corkin, 'Angolan political elites' management of Chinese credit lines', in Power and Alves, eds, *China and Angola*, p. 53.
60. Fernandes, 'China and Angola', p. 72.
61. Bräutigam, *The Dagon's Gift*, is the most comprehensive source.
62. Fernandes, 'China and Angola', p. 73.
63. Hu Fuming, 'Practice is the sole criterion for testing truth', Guangming Daily, 11 May 1978.

representatives is patchy at best and leaves this task to ill-prepared African administrations—or even actively resists demands for involvement in tendering processes for projects by African governments.[64] With regard to planning certainty, by July 2013, three African governments had managed to commit the Chinese side to five-year plans in development cooperation, i.e. provided a list of national priorities from which Beijing should choose its cooperation projects. This was meant as an attempt to switch from supply-driven to demand-driven assistance and rein Beijing into 'aid coordination', despite their official rejection of the discourse.[65]

In other cases, where government officials were less crafty in negotiations—or more interested in kick-backs—raw deals were struck for the African side at the level of project implementation. One such case that hit the news was the purchase of overpriced airport scanners in Namibia, in which corrupt actors flawed a deal; the deal gained prominence not least as it was linked to the son of then Chinese President Hu Jintao. These incidences taint the image of China amongst African civil society actors. Albeit not foreign to Western actors, the problem arguably is a specific one for China, though, as the delivery of state-negotiated projects is conducted by enterprises. In the public debate, little distinction is made between various Chinese actors,[66] while aid agencies from Western countries are usually distinguished from companies.

The prospects of cooperation: the need for policy changes and improved management

The prospects in China–Africa relations are mixed: trade is likely to further increase—if China continues growing—while investments might become more selective. Policies will have to further evolve, including a possible rephrasing of non-interference, in order to improve the Chinese image and to better manage expectations. Issues appear manageable, as the overall reputation of China as a partner for development is positive in Africa, as polls suggest.[67] Overall, the prevailing pragmatism will allow for adjustments in bilateral relations where needed, without major ruptures in policies. Evolution is more likely than revolutionary change, even in cases where doctrines have to be adapted to twenty-first century settings. The biggest unknown in the relationship, however, is the people-to-people dimension.

The need for policy evolution

Challenges will increase with the number of years of substantial engagement by China, particularly in those cases where Chinese investments and finance did not

---

64. Interviews in Eastern Africa, July 2013.
65. Interviews in Eastern Africa, July 2013.
66. Barry Sautman and Hairong Yan, 'Bashing "the Chinese": contextualizing Zambia's Collum Coal Mine shooting', *Journal of Contemporary China* 23(90), (2014), doi: 10.1080/10670564.2014.898897.
67. Pew Research Center, *Global Attitude Project: Obama More Popular Abroad Than at Home, Global Image of US Continues to Benefit*, (17 June 2010), available at: http://www.pewglobal.org/files/2010/06/Pew-Global-Attitudes-Spring-2010-Report-June-17-11AM-EDT.pdf (accessed 15 January 2013).

bring the promised and expected results. For some countries, a development orientation can be expected despite other governance shortcomings, as was the case in Rwanda or Ethiopia over the last decade. Yet, the jury is still out on the development impact in Angola; currently, the construction of much needed infrastructure leads towards a cautiously positive assessment, despite widespread corruption. Trade figures are impressive in a number of cases, including Zambia and South Africa, but it will be crucial—and much more difficult—to change the structure of trade.

In some states across the continent, China has gained an important economic role—often by stepping into breaches that Western countries leave behind when disengaging with what they see as 'rogue regimes'. This, on the one hand, creates international pressure to align with policy stances that Western powers define. China can mostly afford to ignore the pressure for norm internalisation. Zimbabwe and Sudan are two examples often put in the Western media's limelight.

The softening of the orthodox stance on 'non-interference' possibly will require a rephrasing, if not rethinking, of the doctrine. This will mean to square the circle somewhat, as the hands-off attitude is the key selling point to elites. At the moment, pragmatic action can be taken only to some extent in the current rhetorical setting of 'non-interference' and the supposed 'all-weather friendship' that continues even through periods of hardship. Some hardships to come might affect the increasing number of Chinese actors across Africa. Chinese engagement that ignores internal fault lines will ultimately risk—even if involuntarily—getting caught between these lines; ironically, accusations of being 'neo-colonial' are closely linked to the 'non-interference' behaviour if understood as a carelessness.[68] As Aidoo states, 'sections of African populations disagree with the image of China as a non-meddling altruistic partner', which causes troubles for a country that attributes 'paramount importance' to its image.[69] In the case of South Sudan, the government in Beijing seems to have overcome the initial hostility of the newly established country's government.[70] Reputational gains with the government, however, do not automatically result in fostering peace even with well-intended Chinese engagement. As Wheeler argues: 'Perceptions of where, and to whom, the benefits of economic co-operation are distributed matters more for stability than whether it is delivered at all'.[71] Chinese commentators call for the Chinese government to become more proactive in its engagements.[72]

In practice, the perspective of 'mutual benefit' is often narrowed to the partner government. This creates mid- to long-term risks for the Chinese state, as shortcomings or mistakes of partner governments and of (Chinese) implementing enterprises will cause political problems for China's foreign policy. The political

---

68. Suisheng Zhao, 'A neo-colonialist predator or development partner? China's engagement and rebalance in Africa', *Journal of Contemporary China* 23(90), (2014), doi: 10.1080/10670564.2014.898893.
69. Aidoo, 'China's "image" problem in Africa'.
70. Stephen Kuo, 'Not looking to lead—Beijing's view of the crisis between the two Sudans', in Saferworld, ed., *China and South Sudan* (Saferworld Briefing, August 2012).
71. Thomas Wheeler, 'Development through peace—could China's economic co-operation with South Sudan be more conflict-sensitive?', in Saferworld, ed., *China and South Sudan*, p. 10.
72. Zhiqun Zhu, *China's New Diplomacy*, p. 19.

responsibility for ignorance towards social and environmental implications often lies with the African partner government.[73] In the absence of standards and norms (i.e. externally set benchmarks for national or local procedures), the problem is likely to surface repeatedly in a number of contexts and will often be associated with 'China'. This is a new experience for China's government to digest, even more so as it is not used to strong organised pressure groups in its own national context.

The 'mutual benefit' in cooperation with China is particularly questioned in the more advanced African economies, where concern is raised about small industries being outcompeted by Chinese products. Additionally, and potentially more explosive, is the feeling of trade being channelled through Chinese networks and thereby depriving African small traders of their livelihoods.[74] While the majority of African governments presumably do not want to chase away an increasingly important global player and investor and thus rarely criticise Chinese activities openly, the public is less gracious about perceived and real negative implications of Chinese engagement.

For international relations as a whole—and also for the Chinese government—one of the biggest challenges is that 'government is no longer the sole player in China's new diplomacy'.[75] Chinese cooperation policy is deeply rooted in the South–South cooperation discourse, which comes with a state-to-state preference in cooperation, as argued. However, China's government is not in full control of all activities ascribed to 'China', and is unlikely to gain full control over all actors that are considered Chinese. While this diversification of actors in foreign policy can be regarded as a parallel process to Western experiences,[76] it would be wise to reflect the changes in its institutional setting. The very rationale for China's Africa engagement is to involve Chinese business. Yet, quality delivery is crucial for the 'brand China'. State-driven projects—even if implemented by companies—might be better supervised by dedicated state actors, for which a Chinese development agency could be useful.[77] This institutional change would not necessarily compromise the South–South cooperation stance, as the examples of Brazil, India or South Africa illustrate.[78]

Trade unions such as South Africa's congress of unions (COSATU) voice concern about the impacts of Chinese trade on manufacturing—and thus, employment—in some countries,[79] not unlike in historical examples previously. Yet, the scope of

---

73. For the Bui Dam in Ghana, see Oliver Hensengerth, *Interaction of Chinese Institutions with Host Governments in Dam Construction—The Bui Dam in Ghana*, DIE Discussion Paper (Bonn: Deutsches Institut fuer Entwicklungspolitik/German Development Institute 2011); for Angola, see Corkin, 'Angolan political elites' management of Chinese credit lines', pp. 45–67.

74. For Kenya, see Hofmann *et al.*, 'Contrasting perceptions', pp. 75–90; for Zambia, see Gerard van Bracht, 'A survey of Zambian views on Chinese people and their involvement in Zambia', *African East-Asian Affairs* 1, (2012), pp. 54–97.

75. Zhiqun Zhu, *China's New Diplomacy*, p. 10.

76. Hill, *The Changing Politics of Foreign Policy*.

77. One step could be a clearer differentiation between state and corporate/private actors. The establishment of a state agency for the planning and supervision of development cooperation is increasingly recognised as an option by Chinese analysts.

78. Sven Grimm, John Humphrey, Eric Lundsgaarde and Sarah John de Sousa, *European Development Co-operation to 2020: Challenges by New Actors in International Development*, EDC 2020 Working Paper 4 (Bonn: European Association of Development Research and Training Institutes, May 2009).

79. Ian Taylor, *China and Africa: Engagement and Compromise* (Oxon: Routledge, 2006), p. 105.

Chinese citizens' activities on the African continent is of a different nature and has new challenges.

Managing people-to-people relations

We could argue that the world has seen the economic rise of states before, such as Japan in the 1970s and Korea in the 1980s. However, China is of a different quality: it is a much bigger and more powerful country at the global scale. And the Chinese diaspora of traders and investors comes with numbers of Chinese citizens moving to African states that are unprecedented by previously growing Asian economies.

Particularly the growth of the Chinese community in Africa is a politically contested issue with simmering public discontent. Chinese traders' appearance in all parts of the continent should be regarded as the physical arrival of globalisation across Africa; numbers are not much different to Western experiences with migration. Yet, the quality of the experience is different in Africa, where livelihoods are often much more precarious and thus the feeling of being threatened is easy to nurture. Problems with an existing (and still emerging) diaspora will need to be addressed in a cautious, but proactive manner. Populist politicians can exploit these stereotypes and external actors ('the Chinese') easily become scapegoats for flaws in national political processes. Openly racist remarks by some African writers are shockingly similar in structure to some anti-Semitic rambling at the beginning of the twentieth century.[80] While the xenophobic or openly racist drift is (still) somewhat of an exception, the dangers of xenophobia coupled with popular discontent are apparent and need political reactions beyond merely claiming the existence of 'people-to-people' friendship. Also worth noting in this context is the experience of African traders in China. Their increasing number in China, particularly in Guangzhou and—somewhat less numerous—in Yiwu,[81] equally results in some experiences with xenophobia on the Chinese side and is occasionally flaring up in unrest against (perceived or real) police harassment.

A key element in addressing challenges is a better understanding of 'the other'. The Chinese discussion on relations with Africa appears to be far more positive than what is often debated elsewhere. On the Chinese side, academics often still struggle with access to first-hand information on the debates in Africa—a problem exacerbated by still limited 'on the ground' experiences of Chinese researchers and by censorship on the Chinese side. The fact that concerns are hardly voiced in often rosily depicted relations can be (ab)used to dismiss them as purely a 'Western invention' of problems. This, however, is likely to backfire. Open and critical

---

80. George Ayittey, 'The opposition's opening remarks, an online debate', *The Economist*, (15 February 2010), available at: http://www.economist.com/debate/days/view/465 (accessed 16 January 2013). Ayittey claims the existence of a secret 'Chongqing plan' to 'relocate 12 million Chinese to Africa'. This seems to be linked to the Chongqing economic experiment of rapid urban development in that city, which was expected to result in the 'relocation' of 12 million farmers by 2020. Chongqing's then deputy mayor, Zhou Mubing, was reported to have suggested sending these farmers abroad and to Africa, according to the Hong Kong based *South China Morning Post* of 19 September 2007. Somewhat cynical remarks by one official in China, however, do not constitute a 'secret plan' by the Chinese government.

81. Daouda Cissé, *South–South Migration and Trade: African Traders in China*, CCS Policy Briefing (Stellenbosch: Centre for Chinese Studies, June 2013).

dialogue over projects risks slowing down delivery of projects, as well as putting a strain on partner government administrative and political resources. However, more political opening and reform might foster better inclusion of non-state actors in partner countries.

Conclusions

No other emerging economy is as engaged and visible in Africa as China currently is, and none has gained as much global political power. Consequently, no other 'emerging economy' is facing as strong expectations as China. The mere size of China and its power potential makes it a location for hopes and demands; the size of its economy clearly makes a good deal of the demands.

Some expectations towards China, however, are created by political rhetoric that is also cultivated in Beijing. Being different in the engagement is important as a political selling point, but creates its own challenges. 'Winning the hearts and minds' of other developing countries is a challenging task—and the apparent attempt to re-interpret the 'American dream' into a somewhat opaque 'Chinese dream' could, in fact, also be understood as a scarcity of domestic alternatives to the attraction of the USA. It remains to be seen how much a greater domestic orientation of new leaders in Beijing in 2013 and the attempt to brand a 'Chinese dream' lead to greater power of attraction to the outside world.

China's Africa policy has shown some impressive rhetorical continuity—with flexibility in the practical engagement. Structural changes are likely to be needed in the nearer future if China is to successfully manage the expectations and challenges it faces on the African continent. Successes in development are largely dependent on African reforms and structural changes—and, as the Chinese example shows, these are continuous changes and adaptations to changing economic situations. The Chinese government does not prescribe these changes to partners in Africa, but emphasises the importance of reforms in its own development. With this Chinese approach—according to principles of South–South cooperation—success, once it occurs, will not be attributed to external partners, but will be driven by African governments. In this rationale, China can at best present itself as a reliable partner that provided good advice, somewhat reflecting Deng Xiaoping's teaching of 'Lay low, never take the lead, and bide our time'.[82] In this best case scenario, the relationship will, indeed, be one of mutual trust and friendship with numerous economic opportunities on both sides.

Some precautions against political failures on the African side are made, e.g. measures against corruption—despite the rhetoric around meeting at 'eye-level'. Provision of development finance, for instance, often does not result in transfers of funding to African partners, but rather in calls for tender in China for the delivery of services (infrastructure) in the partner country. While one could argue for this being a rather 'Asian approach' (to varying degrees also practised by Japan and South Korea), this is regarded only as second-best by some African governments, who—understandably—would rather be in control of their planning. Overall, in African

---

82. Zhang Chun, 'China engaging into Africa', pp. 130–146.

states with high degrees of agency, Chinese engagement comes with other (reputational and developmental) challenges.

It is unlikely that there is an easy way out of this quagmire of high expectations fanned by South–South rhetoric and persisting preferences towards low-profile political engagement for the Chinese government.

China Goes to Africa: a strategic move?

JIANWEI WANG and JING ZOU

Entering the twenty-first century, particularly under the reign of Hu Jintao, China began to pursue an increasingly pro-active diplomacy in Africa. Most analysis on China's offensive diplomacy in Africa focuses on Beijing's thirst for energy and raw materials, and for economic profits and benefits. That is why it is often called 'energy diplomacy' or 'economic diplomacy' as if China, just like Japan in the 1980s, became another 'economic animal'. But if one looks at the history of the PRC's foreign policy, Beijing has seldom pursued its diplomacy from purely economic considerations. Is this time any different? This article exams China's diplomacy in Africa from a strategic and political perspective such as its geo-strategic calculations, political and security ties with African countries, peacekeeping and anti-piracy efforts in the region, support for African regionalism, etc. It argues that China's diplomatic expansion in Africa, while partially driven by its need for economic growth, cannot be fully understood without taking into consideration its strategic impulse accompanying its accelerating emergence as a global power. Africa is one of China's diplomatic 'new frontiers' as exemplified by new Chinese leader Xi Jinping's maiden foreign trip to Africa in 2013.

Symptom of China's rise

The most common and prevailing explanation is to attribute the surge of China's African diplomacy to its fast-growing economy and the related thirsty for energy and natural resources. Therefore China's intensified diplomatic activities in Africa are often called 'energy diplomacy' or even 'oil diplomacy'. There is certainly a lot of truth in this.[1] As we all know, China has the fastest developing economy in the world, with an average annual 9.8% growth between 1979 and 2012.[2] As the world's second largest economy, China is already a risen power. It is common sense that with the rapid growing scale of China's economy, it faces a shortage of oil and many other raw materials requiring a steady and massive supply of these resources to sustain its

---

* Jianwei Wang is professor and head of the Department of Government and Public Administration, University of Macau; Jing Zou is a Ph.D. student majoring in international relations at the Department of Government and Public Administration, University of Macau.

1. For more detailed discussion about China–Africa economic relations, please refer to Suisheng Zhao, 'A neo-colonialist predator or development partner? China's engagement and rebalance in Africa', *Journal of Contemporary China* 23(90), (2014), doi: 10.1080/10670564.2014.898893; Fei-Ling Wang and Esi A. Elliot, 'China in Africa: presence, perceptions and prospects', *Journal of Contemporary China* 23(90), (2014), doi: 10.1080/10670564.2014.898888; and Sven Grimm, 'China–Africa cooperation: promises, practice and prospects', *Journal of Contemporary China* 23(90), (2014), doi: 10.1080/10670564.2014.898886.

2. National Bureau of Statistics of the People's Republic of China, 'The great socio-economic changes in China since 1978', *People's Daily*, (6 November 2013).

growth. China's demand for oil has been so rapid that it became an energy importer instead of a net exporter in 1993 and the second largest oil importer in the world behind the US in 2004. China's dependency on foreign oil exceeds 55%[3] and its per-capita proven remaining recoverable reserves of oil and gas are only 7.7% and 7.1% of the world average level, respectively.[4] Chinese analysts never deny the urgency for China to secure new sources of energy and other raw materials. Indeed they think it is a legitimate reason for China to be diplomatically pro-active in regions with rich natural resources. Thus, as early as in the 1990s, Africa was integrated into China's energy strategy—China began to import oil from Africa in 1992, and the quantity increased from 500,000 tons from that year to 708.5 million tons in 2010. China imports an estimated one-third of its oil from Africa.[5]

Besides oil and natural gas, China also needs other natural resources such as timber and raw materials such as copper, nickel, bauxite and so on that are abundant in Africa. According to conservationists and some fairly compelling statistics, China is already importing vast quantities of timber, and imports of industrial wood—used in construction, furniture-making and pulp mills—have more than tripled since 1993.[6] Also because of the demand of economic development, China has invested US$170 million in copper mines in Zambia; it has also invested in cobalt and copper mines in the Democratic Republic of Congo, titanium mines in Kenya and so on.[7] In addition, with a projected increase in population, the loss of vital agricultural land to industry and increasing consumption amongst urbanizing people, Beijing perceives a need to obtain stable sources of key foodstuffs. One solution to this problem is the wholesale embrace of genetically modified crops and investment in agriculture, fisheries and related secondary production facilities in Africa.[8]

Needless to say, the economic motivation behind China's African diplomacy is not just limited to energy and raw materials. China also needs Africa as largely unexplored markets for its surplus goods and capitals. As the Chinese trade analyst Alden put it, 'Chinese products are well-suited to the African market. At the moment, China is in a position to manufacture basic products at very low prices and of satisfactory quality'.[9] Besides being rich in natural resources and raw materials, Africa is also a huge potential market for Chinese products. Trade between Africa and China has been growing dramatically, having increased from US$10.6 billion in 2000 to US$166.3 billion in 2011.[10] China has replaced the US as Africa's No. 1 trading partner in

---

3. 'China's dependency on foreign oil exceeds 55%', *People's Daily Online*, (11 August 2010), available at: http://english.people.com.cn/90001/90778/90862/7100858.html (accessed 12 January 2013).

4. Yangzhe Li, 'Energy conservation in China, philosophy, measures, achievements & cooperation', (12 September 2012), available at: http://annualmeeting.naseo.org/presentations/China.pdf (accessed 13 January 2013).

5. Shelly Zhao, 'The geopolitics of China–African oil', *China Briefing*, (13 April 2011), available at: http://www.china-briefing.com/news/2011/04/13/the-geopolitics-of-china-african-oil.html (accessed 10 January 2013).

6. Judith van de Looy, 'Africa and China: A Strategic Partnership?', ASC Working Paper 67/2006, African Studies Centre, Leiden, The Netherlands, available at: https://openaccess.leidenuniv.nl/bitstream/handle/1887/12883/asc-075287668-172-01.pdf?sequence=2 (accessed January 13, 2003).

7. Piet Konings, 'China and Africa: building a strategic partnership', *Journal of Developing Societies* 23(3), (2007), p. 354.

8. Chris Alden, 'China in Africa', *Survival* 47(3), (2005), p. 149.

9. *Ibid.*, p. 150.

10. 'Sino–Africa energy cooperation is indispensible', *China Daily Online*, (25 July 2012), available at: http://www.chinadaily.com.cn/hqcj/zgjj/2012-07-25/content_6532270.html (accessed 13 January 2013).

2009.[11] In particular, Africa opens opportunities for trade and investment for those Chinese enterprises and companies who are struggling in the domestic market. In other words, the lower level of economic development in some African countries provides opportunities to keep Chinese companies afloat. While most Chinese companies are still too weak to compete with enterprises in more advanced countries, they nevertheless have comparative advantages in market competition in African countries. That is why the Chinese government has taken a lot of measures to encourage Chinese companies to 'go out'. Direct Chinese investment in Africa has risen over the past few years, especially since the China–Africa Cooperation Forum in 2000. And the most recent example is China's decision to set up a China–Africa development fund of US$5 billion to encourage Chinese companies to invest in Africa.[12]

China's economic agenda of its diplomacy in Africa therefore is much broader and deeper than just 'energy diplomacy'. Needless to say, China's economic interests will remain the main focus of its policy towards Africa. Yet economic motivations alone cannot fully explain the surge of this new component in China's overall foreign policy in recent years. China's African diplomacy has many non-economic dimensions that are also important and consequential. In a nutshell, the expansion of China's diplomatic horizon in a large measure is a symptom of China's transformation from a regional power to a global power. For a long time China often took an indifferent attitude towards what was going on in remote areas unless China's narrowly-defined national interests were directly affected. That was one of the reasons why China often abstained from voting in the UN Security Council in the 1970s and 1980s. But with China's growing economic and military power and the resulting enlargement of its diplomatic scale, China's definition of national interest has been evolving. The expectation of the international community on China has also been rising. Now what is taking place in the Middle East, Africa and Latin America could also impact in various ways upon China's economic and security interests. China simply cannot implement its classic principle of 'non-intervention' in its strictest sense any more. The emerging global nature of China's diplomacy is reflected in China's more active participation in activities such as the UN peacekeeping operations and anti-piracy efforts in the region. China's interest in influencing regional and international affairs, including Africa, has also been growing. Just like other major powers, China began to appoint special envoys to handle some hot-bottom regional and international issues. For instance, in 2007 China appointed special envoys for the Darfur issue in Sudan in particular and for African affairs in general. The institutionalization of the special envoy system in China's foreign affairs points to China's transition from a regional power to a global one although China still shies away from publicly recognizing this.

Another factor contributing to and stimulating China's pro-active diplomacy in peripheral areas like Africa is the 9/11 attack and the consequent change in the geo-strategic position of the United States in world affairs. On the one hand, under the

---

11. Information Office of the State Council of the People's Republic of China, *China–Africa Economic and Trade Cooperation (2013)*, available at: http://www.scio.gov.cn/ztk/dtzt/2013/9329142/329145/Document/1345037/1345037.htm (accessed 13 September 2013).

12. 'Chinese President Hu Jintao delivers speech at Beijing Summit of FOCAC', 4 November 2006, available at: http://english.focacsummit.org/2006-11/04/content_4951.htm (accessed 13 January 2013).

heavy influence of neo-conservative foreign policy ideology, the Bush administration took the opportunity to expand its sphere of influence globally, particularly in the Middle East. On the other hand, the war in Iraq greatly strained US relations with its allies and other major powers, alienated America from world public opinion, preoccupied the strategic attention of American policy makers and the bulk of the American military force, and severely weakened US 'soft power' in the world. This situation offered new possibilities for China's diplomacy. During the Jiang Zemin era, Beijing's diplomacy was pretty much America-centered. With 9/11 dramatically changing the strategic landscape and the United States being bogged down in Iraq and Afghanistan and its influence in other geopolitical regions declining, Beijing decided to take advantage of this unexpected historical opportunity to expand its scope of diplomacy so as to add to China's so-called 'strategic space'.[13] Chinese analysts believe that a strong partnership with peripheral regions such as Africa could increase China's room for diplomatic maneuver in dealing with the United States and other Western powers. Some Chinese foreign policy analysts do try to use the 'new frontier' diplomacy to indirectly improve China's posture vis-à-vis the United States and other major powers, 'softening' Washington's policy on China.[14] In other words, they argue that a successful 'new frontier' diplomacy in areas such as Africa is actually conducive to China's relations with major powers such as the United States at the 'center' and may effectively reduce China's strategic dependency and risk in dealing with the United States. From that perspective we should understand China's 'new frontier' diplomacy, including the one in Africa, as a strategic move rather than just an opportunistic tactic.

While China's pro-active diplomacy in Africa reflects its growing interest and power, it also in a way indicates its uniqueness as a rising power. In many aspects, China's rise is asymmetric and strength is often paralleled with weakness. Among other things, from the Chinese perspective, with Taiwan still politically separated from the mainland, China is not a unified country yet. One constant aspiration of all PRC leaders is to realize China's complete unification. Therefore for a long time, one major foreign policy objective of China in Africa was to gain support for the 'one-China principle' and isolate Taiwan diplomatically. This strategic priority, however, has become less important in recent years with the ease of tension between Taiwan and the mainland. Both sides followed a tacit 'diplomatic truce'. With only three African countries still remaining in Taiwan's court, Africa is much less a diplomatic battleground between Taiwan and China. However, the flame could be rekindled with the possible 'regime change' in the 2016 election in Taiwan. Another challenge for China's rise is that its domestic political ideology and practice are often at odds with the West-dominated international mainstream. In the struggle for normative legitimacy in international affairs, Beijing needs Africa's support. Among other things, African countries played a key role in Beijing's successful campaign to avoid UNCHR (United Nations Commission on Human Rights) censure for its human rights record.[15]

---

13. Yiwei Wang, 'Economic diplomacy shows its charm', *Global Times*, (3 December 2004).
14. Fu Chen, 'Latin America: "new land" of Chinese diplomacy', *South Weekend*, (2 December 2004).
15. Joshua Eisenman, 'China's post-Cold War strategy in Africa: examining Beijing's methods and objectives', in Joshua Eisenman, Eric Heginbotham and Derek Mitchell, eds, *China and the Developing World: Beijing's Strategy for the Twenty-first Century* (New York: M.E. Sharpe, 2007), p. 35.

Cultivating more comprehensive relationships with Africa

During much of the Cold War period, China's diplomacy in Africa was characterized by 'politics or ideology in command'. It was largely sustained by the ideological doctrine of supporting the movement of national liberation and promoting world revolution. For that purpose, China's African policy was often dominated by the pure ideological considerations rather than economic rationales. Seeking economic profits and benefits was never the main consideration. Since China adopted its reform and openness strategy in the late 1970s, 'politics takes command' has been gradually replaced by 'economics takes command'. Revolutionary ideology has completely lost its relevance in China's relations with African countries. As analyzed above, one major goal of China's diplomacy is to sustain China's domestic economic development. For that reason, the term 'economic diplomacy' has increasingly obtained currency in China's dealing with African countries. The salience of the 'economic diplomacy' can be seen from the fact that the State Council held a special working conference on China's economic diplomacy towards developing countries in 2004. This was the first time in the history of the PRC.[16]

However, in recent years the Chinese government has begun to realize that economic diplomacy alone is not sufficient to safeguard China's interests in African countries. The chronicle of political instability and military conflicts in African countries imposes constant threat to China's economic and other interests there. To address these issues and minimize their adverse impact, China needs to modify its single-dimensional diplomacy and cultivate more comprehensive relationships with African countries. This can be seen from the *Beijing Action Plan* of the fifth ministerial conference of the FOCAC. In this document, the areas of China–Africa cooperation include political affairs and regional peace and security, cooperation in international affairs, economic cooperation, cooperation in the field of development, and cultural and people-to-people exchanges with each field including many concrete projects.[17] While 'economic diplomacy' is the key, it alone cannot guarantee stable relationships. Beneficial economic relations need to be cushioned by an ever-thickening and comprehensive network of interdependence. One important component of this network is human resources. In recent years, Beijing has invested considerable resources to exercise its 'soft power' by training and influencing the elites in African countries. For example, at the fifth ministerial conference of the FOCAC in 2012, Chinese President Hu Jintao pledged to train 30,000 African professionals in various sectors, offer 18,000 government scholarships, and take measures to improve the content and quality of the training programs.[18] Over the years China has also put a lot of resources into training young African diplomats. While the expectation is not necessarily to produce a 'pro-China elite' in those countries, Beijing does want the African political and economic elites to understand China's situation and policy better to minimize misconceptions and to nurture a more

---

16. Wang, 'Economic diplomacy shows its charm'.
17. 'The Fifth Ministerial Conference of the Forum on China–Africa Cooperation, Beijing Action Plan (2012–2015)', (23 July 2012), available at: http://www.focac.org/eng/ltda/dwjbzjjhys/hywj/t954620.htm (accessed 11 August 2013).
18. *Ibid.*

positive view of China. What's more, to meet the needs of many African countries, China has organized various seminars to train African officials on how to lift the population out of poverty by introducing China's experience of poverty relief and elimination.[19] In addition, cultural exchanges between China and Africa have also flourished. As early as 2000, China signed bilateral cultural cooperation agreements and government annual implementation plans with all the African countries that have diplomatic relations with China. And from 2007 to 2010, China and Africa successfully held five sessions of the 'China–Africa Cultural Focus', three sessions of the 'African Cultural Visitors Program', two sessions of the 'Exchange Program of Chinese and African Cultural Visitors', etc.[20]

In weaving a comprehensive network with African countries, military and security ties stand out as becoming more active and dynamic in recent years. Military exchanges and cooperation between China and African countries involve mutual visits of high-level military officers, military training programs, financial aid and joint military exercises. Since 2000, Chinese military leaders have paid several visits to many African countries, such as Algeria, Nigeria, Egypt, South Africa and so on; in the meantime, lots of African military delegations have also visited China.[21] The military training program is divided into two forms: one is that African military personnel come to China to study and accept training; the other is the Chinese military officers go to African countries to train their solders.[22] Referring to financial aids, among other things, in 2001, China supplied US$1 million to Nigeria to upgrade its military equipment; in 2005, China granted US$600,000 to Liberia for military capability building; in 2007, China Export–Import Bank granted a US$30 million loan to Ghana to buy military equipment and build a communication system between national security agencies; in 2010, China donated US$1.5 million to Mauritania to procure military engineering equipment.[23] In addition, China conducted joint military exercises with several African countries including bilateral joint maritime exercises with South Africa and Egypt, and a humanitarian and medical training operation with Gabon in 2009 named Operation Peace Angel. China also invited the military officers from South Africa and Egypt to observe the Chinese military exercise within China. These two countries also received Chinese navy visits. For example, in 2002 a Chinese destroyer passed through the Suez Canal and docked in Alexandria during the PLA navy's first around-the-world trip. In 2000, during the first ever cruise to Africa, Chinese navy vessels docked in Tanzania and South Africa.[24] As the PLA is an actor under the direct control of the Chinese Central

---

19. The Central People's Government of the People's Republic of China, 'Office of Poverty Alleviation of the State Council held seminar on poverty alleviation for officials from African countries', (5 July 2006), available at: http://www.gov.cn/gzdt/2006-07/05/content_328103.htm (accessed 11 August 2013).
20. Yongpeng Zhang, 'China and Africa should pay more attention to cultural exchange', (8 April 2013), available at: http://ezheng.people.com.cn/proposalPostDetail.do?id=760958 (accessed 12 October 2013).
21. Xuejun Wang, 'Review on China's engagement in Africa peace and security', *China International Studies*, (January/February 2012), p. 78.
22. 'China's growing role in African peace and security', *Saferworld Report*, (January 2011), p. 40, available at: http://www.saferworld.org.uk/downloads/pubdocs/Chinas%20Growing%20Role%20in%20African%20Peace%20and%20Security.pdf (accessed 30 August 2013).
23. *Ibid.*, p. 39.
24. *Ibid.*, p. 40.

Military Commission, these military and security activities with African countries are of a strategic nature in support of the larger foreign, diplomatic, economic and security agenda set by China's leadership.[25] While the growth of Sino–African security and military cooperation has been noticeable, the depth and width of such relations is still unknown due to a lack of transparency.[26]

Related to the increasing comprehensiveness of China's diplomacy in Africa, another new feature in recent years is a transition from the pure bilateral diplomacy to increasingly conscious application of multilateral diplomacy and stronger support for African multilateralism and regionalism. While bilateral relations are still important, Beijing has gradually learnt how to use the multilateral setting to achieve collective results which are not always attainable in traditional bilateral diplomacy. The most interesting example in this regard is the creation of the Forum of China–African Cooperation (FOCAC). FOCAC was established in 2000 with the purpose to 'conform to the changing international situation, meet the requirements of economic globalization and see co-development through negotiation and cooperation'.[27] The forum is convened every three years and China and African countries take turns in hosting the event. The event which drew a lot of the world's attention was the Beijing summit and third ministerial conference of the FOCAC in November 2006. The grand scale of the event visibly raised China's diplomatic profile in the world as well as in Africa. It was the first time that China had received so many state leaders simultaneously and a selective vehicle ban had to be imposed to ensure smooth traffic for the African leaders in Beijing. The Beijing Declaration and Action Plan adopted by the summit elevated the China–Africa relations to a new level of a 'new type of strategic partnership'.

China's interest in the multilateral approach is also reflected in its desire to become involved in regional organizations in Africa and its support for African regionalism. In recent years China has taken action to enhance its relations with regional organizations in Africa, such as the African Union, as these organizations have 'a unique political, moral, and geographical advantage in handling conflict prevention and solution' in their particular region.[28] In 2005, China appointed representatives to the African Union (AU), the Southern African Development Community (SADC), the Economic Community of West African States (ECOWAS) and the Common Market for Eastern and Southern Africa (COMESA), and in November 2008 the first annual AU–China Strategic Dialogue was held in Addis Ababa.[29] Among other things, in 2006, Hu Jintao was committed to building a new conference center for the African Union to 'support African countries in their efforts to strengthen themselves through unity and support the process of African integration'.[30] China also repeatedly appealed to the international community to support African regional and sub-regional organizations in solving regional problems. In addition China also made efforts to

---

25. *Ibid.*, p. 38.
26. *Ibid.*, p. 37.
27. 'Creation of the forum', *China.org.cn*, (10 December 2003), available at: http://www.china.org.cn/english/features/China-Africa/82047.htm# (accessed 13 August 2013).
28. 'China's growing role in African peace and security', *Saferworld Report*, p. 58.
29. *Ibid.*, p. 57.
30. 'Chinese President Hu Jintao delivers speech at Beijing Summit of FOCAC'.

develop its relations with African NGOs. For example, in the framework of the FOCAC, Chinese Young Volunteers Overseas Service Plan started in 2002 and Chinese young volunteers went to Africa in 2005 to provide volunteer service. And in 2007, China and African NGOs held their first direct dialogue meeting in Shanghai.

Contributing to conflict resolution and peace building in Africa

Compared to China's diplomacy in Africa in the past, nowadays Beijing has become more pro-active in facilitating conflict-resolution in Africa. One salient example is China's increasingly significant contributions to peacekeeping operations in Africa. For a long time since China returned to the United Nations in the 1970s, it opposed UN peacekeeping operations regarding them as a manifestation of big power hegemony and imperialist intervention in the domestic affairs of sovereign nation states. This attitude began to change in the late 1980s and early 1990s. It was in Africa that China started getting involved in UN peacekeeping operations. In 1989, China deployed a team of 20 civilian observers to join the 'UN Transition Assistance Group' in support of Namibian independence from South Africa. This was the first time that had China participated in a UN peacekeeping mission.[31] And from then on, China has gradually become the most active of the UN Security Council Big Five in participating in UN peacekeeping operations. Particularly in the twenty-first century, China has significantly increased its contribution in terms of the size of its troops, budget-sharing and improvement of peacekeeping operations.[32] For example, in 2000, China deployed fewer than 100 peacekeepers, but the years thereafter have seen a 20-fold increase.[33] China's contribution to the peacekeeping budget grew from around 0.9% throughout the 1990s, to 1.5% by December 2000, and was above 3% by 2008.[34] China now ranks as the largest personnel contributor to UN peacekeeping operations among the five permanent members of the UNSC and the seventh top provider of financial contributions.[35] As a result of China's growing involvement in peacekeeping operations, China has been paying more attention to the training of peacekeeping personnel. In August 2000, the Chinese Civilian Peacekeeping Police Training Center was established in Langfang, Hebei Province, which normally takes responsibility for the selection and training of the peacekeeping Civpol and Formed Police Unit.[36] Also in June 2009, another peacekeeping center for the Ministry of National Defense was founded in Huairou, Beijing, the first organization of the Chinese People's Liberation Army conducting professional peacekeeping training

---

31. Wang, 'Review on China's engagement in Africa peace and security', p. 73.
32. Ziwen Deng and Cuiwen Wang, 'Why China participated in peacekeeping operations in Africa after the Cold War', *Quarterly Journal of International Politics* 2, (2012), pp. 3–6.
33. International Crisis Group, *China's Growing Role in UN Peacekeeping*, Asia Report No. 166, (17 April 2009), p. 1, available at: http://www.crisisgroup.org/en/regions/asia/north-east-asia/china/166-chinas-growing-role-in-un-peacekeeping.aspx (accessed 30 August 2013).
34. *Ibid.*, p. 8.
35. 'China's growing role in African peace and security', *Saferworld Report*, p. 73.
36. 'A brief introduction to China peacekeeping Civpol Training Center', available at: http://www.mps.gov.cn/n16/n983040/n1372264/n1372567/1501154.html (accessed 30 August 2013).

and international exchanges. Additionally, numerous Chinese officers have taken part in international training courses and exchanges with other peacekeeping countries.[37]

Among China's overall involvement in UN peacekeeping operations, the great majority of peacekeepers have been deployed in Africa. By 2012, it had accumulated to 1,622 persons and the countries involved were Mozambique, Sierra Leone, Congo (DRC), Liberia, Cote d'Ivoire, Burundi, Sudan, Western Sahara, Ethiopia, Eritrea, and so on.[38] By the end of October 2013, China had participated in ten UN peacekeeping operations in Africa with a total manpower of 1,919.[39] It was in Africa where China upgraded its involvement in peacekeeping operations in various ways and from time to time. Actually, China for long time declined to contribute any combat troops to UN peacekeeping operations although its provision of civilian police, military observers and enablers filled a key gap and was important to the viability and success of the missions.[40] However in June 2013, China broke this taboo by announcing that it would send almost 400 medical, engineer and security troops to Mali as part of the UN's stabilization mission there. This is the 24th peacekeeping mission that China has been involved in since 1990 and the first time China had sent security troops.[41] As a result, a total of 135 security forces arrived in Mali on 2 December 2013, and recently a second batch of 245 Chinese officers and soldiers also reached there on 16 January 2014.[42] Another special contribution China has made to the peacekeeping operation in Africa is mine sweeping. In 2007, China launched a de-mining assistance program for Africa, providing training for de-mining personnel and landmine removal devices. China provided minesweeping equipment and funds and training to many African countries including Angola, Mozambique, Chad, Burundi, Guinea Bissau, Sudan, Ethiopia and Egypt. In 2010 the Engineer Command College of the PLA also gave a six-week training course to 21 Sudanese de-miners in China. In fact, although China is not a signatory to the 1997 Ottawa Treaty banning the use, stockpiling, production and transfer of anti-personnel mines, the PLA has been very instrumental to de-mining in African countries.[43]

Because of all these efforts, Chinese peacekeepers are always considered highly professional, well trained and able to work effectively in difficult operational environments.[44] The chairman of UNTAC Akashi Asahi said on one occasion 'without the participation of the Chinese engineers, the UNTAC could not have achieved so much'.[45] With regard to the Chinese contingents in MONUC, UN Under-Secretary-General for Peacekeeping Affairs Guehenno said that China has set a brilliant example

---

37. 'China's growing role in African peace and security', *Saferworld Report*, p. 75.
38. Wang, 'Review on China's engagement in Africa peace and security', pp. 73–74.
39. Statistics from Resources of United Nations Peacekeeping, *Troop and Police Contributors*, available at: http://www.un.org/en/peacekeeping/resources/statistics/contributors.shtml (accessed 11 November 2013).
40. International Crisis Group, *China's Growing Role in UN Peacekeeping*, Introduction, p. i.
41. 'China will send UN peacekeepers to Mali', *Xinhua*, (27 June 2013), available at: http://news.xinhuanet.com/politics/2013-06/27/c_116319215.htm (accessed 18 September 2013).
42. 'China sends 245 peacekeepers to Mali', *Xinhua*, (15 January 2014), available at: http://news.xinhuanet.com/english/china/2014-01/15/c_133047867.htm (accessed 18 January 2014).
43. 'China's growing role in African peace and security', *Saferworld Report*, p. 39.
44. International Crisis Group, *China's Growing Role in UN Peacekeeping*, p. 7.
45. Ping Zhang, 'Remarks on the Chinese People's Liberation Army's participation in UN peacekeeping operations', (March 2007), available at: http://www.docin.com/p-635631031.html (accessed 18 August 2013).

for holding the UN Charter.[46] Also the Liberian President Allen Johnson Sirleaf praised the Chinese contingents as a strong, well-trained, highly-disciplined and professional force and a friendly emissary of the great Chinese people and Army.[47] And when UN Secretary-General Ban Ki-moon was on his sixth visit to China, he said thanks many times at the peacekeeping center of the Ministry of National Defense, noting that China has made great contributions to world peace and security.[48]

Besides increasingly putting boots on the ground in peacekeeping operations, China has often exercised its political influence in the region to facilitate the operation of the peacekeeping missions. For instance, in the case of Sudan, China was able to persuade the reluctant Sudan Bashir regime to accept the UN peacekeeping forces.[49] China also supports efforts to localize peacekeeping operations. The Chinese government consistently endorses the concept and practice of 'solving African problems by Africans'[50] by providing financial support to Africa's independent peacekeeping operations. In 2005 and 2006, China contributed US $400,000 respectively as special donations to the African Union to help with its peacekeeping mission in Darfur. In 2008, China contributed US$300,000 to the AU's peacekeeping mission in Somalia. And in August 2009, besides the US$400,000 donated directly to the AU for its Somali mission, China contributed a separate RMB5 million as logistic support to Uganda and Burundi—the two major troop providers for the AU mission in Somalia (AMIMO).[51]

China's efforts to build peace in Africa do not stop at conflict-resolution such as peacekeeping. The Chinese government realized that long-term peace and stability in Africa is a guarantee of Chinese interests there and a prerequisite to promote the development of Sino–Africa relations, and thus China attached great importance to post-conflict(war) reconstruction in Africa. The Chinese government, large-scaled state-owned enterprises, various banks, and small-scaled and medium-scaled enterprises have participated in the post-conflict reconstruction of many African countries. For example, after the civil war in Sierra Leone, China cancelled several debts and signed at least eight separate agreements between 2001 and 2007 to help its post-war economic recovery. China also donated US$3 million to Liberia in 2004 and a further US$1.5 million in 2006 to support its post-conflict government. Chinese companies helped countries like Sudan, DRC, Sierra Leone, etc., construct transport links, power generation and basic structures such as schools and hospitals for resuming production and development, and to create employment opportunities for young, unemployed men who have little other experience beyond war.[52]

China's active participation in African peacekeeping operations has obviously increased its military presence in the region, but with China's material and strategic interest in the region growing, a more high-profile and direct deployment of military

---

46. *Ibid.*
47. *Ibid.*
48. 'UN chief hails China's peacekeepers', *China Daily Online*, (20 June 2013), available at: http://africa.chinadaily.com.cn/china/2013-06/20/content_16638726.htm (accessed 11 October 2013).
49. International Crisis Group, *China's Growing Role in UN Peacekeeping*, Introduction, p. i.
50. 'China's growing role in African peace and security', *Saferworld Report*, p. 34.
51. Wang, 'Review on China's engagement in Africa peace and security', pp. 74–75.
52. 'China's growing role in African peace and security', *Saferworld Report*, p. 85.

assets is needed to protect China's interest as well as contribute to the public good as a responsible great power. China's decision to join the efforts of the international community to fight pirates in the Horn of Africa is one example in point. In December 2008, the Chinese government deployed two warships and a vessel to the Gulf of Aden for the first time to escort and protect Chinese ships and crews as well as international vessels against possible attack from the Somalia pirates. In addition China joined the Convention for the Suppression of Unlawful Acts against the Safety of Maritime Navigation and ratified the International Code for the Security of Ships and Port Facilities, actively participating in various bilateral and multilateral cooperation to prevent and fight pirates, including launching international conferences to discuss the issue and come up with anti-piracy policy recommendations.[53] By February 2013, Chinese naval escort fleets had escorted 5,046 Chinese and foreign vessels and rescued more than 50 foreign and domestic ships during the previous four years, and by August 2013, China had dispatched 15 teams to carry out these escort missions.[54] The Chinese navy in the Gulf of Aden and off the coast of Somalia conducted some combat operations by more directly using force than the peacekeeping operations.

The anti-piracy mission has also provided the Chinese navy with an excellent opportunity to gain valuable experience of modernization—protecting commercial and national security interests on the sea, learning advanced technology and experiences from other countries through cooperation, securing its commercial and energy lines of supply with the Middle East and Africa[55] and acquiring capabilities to operate in the far seas (such as remote recharging, information transmission, interdepartmental coordination and so on). For some scholars, China's anti-piracy operations have supplied an 'international public good' to benefit not just Chinese ships, but all the ships operating in the gulf, which has in turn safeguarded international free trade and upheld international laws governing the navigation of commercial ships.[56] Here it is important to point out that a shift by the Chinese navy in strategy from 'near-seas active defense' to 'far-seas operations'[57] through anti-piracy in Africa has indeed been improving Chinese maritime capabilities, which have always been worried, doubted and criticized before. The anti-piracy operation off the coast of Somalia is the first known operational deployment of the Chinese military in Africa other than its peacekeeping troops: 'Not since Admiral Zheng He's 15th century mission to Africa has China's navy operated in such far away waters'.[58] The Chinese navy's ongoing anti-piracy operations in the Gulf of Aden have also

---

53. *Ibid.*, p. 36.
54. The Central People's Government of the People's Republic of China, 'Chinese navy escort fleet of the fourteenth batch taking over the escort mission', available at: http://www.gov.cn/jrzg/2013-02/16/content_2332869.htm (accessed 13 August 2013).
55. 'China's growing role in African peace and security', *Saferworld Report*, p. 66.
56. *Ibid.*, p. 67.
57. James M. Pendergast, 'China's anti-piracy mission: turning blue-water theory into practice and the implications for the US navy', a paper submitted to the Faculty to the Naval College in partial satisfaction of the requirements of the Department of Joint Military Operations, 11 May 2010, p. 6, available at: http://www.google.com.hk/url?sa=t&rct = j&q = &esrc = s&frm = 1&source = web&cd = 1&ved = 0CCsQFjAA&url = http%3A%2F%2Fwww.dtic.mil%2Fcgi-bin%2FGetTRDoc%3FAD%3DADA525093&ei = e8A6UpDaEI6MiQejn4FA&usg = AFQjCNGDDIRS2fErm1Z5y_19HZRdDNY1Zw (accessed 30 August 2013).
58. 'China's growing role in African peace and security', *Saferworld Report*, p. 66.

enhanced Beijing's foreign policy capability in crises. During the Libya crisis in 2011, the Chinese missile destroyer *Xuzhou* in the area was easily diverted to escort commercial vessels to carry out the Libya evacuation.[59] These operations also raised the question of whether China should seek a more permanent military presence in the region to support its navy operation. Some Chinese military officials argue that a long-term base in the region would be necessary to avoid the long journeys required to re-supply the Chinese navy.[60]

Strategic competition with the West?

The above cursory analysis indicates that China's diplomacy in Africa is omnibearing rather than one-dimensional. It is much broader than 'economic diplomacy' or 'energy diplomacy'. As a whole, China's 'new frontier' diplomacy in Africa has been successful. Not only has China benefited economically from this new endeavor of diplomacy, politically and strategically China has also obtained more flexibility and maneuverability by new allies and new markets in international political and economic affairs. This new geopolitical reality shaped by the increasing Chinese presence in the region has caused alarms and backlashes that could challenge China's strategic position in Africa in the long run. Beijing is often ill-prepared for dealing with the global repercussion of its 'new frontier' diplomacy in Africa. Very often Beijing failed to foresee such consequences and was slow in reacting and taking effective measures to address the problems.

One typical criticism regarding China's African diplomacy is the charge of Chinese 'neocolonialism'. Complaints about China's actions resembling very much the old Western pursuit of resource extraction in Africa that damage the local business environment and social fabric, especially undermining labor rights and employment opportunities,[61] can be heard in both 'frontier' and 'center' countries. China's aggressive efforts at securing the supply of oil and other natural resources directly from African countries, bypassing the world market and 'locking up' monopoly rights to explore oil, natural gas and other minerals in those countries, have convinced many Westerners that China is pursuing a strategy of 'robbing' natural resources from third world countries. The flood of cheap labor-intensive Chinese products in the markets of African countries has also caused the bankruptcy of local companies and a soaring trade deficit in these countries. And in contrast with Western companies that employ mainly local workers, Chinese companies tend to keep local hiring to a minimum and hire mostly their own professionals and workers, so the local workers always grumble about labor discrimination, despotic treatment, poor work conditions and low wages.[62] As a result, anti-Chinese riots erupt from time to time with Chinese merchants often victimized and the anti-dumping measures against Chinese products have mushroomed.

---

59. Shaio Zerba, 'China's Libya evacuation operation: a new diplomatic imperative—overseas citizen protection', *Journal of Contemporary China* 23(90), (2014), doi: 10.1080/10670564.2014.898900.
60. 'China's growing role in African peace and security', *Saferworld Report*, p. 40.
61. Wang and Elliot, 'China in Africa'.
62. Zhao, 'A neo-colonialist predator or development partner?'.

Chinese officials and analysts understandably tend to deny that China's policy has anything to do with 'neocolonialism'. Instead they argue that China's 'new frontier' diplomacy aims at helping developing countries develop and nurture a mutually beneficial win–win situation. For example, they point out that the economic woes of many small–medium size enterprises in African countries started long before China increased its exports there and such woes are largely a result of economic globalization and China should not be blamed for this. Actually China's high demands for raw materials and primary products have enabled these products to maintain relatively high prices, which is beneficial to African countries. At the same time, because of China's cheap labor, the manufactured goods these countries import from China are also relatively inexpensive compared to the same products from more developed countries. Therefore the so-called 'scissors' effect between primary goods and manufactured goods is not that big compared to the situation between developing countries and advanced countries as existed in the 1950s and 1960s. In addition, China has purchased oil and other natural resources from those countries according to the world market price, which is totally different from the practice of exploiting colonies followed by old colonial powers in the past.[63]

While rebuking the accusation of 'neocolonialism', China also takes concrete actions to demonstrate its good intentions and to reduce suspicion and resentment among the populations in African countries. For example, by 2007, China had written off total debts of RMB16.6 billion (US$2.13 billion) for 44 recipient developing countries, 31 of which are African countries with a total write-off value of RMB10.9 billion (US$1.40 billion). Another debt cancellation of RMB10 billion (US$1.28 billion) for African countries is currently under negotiation. The write-offs in total will cumulate to 60% of all debt obligations to China.[64] At the Beijing Summit of the FOCAC in 2006, Chinese President Hu Jintao offered another package of goodies to African countries, including doubling the 2006 foreign economic assistance to Africa by 2009; providing US$3 billion of preferential loans and US$2 billion of preferential buyer's credits to Africa in the next three years; cancelling debt in the form of all the interest-free government loans that matured at the end of 2005 owed by the heavily indebted poor countries and the least developed countries in Africa that have diplomatic relations with China; and further opening up China's market to Africa by increasing from 190 to over 440 the number of export items to China receiving zero-tariff treatment from the least developed countries in Africa having diplomatic ties with China. He also promised to build more hospitals, schools and funds for providing artemisinin and building malaria preventive and treatment centers.[65] These substantial measures were naturally applauded by the African leaders.

---

63. 'West's notion that "China entered Latin America" is misguided', *China.com.cn*, (16 March 2006), available at: http://www.china.com.cn/zhuanti/115/hpdl/txt/2006-03/16/content_6156484.htm (accessed 21 August 2013). This line of argument is largely supported by By Deborah Brautigam, a leading American expert on China-African relations, in her book, The Dragon's Gift: The Real Story of China in Africa (New York: Oxford University Press, 2009).

64. Helmut Reisen, 'Is China actually helping improve debt sustainability in Africa?', Preliminary draft, p. 1, available at: http://www.iddri.org/Evenements/Conferences-internationales/070706_PaperReisen_confpaysemergents.pdf (accessed 13 August 2013).

65. 'Chinese President Hu Jintao delivers speech at Beijing Summit of FOCAC'.

As part of the criticism of 'neocolonialism', China's practice of not attaching any political conditions to its economic assistance and favorable loans to African countries was criticized by Western countries as helping sustain political repression and economic mismanagement in some of those countries. As reflected in the different approaches towards peace building in some African countries with 'rogue regimes' between China and the Western countries, China pays more attention to economic development such as reducing poverty and resolving unemployment rather than the 'color' of the regime in these countries. But for Beijing, this kind of practice without any additional conditions of course is not something new. China has been consistent on this issue since the 1960s. Indeed, Beijing boasts that this is the major difference between its policy and Western policies towards third world countries and takes it as a main reason why China's economic assistance is more welcomed in African countries. This practice comes from the deep-rooted principle of 'non-intervention' in Chinese foreign policy philosophy. First, it is morally wrong to interfere with the domestic affairs of another sovereignty country. Second, historical experience has shown that such interventions configured in political and economic conditions seldom work. Third, since China's own domestic political system shares some similarities with that of many African countries, Beijing is in no position to criticize others. In this regard, Beijing understands that its practice is creating some new norms which are at odds with Western norms. Although Beijing is not pushing it, it is certainly happy to see the gradual formation of the so-called 'Beijing consensus' based on these norms in contrast with the so-called 'Washington consensus'.[66]

However, with China's economic stakes in African countries getting higher, Beijing is forced to modify its adherence to the 'non-intervention' principle. In other words, China can no longer take an indifferent attitude towards domestic development in African countries as such development could have an adverse impact on China's interests. Chinese scholars point out that the political and economic instability in some African countries constitute major obstacles in deepening China's relations with them. Therefore domestic governance in those countries with extensive economic ties with China should be a concern for Beijing. Of course, Beijing's first and foremost concern about domestic governance, just as in its own domestic situation, is stability rather than democratization. As Hu Jintao told his African counterparts, peace and stability are the preconditions for Africa's development and prosperity and China will spare no effort to support the course of peace and stability in Africa.[67] That is also why China is willing to send more peacekeeping troops to Africa as it is related to China's broad national interest.[68]

Another example illustrating China's dilemma on this 'no condition' and 'non-intervention' issue is the Darfur crisis in Sudan. For a long time Beijing tended to view the Darfur issue from China's narrow economic interests in the country and paid

---

66. 'African countries are forming a "Beijing consensus" foreign assistance with no political and economic conditions', *China Daily Online*, (4 June 2007), available at: http://www.chinadaily.com.cn/hqzg/2007-06/04/content_886772.htm (accessed 11 August 2013).

67. 'President Hu: wide-ranging consensus reached during Beijing Summit', *Xinhua*, (5 November 2006), available at: http://news.xinhuanet.com/english/2006-11/05/content_5293311.htm (accessed 13 January 2013).

68. 'Chinese Foreign Ministry rebukes the criticism that "China's arms flame conflict in Africa"', *nen.com.cn*, (19 December 2006), available at: http://news.nen.com.cn (accessed 11 August 2013).

little attention to world opinion on the crisis. China firmly believed that Sudan's sovereignty should be respected and no UN peacekeeping troops should be sent without the consent of the Sudanese government, and secondly that pressure or intervention would not produce the desired results. But the campaign launched by some American celebrities and NGOs to link China's close economic and military ties with the Khartoum government with its hosting of the 2008 Olympics as a leverage point to put the heat on China caught Beijing off guard. The otherwise glorious '2008 Beijing Olympics' was now associated with the word 'genocide'. Beijing suddenly realized that the domestic development in Sudan could jeopardize broader Chinese foreign policy interest. Here, whether China really had the ability to stop the tragedy in Darfur does not matter; what matters is the perception that holds China responsible for what was happening in Darfur. Facing mounting pressure, Beijing had to compromise its stand on 'non-intervention' to control the damage. Chinese President Hu Jintao personally put pressure on the Sudanese President to cooperate with the United Nations during his visit in February 2007. Realizing the seriousness of the problem, Beijing also appointed a full-time special envoy to handle the Darfur issue. This episode evidently taught Chinese leaders a lesson: China can no longer just do business as usual with African countries while totally ignoring the domestic situation in those countries. 'Domestic affairs' could have an impact on China's tangible and intangible foreign policy interest. For China's own long-term strategic interest, China needs to promote at least 'good and humane governance' in African countries.[69]

What's more, with China's increasingly robust economic presence and pro-active diplomacy in Africa, more and more Chinese citizens work and live in African countries, many of which are in volatile situations. Beijing is obliged to develop its diplomatic and military capabilities to protect its citizens when necessary. This makes China's otherwise rigid doctrine of non-interference even more difficult to implement if the Chinese government does not want to put its citizens in harm's way. The most noteworthy example was the mass evacuation of more than 35,000 Chinese citizens from Libya during the 2011 civil war. At that time the Chinese leadership asked to spare no efforts to ensure the safety of Chinese citizens there. Such a strong posture obviously is a departure from the strict adherence to the principle of non-intervention.[70] Once again the growing presence of Chinese interest in the continent put this traditional principle of Chinese foreign policy under test.

This dilemma can also be seen from another practice of China's diplomacy in Africa—the transfer of Chinese arms to African countries. This has been one of the most controversial aspects of China's engagement in Africa. From 1996 to 2003, Chinese arms sales to Africa were second only to Russia's, making up about 14% (US$900 million) of Africa's total conventional arms imports.[71] However, between

---

69. The action plan adopted at the China–Africa summit did mention that both sides should 'share experience on governance to pursue common development and progress'; see 'Action Plan adopted at China–Africa Summit, mapping cooperation', *Xinhua*, (5 November 2006), available at: http://news.xinhuanet.com/english/2006-11/05/content_5292285.htm (accessed 15 August 2013).

70. Grimm, 'China–Africa cooperation'. doi: 10.1080/10670564.2014.898886

71. Richard F. Grimmett, 'Conventional arms transfers to developing nations, 1996–2003', *CRS Report for Congress*, (26 August 2004), p. 39, available at: http://www.fas.org/man/crs/RL32547.pdf (accessed 15 August 2013).

2004 and 2011, the Chinese arms sales to Africa surpassed Russia, doubled, and increased to US$1,800 million.[72] Although Beijing asserts that its arms exports should be conducive for the importing state's legitimate self-defense, should not undermine peace, security or stability in the recipient's region or the whole world, and should not be used as a means of interfering in the internal affairs of the recipient country,[73] in reality Chinese arms sales do affect the peace and stability in and between many African countries. Arms sales are a result of mutual needs between China and some African nations. China sells weapons to African countries for multiple purposes. Its arms industry needs Africa's arms market for profits. China also uses the arms sales to support some African regimes for its own political and economic interests. Finally Chinese workers in some violent and chaotic African countries also need Chinese weapons to protect themselves. However by doing so, China often indirectly interferes in the domestic affairs of African countries. For example, China helped establish three weapons factories in Sudan. Chinese-made tanks, fighter planes, bombers, helicopters, machine guns and rocket-propelled grenades have intensified Sudan's two-decade-old north–south civil war.[74] Chinese weapons, such as Kalashnikov rifles, rocket-propelled grenades and artillery pieces mounted on Toyota Land Cruisers, also found their way into Chad and were used by Chadian rebels to almost overthrow the government there.[75] It is no secret that China often sells arms to countries with poor human rights records. Some analysts also point out that China does not necessarily sell arms to Africa for the economic benefits as the profits from small arms and light weapons transferred to Africa are very limited. China likely views such sales as one means of increasing its ability to obtain access to significant natural resources, especially oil, and of enhancing its status as an international political power.[76]

Much of these criticisms of China's Africa diplomacy come from Western countries. Here the question is raised about how China's diplomatic offenses in Africa have affected China's relations with major powers, as implied already in some Chinese calculations of the pros and cons found in conducting 'frontier' diplomacy. One interesting byproduct of China's 'new frontier' diplomacy is to stimulate other major powers to be more pro-active in Africa too. High-level government leaders and officials from the United States, Japan, Russia, South Korea and some other countries have begun to show up in Africa more often than before. For example, after Chinese President Hu Jintao visited Africa in May 2006, Japanese Prime Minister Koizumi immediately followed suit.[77] The Russian president visited Africa for the first time in

---

72. Richard F. Grimmett and Paul K. Kerr, 'Conventional arms transfers to developing nations, 2004–2011', *CRS Report for Congress*, (24 August 2012), p. 52, available at: http://www.fas.org/sgp/crs/weapons/R42678.pdf (accessed 15 August 2013).

73. Ministry of Foreign Affairs of the People's Republic of China, *China's Policy and Regulation on Arms Trade*, available at: http://www.fmprc.gov.cn/eng/wjb/zzjg/jks/kjlc/cgjkwt/t410766.htm.

74. Peter S. Goodman, 'China invests heavily in Sudan's oil industry', *Washington Post*, (23 December 2004), available at: http://www.washingtonpost.com/wp-dyn/articles/A21143-2004Dec22.html (accessed 15 August 2013).

75. "Rebels' arms made in China', *The Washington Times*, (27 April 2006), available at: http://www.washingtontimes.com/news/2006/apr/27/20060427-101338-2606r/ (accessed 15 August 2013).

76. Grimmett and Kerr, 'Conventional arms transfers to developing nations, 2004–2011', p. 11.

77. 'In its diplomatic competition in Africa, China won over Japan', *ido.3mt.com.cn*, (9 May 2006), available at: http://ido.3mt.com.cn/Article/200605/show359071c30p1.html (accessed 11 February 2013).

2006. This ice-breaking visit was attempted to restore Russian influence in Africa, as from the 1990s to the first few years of the twenty-first century, there was almost no diplomatic intercourse between Russia and Africa.[78] The Russians concluded that Russia could learn from China's successful African diplomacy.[79] The Korean president also visited Africa in 2006 for the first time in 24 years and then, in the same year, Korea launched the Korea–Africa Forum, similar to the FOCAC.[80] Additionally, major Western countries such as the US, the UK, France and Germany, almost without exception, have increased their economic aid to African countries in recent years.[81] While it cannot be said that these diplomatic initiatives are exclusively driven by China, it is safe to say that the 'China factor' was one of the motivations.

Understandably the apprehension over strategic competition mainly comes from the West which historically was dominant in Africa. In the United States, some Congressmen, foreign policy elites and media were evidently agitated by China's active diplomacy in Africa. They worried that China's gain might be America's loss and because of the cost of war in Iraq and Afghanistan, the United States simply did not have the energy and resources to compete with the Chinese in the region. Actually there was a debate within the administration about China's role in Africa and its implications for American interests. Some officials look favorably at China's growing role in Africa, especially developing infrastructure such as roads, power plants and telecommunications, and think these projects are good for Africa's economic development.[82] They were also not particularly bothered by China's efforts to seek oil and raw materials in Africa.[83] But some American officials were disturbed by China's involvement with 'troublesome states' simply for energy and other economic benefits.[84] Still some cool-headed policy analysts argue that 'like other growing economies, China is a legitimate competitor for natural resources' and 'it is necessary to recognize that the rise of China, India, and other Asian countries changes the strategic and economic environment in Africa' and to adapt to the new situation, the United States should develop a new, more comprehensive and strategic approach towards Africa.[85]

---

78. 'Russian president visited Africa for the first time for expanding its strategic space' *www.ce.cn*, (7 September 2006), available at: http://www.ce.cn/xwzx/gjss/gdxw/200609/07/t20060907_8462513.shtml (accessed 25 December 2013).

79. 'Reading Russia's new diplomacy thrust', *CRI Online*, (23 March 2007), available at: http://gb.cri.cn/12764/2007/03/23/1865@1512125.htm (accessed 11 February 2013).

80. Debin Zhan, 'Korea–Africa Forum will be held in Seoul from November 7 to 9', *Global Times*, (8 November 2006).

81. Chih-heng Yang, 'Japan and China's assistant diplomacy in Africa: a comparative view', *Review of Global Politics* 24, (2008), p. 63.

82. 'Darfur adds to US doubts over Beijing's foreign policy', *Financial Times*, (14 June 2007).

83. 'US State Department officials: China's effort to find oil in Africa does not threaten the United States', (18 January 2006), available at: http://www.china-embassy.org/chn/zmgx/t231605.htm (accessed 11 February 2013).

84. Robert Zoellick, *Whither China: From Membership to Responsibility?*, National Committee on US–China Relations, (21 September 2005), available at: http://usinfo.state.gov/eap/Archive/2005/Sep/22-290478.html (accessed 11 February 2013).

85. Council on Foreign Relations, *More than Humanitarianism: A Strategic US Approach toward Africa*, Independent Task Force Report No.56 (January 2006), p. 52, available at: http://www.cfr.org/world/more-than-humanitarianism/p9302?breadcrumb=%2Fpublication%2Fby_type%2Ftask_force_report (accessed 20 August 2013).

It is interesting to note that sometime Western European countries turned out to be more vocal in criticizing China's 'new frontier' diplomacy given their own colonial history in these areas. France seldom criticized China's foreign policy in the past and has the closest political relations with Beijing, but on China's diplomacy in Africa, senior French government officials became very critical. The French defense minister reportedly accused China of selling too many arms to African countries causing an escalation of military conflicts there. He also charged that the Chinese capital was penetrating many parts of the African economy and China had the evil intention of carving up the mineral resources of Africa.[86] A French presidential candidate was among the first to advocate the boycott of the 2008 Beijing Olympics. China's diplomacy in Africa has also become one of the most discussed issues in the European Parliament. Some EU officials alleged that China's activities in the region broke the traditional balance. Historically Africa was within the European sphere of influence and the EU needed to take more forceful action to prevent the continent from slipping away into Chinese hands. Partly as a response to China, the European Union put forward its 'Africa strategy' in December 2005.[87]

More recently China has encountered another unexpected competitor in Africa—Japan. China–Japan relations have steadily deteriorated since 2010 when China surpassed Japan to become the second largest economy in the world. This power transition, as predicted by the realist IR theories, caused heightened tension and the possibility of conflict between these two Asian neighboring giants. This does not just apply to the traditional thorny issues such as territorial disputes and the war history, but also spills over into other domains and regions such as Africa. Imitating China's grand FOCAC in 2006, Japan began to raise the profile and scale of the TICAD (Tokyo International Conference on African Development) in 2008. Particularly since Shinzo Abe came to power in 2013, he has intensified Japan's diplomatic offensive in Africa largely in competition with China. At the fifth Tokyo International Conference on African Development in 2013, with 40 African heads of state attending, Japan pledged US$32 billion economic assistance to Africa over the following five years. Japan made a direct intervention in the crisis in South Sudan by sending 400 military personnel, a rare move in the traditionally pacifist Japanese diplomacy.[88] In January 2014, Abe made a high profile trip to Africa, the first one by a Japanese prime minister since 2006. During that trip, he declared that 'Africa is the frontier for Japanese diplomacy'.[89] His trip coincided with Chinese Foreign Minister Wang Yi's African tour. They exchanged jabs with each other. The world media widely viewed Abe's trip as an attempt to counter China's rise in Africa.[90] A Chinese diplomat bluntly characterized Abe's trip as part of a 'China containment policy'.[91]

---

86. 'Chinese Foreign Ministry rebukes the criticism that "China's arms flame conflict in Africa"', *nen.com.cn*.

87. 'Watch out of "China Factor"', available at: http://www.globalbizfin.com/staticWebs%5CArticle%5C633068960887656250s2.aspx (accessed 11 February 2013).

88. John Watanabe, 'Japanese PM's aggressive diplomatic push in Africa', *World Socialist Web Site*, (18 January 2014), available at: https://www.wsws.org/en/articles/2014/01/16/japa-j16.html (accessed 3 February 2014).

89. 'Abe pitches "business diplomacy" in Africa to counter China', *The Asahi Shimbun*, (13 January 2014), available at: https://ajw.com/article/behind_news/politics/AJ201401130065 (accessed 3 February 2014).

90. Zhen Li *et al.*, 'Africa becomes the new frontline of Japan's diplomatic offensive', *Global Times*, (10 January 2014).

91. Watanabe, 'Japanese PM's aggressive diplomatic push in Africa'.

As one African analyst put it, Japan's arrival 'signals the end of an era. The era of safe Chinese investment and cheap exports on the continent is over'.[92]

While initially overlooking the global ramifications of China's 'new frontier' diplomacy, Chinese foreign policy elites now are fully aware of the aftermath of this diplomatic initiative and have begun to discuss whether China's new-type strategic partnership with African countries could lead to conflict with Western countries. Some Chinese scholars conclude that if China only engages in trade and economic activities without seeking military bases and territorial concession, it could avoid a strategic rivalry with the United States and other Western powers. China should pursue a strategy of so-called 'non-military mercantilism', doing business while avoiding forming any military or security alliances with African countries.[93] Others suggest that in economic relations, China should follow the principle of 'competitive cooperation'[94] and refrain from establishing exclusive economic 'spheres of influence' in Africa. Indeed, China's diplomacy in Africa is no longer just a bilateral business between China and African countries; it has entered the formal agenda of strategic consultation between China and other major powers. For example, China's diplomacy in Africa has become a topic of regular consultation on African affairs under the framework of the Sino–American Strategic and Economic Dialogue. By the same token, it has become an integral part of EU official documents on EU China policy and EU–China relations. Likewise, the EU and China also conduct strategic dialogue on African issues. All these indicate that China's diplomacy in Africa has obtained strategic importance in world politics.

Conclusion

From the 1950s to the present, there have been not just continuities but also shifts in China's African policy. The importance of Africa in China's overall foreign policy map also experienced ups and downs. Since the 1990s, China has gradually 'rediscovered' Africa. China's African diplomacy became more pro-active, particularly under the reign of Hu Jintao, and has borne some fruit. Also the remarkable strategic position of Africa in China's diplomacy can be seen in the newly elected President Xi Jinping choosing three African countries as his first four-state trip. While this diplomatic offensive was immediately and directly motivated by the economic consideration to meet China's increasing appetite for energy and natural resources, it nevertheless has also brought some political and strategic benefits to Beijing. And it reflects China's growing diplomatic horizon and expanding definition of its national interest while it has been transforming from a strictly regional power to an increasingly global player. However, different from the Soviet Union during the Cold War, China does not have a clear and well-defined strategic blueprint to turn Africa into its sphere of influence. In other words, potential competition or even

---

92. Narcisse Jean Alcide Nana, 'China's oil safari crosses Japan's energy diplomacy in Africa', *The Ghanaian Chronicle*, (15 January 2014), available at: http://thechronicle.com.gh/chinas-oil-safari-crosses-japans-energy-diplomacy-in-africa/ (accessed 3 February 2014).

93. 'View China's African diplomacy from the sudden "loss of Chad"', available at: http://blog.phoenixtv.com/index.php/uid_626063_action_viewspace_itemid_264834 (accessed 11 February 2013).

94. 'Watch out of "China Factor"'.

strategic rivalry with the United States and other Western countries is not a result by design, but by default. The success of China's African diplomacy has alarmed other major powers and caused strong reactions from them. In that way it can be argued that Beijing may have triggered another round of strategic competition among the major powers in the continent. How to avoid possible strategic rivalry with the United States, European powers and even Japan while securing its own new interest and gains in the region is a challenge to the Chinese government. In the meantime, China's success in Africa has also brought its own problems and pitfalls. China's diplomatic fruits are often considered as being purchased with huge sums of money and hence are increasingly costly to Beijing, and in the meantime negative attitudes towards the Chinese presence in Africa are increasingly entangled.[95] Even President Xi acknowledged candidly that the Sino–Africa relationship faced many strains.[96] How to avoid being perceived and treated as a new colonial power by the African people is another challenge to Beijing. In the final analysis, China needs to play a two-level game in Africa: managing its relations with local elites and publics as well as with other major players simultaneously. In other words, China needs to avoid being viewed as both another Soviet Union by other major powers and another colonial power by the local population. For such a daunting task, Beijing does not seem to be fully prepared.

Acknowledgement

This work is supported by a research grant from University of Macau.

---

95. Wang and Elliot, 'China in Africa'.
96. Zhao, 'A neo-colonialist predator or development partner?', citing Chris Buckley, 'China's leader tries to calm African fears of his country's economic power', *New York Times*, (26 March 2013), available at: http://www.nytimes.com/2013/03/26/world/asia/chinese-leader-xi-jinping-offers-africa-assurance-and-aid.html?_r=0#h.

China's Libya Evacuation Operation: a new diplomatic imperative—overseas citizen protection

SHAIO H. ZERBA

This article examines China's response to the 2011 Libya crisis and the emergence of a new diplomatic imperative: overseas citizen protection. Over a 12-day period in February and March 2011, China evacuated more than 35,000 Chinese nationals from civil war torn Libya, testing the overseas crisis management capacity of the Chinese government. Because of increasing domestic pressure to protect the growing population of Chinese citizens abroad, Beijing is developing its diplomatic and military capabilities to manage and mitigate crises overseas. The large-scale Libya evacuation served as a wake-up call for Beijing, that conducting business in high risk countries around the world comes at a price; as a consequence, Beijing will be compelled to reassess its global strategic posture and foreign policy principles.

Introduction

The People's Republic of China protects the legitimate rights and interests of Chinese nationals residing abroad and protects the lawful rights and interests of returned overseas Chinese and of the family members of Chinese nationals residing abroad. [Article 50 of the Constitution of the People's Republic of China (PRC)][1]

Anti-government protests, which eventually deposed the Gaddafi regime, began in Libya on 15 February 2011, soon engulfing the two largest cities of Tripoli and Benghazi. Before the eruption of the conflict, Chinese companies were extensively engaged in labor-intensive projects in Libya such as construction, railways, oil exploration and telecommunication. According to the Chinese Ministry of Commerce (MOFCOM), during the time of the crisis, altogether 75 Chinese

---

* Shaio Hui Zerba is an Assistant Professor of Political Science at the United States Air Force Academy, Colorado Springs, Colorado. Professor Zerba is a Lieutenant Colonel in the United States Air Force with over 16 years of service. Her Air Force assignments include Germany, Japan, New Mexico, California and Washington, DC. The views presented here are solely those of the author and do not represent the United States Department of Defense or the United States Air Force Academy.

1. 'Constitution of the People's Republic of China', *People's Daily*, (4 December 1982), available at: http://english.peopledaily.com.cn/constitution/constitution.html (accessed 21 December 2012).

enterprises reported having a total of 50 large projects in the country, with a contract value of US$20 billion and more than 36,000 Chinese employees.[2] A week into the crisis, many Chinese enterprises in Libya were raided and robbed and dozens of Chinese employees seriously injured. Looters robbed Chinese workers of their cars, computers, cellphones, cash, construction machinery, and set fires to office facilities. MOFCOM estimated the losses exceeded US$1.5 billion.[3] The Chinese leadership quickly moved to protect Chinese citizens in Libya and by 2 March, a total of 35,860 Chinese citizens had been evacuated from Libya.

In the early 2000s, China initiated the 'Go Global' policy to encourage economic growth and promote Chinese investments abroad.[4] As a consequence of this economic strategy, Chinese nationals have increasingly begun to span the globe and many work and live in volatile regions, facing dangerous conditions such as harassment, robbery, kidnapping and death. With the increasing number and frequency of violence against Chinese nationals overseas, public outcry is reverberating for the Chinese government to do more. In order to maintain the popular support of its citizens, the Chinese government is under significant domestic pressure to protect the growing number of Chinese citizens living and working abroad. Thus, the safety of Chinese citizens overseas has become a new diplomatic imperative for Beijing.

'Overseas citizen protection' (海外公民保护)[5] reflects the Chinese government's motto of 'putting people first and running the government in the interest of the people'[6] and is a new diplomatic imperative for China.[7] Overseas citizen protection first came to prominence in 2004 following a deadly attack on 14 Chinese workers in Afghanistan and Pakistan. Since this shocking event, the Chinese government has reinvigorated the overseas citizen protective system and rolled out dozens of overseas evacuation operations, rescuing tens of thousands of Chinese nationals trapped in dangerous conditions such as the riots in East Timor in 2006.[8] China's latest major evacuations occurred in 2011, during the Arab Spring uprisings in Egypt and Libya, rescuing 1,800 and 35,000 citizens, respectively.[9]

---

2. Ding Ying, 'Out of Libya', *Beijing Review*, (10 March 2011), p. 14.

3. *Ibid.*, pp. 11–14.

4. 'China's outward direct investment', in *OECD Investment Policy Reviews: China 2008* (Paris: OECD, 2008), p. 84, available at: http://www.oecd.org/investment/investmentfordevelopment/41792683.pdf (accessed 17 January 2013).

5. The term 'overseas citizen protection' (海外公民保护), refers to efforts by a range of Chinese institutions to assist or evacuate Chinese citizens working abroad. Mathieu Duchâtel and Bates Gill, *Overseas Citizen Protection: A Growing Challenge for China* (Stockholm: Stockholm International Peace Research Institute, February 2013), available at: http://www.sipri.org/media/newsletter/essay/february12 (accessed 1 January 2013).

6. 'China's Libya evacuation highlights people-first nature of government', *Xinhua*, (3 March 2011), available at: http://news.xinhuanet.com/english2010/indepth/2011-03/03/c_13759953.htm (accessed 21 December 2012).

7. China's new diplomatic imperative raises the question of whether protecting Chinese abroad encompasses the overseas Chinese or solely Chinese citizens. Beijing often glosses over the distinctions between citizens and non-citizens by using the term 'overseas Chinese' (华侨) to refer to all ethnic Chinese living abroad. While covering the Libya evacuation, the Chinese controlled media carefully described the evacuation of 'Chinese citizens'. Notably, Chinese media avoided using the ambiguous term 'overseas Chinese' which suggests Beijing does not intend for its new diplomatic imperative to encompass the overseas Chinese.

8. 'China's Libya evacuation highlights people-first nature of government', *Xinhua*.

9. Ding Ying, 'Out of Libya', p. 14.

China's high profile evacuation of Chinese nationals from war torn Libya in 2011 exemplifies Beijing's new diplomatic imperative. To protect Chinese citizens abroad, Beijing is developing its diplomatic and military capabilities to manage and mitigate crises overseas. These expanding capabilities may signal a possible shift in Beijing's long-standing low profile, noninterventionist policies in favor of more proactive and assertive policies to protect Chinese interests abroad. This article explores Beijing's new diplomatic imperative through the lens of the Libya evacuation case by examining: (1) China's expanding interests in Africa and abroad; (2) the growing domestic pressure placing the safety of Chinese citizens overseas on the political agenda; (3) features of China's Libya evacuation; (4) the development of China's diplomatic and military capabilities to manage crises overseas; and finally (5) Beijing's options for protecting Chinese interests and citizens overseas and the possible shift in Beijing's global strategic posture to accommodate these options.

China's expanding interests in Africa

Since the late 1990s, the Chinese government has begun encouraging Chinese businesses to invest abroad in the 'going out' strategy (走出去). In 2000, the 'Go Global' policy was officially formulated by Premier Zhu Rongji in his annual policy address to the National People's Congress (NPC) as a platform for Chinese firms to become more competitive in the world economy.[10] Accordingly, the Tenth Five Year Plan (2001–2005) listed overseas investment by Chinese enterprises as one of the four key components to enable the Chinese economy to 'adjust itself to the globalization trend'.[11] The 'Go Global' policy led to an increase in Chinese trade with the resource abundant regions of Southeast Asia, Latin America and Africa.[12] Consequently, Africa now stands as China's fourth largest investment destination and China has become Africa's largest trading partner.[13] According to statistics from the Forum on China–Africa Cooperation, the trade volume between China and Africa was US$10.6 billion in 2000 and jumped to US$130 billion in 2010.[14] Furthermore, Wang Chengan, administrative associate director of the China Society of African Issue Studies, claims that the total trade volume between China and Africa for 2011 is expected to reach between US$150 and US$160 billion and annual trade volume is expected to exceed US$300 billion in three–five years.[15]

According to incomplete statistics from the Forum on China–Africa Cooperation, there are over 2,000 Chinese companies actively operating in Africa, covering sectors

---

10. 'China's outward direct investment', in *OECD Investment Policy Reviews*.
11. *Ibid*.
12. David Zweig, *China and the World Economy: The Rise of a New Trading Nation* (World International Studies Association, 24 July 2008), p. 9, available at: http://www.cctr.ust.hk/materials/working_papers/WorkingPaper25_DZ_China_World.pdf (accessed 17 January 2013).
13. 'YEARENDER: China–Africa cooperation: unlimited opportunity with vitality', *Forum on China–Africa Cooperation*, (20 December 2011), available at: http://www.focac.org/eng/zxxx/t888902.htm (accessed 21 December 2012).
14. 'China is committed to promoting economic and trade cooperation through cross-cultural dialogues', *Forum on China–Africa Cooperation*, (27 May 2011), available at: http://www.focac.org/eng/zfgx/rwjl/t825891.htm (accessed 21 December 2012).
15. 'China, Africa trade volume to hit 150 billion', *Forum on China–Africa Cooperation*, (1 December 2011), available at: http://www.focac.org/eng/zfgx/jmhz/t883297.htm (accessed 21 December 2012).

such as agriculture, energy, construction and infrastructure.[16] New construction contracts in Africa signed by Chinese companies reached US$25.2 billion and accomplished a turnover of US$23.7 billion.[17] Along with heavy investment and large infrastructure projects funded by Chinese banks, these projects require skilled workers. Chinese companies rely on Chinese workers to man these large infrastructure projects. This has resulted in a marked increase of Chinese expatriates living in Africa. The population of Chinese expatriates has exploded exponentially in the last decade. The statistical data on the actual population of Chinese in Africa is unreliable and varies in range from one million to half a million. Huang Zequan, Vice Chairman of the Chinese–African People's Friendship Association, declared there were approximately 550,000 Chinese nationals living in Africa in 2008.[18] The Asahi Shimbun reports an estimated one million Chinese live in Africa as of 2012.[19]

Domestic pressure for protection of Chinese citizens overseas

While economically successful, the 'going out' strategy increasingly exposes the Chinese government to the inherent risks associated with economic ventures abroad, especially in unstable regions, as the Libya crisis demonstrated. By the end of 2011, according to the Ministry of Commerce, over 18,000 companies have operations overseas, employing 1.2 million Chinese and holding assets worth a combined US$1.5 trillion.[20] A large portion of these citizens and firms work in dangerous countries and regions with weak security, frequent wars, rampant terrorism and political instability. According to the *Asia Times*, in the past five years, over 100 Chinese citizens have been kidnapped or attacked, with 14 killed, in ten countries such as Afghanistan, Cameron, Colombia, Ethiopia, Myanmar, Pakistan, Nigeria, Thailand and Yemen.[21] Thus, the protection of Chinese businesses and citizens overseas, especially in high risk areas, is not only a new foreign diplomacy concern, it is also a domestic policy issue.

Compounding the security risk for Chinese nationals is the negative perception of Chinese activities in these countries. Poor labor practices linked to Chinese projects is one of the most damaging factors to China's image and reputation.[22] Recriminations of Chinese business practices include low wages, labor discrimination and inhumane working conditions. Field research conducted in eight African countries on perceptions of China in Africa found that Chinese business activities and personnel were viewed as opportunistic, even predatory, causing resentment among African

---

16. 'China is committed to promoting economic and trade cooperation through cross-cultural dialogues', *Forum on China–Africa Cooperation*.
17. 'YEARENDER', *Forum on China–Africa Cooperation*.
18. Serge Michel, 'When China met Africa', *Foreign Policy* 166, (May/June 2008), p. 41.
19. 'China rapidly expanding its presence in Africa', *The Asahi Shimbun*, (19 July 2012), available at: http://ajw.asahi.com/article/asia/china/AJ201207190100 (accessed 24 August 2013).
20. Zhang Haizhou, 'Better protection of overseas citizens and assets proposed', *China Daily Asia Pacific*, (9 March 2012), available at: http://www.chinadailyapac.com/article/better-protection-overseas-citizens-and-assets-proposed (accessed 21 December 2012).
21. Jian Junbo, 'Beijing's new overseas imperative', *Asia Times*, (17 February 2012), available at: http://www.atimes.com/atimes/China/NB17Ad01.html (accessed 17 January 2013).
22. Suisheng Zhao, 'A neo-colonialist predator or development partner? China's engagement and rebalance in Africa', *Journal of Contemporary China* 23(90), (2014), doi: 10.1080/10670564.2014.898893.

locals.[23] Resentment often extends to non-Chinese nationals and entities with a 'Chinese face', complicating Chinese diplomacy. For instance, Zambia's Collum Coal Mine shooting incident in 2010, in which Chinese supervisors shot and wounded 13 protesting Zambian miners, became an indictment on Chinese business practices in Africa. Although the Collum Coal Mine is not a Chinese State Owned Enterprise (SOE) and its owner not a Chinese citizen, but an Australian of Chinese descent, the Chinese Embassy, nonetheless, felt compelled to intervene. Despite the tenuous link between the Collum Coal Mine and the Chinese government, the Chinese Embassy demanded the Collum Coal Mine compensate the victims, apologize to both its Zambian workers and to Chinese-owned firms in Zambia.[24]

Due to the rising concern about the safety of Chinese citizens overseas, the Chinese government finds itself increasingly called to manage crises abroad. The most notable example was the mass evacuation of Chinese nationals from Libya in 2011. Against the backdrop of the Arab Spring, domestic political pressure compelled Beijing to implement strategies to bolster its authority at home. A beneficial byproduct of the prompt evacuation of citizens in Egypt and Libya was the rallying of popular support for the Chinese leadership. Unsurprisingly, the Chinese media touted the Libya evacuation as a great national achievement showcasing how far China has advanced economically, militarily and diplomatically since its Century of Humiliation (1839–1949). The 'massive, orderly and extraordinarily efficient evacuation' was widely regarded by Chinese controlled media as a reflection of the Chinese government's motto of 'putting people first and running the government in the interest of the people'.[25] While the manipulation of nationalistic passion can strengthen the authority of the regime, it could easily become a catch-22, or no-win situation, for Beijing because it also raises public expectations.

Like many governments around the world, China's leaders are sensitive to the public mood. According to Pew Global Attitudes Project, since 2002, a majority of Chinese have consistently predicted that they would have a better life in five years. In spring 2010, 74% believed their lives would be better in five years.[26] Despite the optimism the Chinese show for the future, there are growing concerns about the challenges the country faces. Arguably, widespread optimism in China could inflate popular expectations, which if unmet could lead to personal or social frustration. In another poll conducted by Pew in 2012, the Chinese people believed their country faced serious and growing challenges. They were concerned about the side effects of rapid economic growth—including the gap between rich and poor, rising prices, pollution and political corruption.[27] A disquieting trend for the Communist

---

23. Fei-Ling Wang and Esi A. Elliot, 'China in Africa: presence, perceptions and prospects', *Journal of Contemporary China* 23(90), (2014), doi: 10.1080/10670564.2014.898888.
24. Barry Sautman and Yan Hairong, 'Bashing "the Chinese": contextualizing Zambia's Collum Coal Mine shooting', *Journal of Contemporary China* 23(90), (2014), doi: 10.1080/10670564.2014.898897.
25. 'China's Libya evacuation highlights people-first nature of government', *Xinhua*.
26. Pew Global Attitudes Project, *Upbeat Chinese Public May Not Be Primed for a Jasmine Revolution* (Pew Research Center, 31 March 2011), available at: http://www.pewglobal.org/2011/03/31/upbeat-chinese-public-may-not-be-primed-for-a-jasmine-revolution/ (accessed 21 December 2012).
27. Pew Global Attitudes Project, *Growing Concerns in China about Inequality, Corruption* (Pew Research Center, 16 October 2012), pp. 2–4, available at: http://www.pewglobal.org/files/2012/10/Pew-Global-Attitudes-China-Report-FINAL-October-10-2012.pdf (accessed 21 December 2012).

Chinese Party (CCP), there was growing concerns about corruption—half said corrupt officials are a very big problem, up 11 percentage points since 2008.[28] Although the Chinese have consistently rated their national and personal economic situations positively over the last few years, they are now grappling with the concerns of a modern, increasingly wealthy society.

Domestic unrest, cynicism and corruption are several challenges to the CCP authority and legitimacy. As a consequence, the Chinese government carefully manages the expectations of its citizens and public demand for more accountability. One factor in the mounting domestic pressure for more accountability is China's active blogosphere and microblogging community. According to a China Internet Network Information Center (CNNIC) January 2011 report, China's total Internet users reached 457 million, up 19.1% compared to 2009.[29] While the growth of Internet users is statistically significant, more striking is the rapid increase in microblogs. Since launching in 2009, Weibo, China's version of Twitter, has gained more than 350 million users (see Figure 1).[30] A virtual public square, Weibo is changing the way Chinese citizens interact with the state by helping to set the national agenda. 'This is unprecedented in Chinese history', says Kaiser Kuo, the director of Corporate Communications at Baidu.com, the leading Chinese search engine. 'There's never been a time when there's been a comparably large and impactful public sphere. It's now driving, in many ways, the entire national dialogue.'[31]

Social network sites such as Weibo have been widely used by Chinese nationals abroad seeking assistance and consular protection. A testament to the popularity of Weibo with Chinese traveling abroad, during an online forum between *China Daily* readers and Huang Ping, Director General of the Foreign Ministry's Department of Consular Affairs in January 2012, readers asked if the Department of Consular Affairs would open an official Weibo account. Huang Ping explained,

> Owing to staff limits we haven't opened an official Weibo. But we always post information about consular affairs through the Foreign Ministry's official Weibo which already has millions of fans. We will try to make better use of the new technology to improve the consular service work in the future.[32]

Further demonstrating the appeal of Weibo, *Global Times*, a pro-Chinese media outlet, describes several anecdotal stories of Chinese youth using Weibo to request consular assistance. In their first account, a Chinese student was detained and harassed at Manila airport. After some fellow Chinese sought help on Weibo through the foreign ministry's official account, staff members at the Chinese embassy in the Philippines soon went to the airport to settle the issue. In another example, a Chinese girl vacationing in the Aegean Sea had her wallet stolen while traveling in Athens.

---

28. Ibid.
29. China Internet Network Information Center (CNNIC), *China Internet Statistics Whitepaper* (China Internet Watch, January 2011), available at: http://www.chinainternetwatch.com/whitepaper/china-internet-statistics/ (accessed 21 December 2012).
30. Mary Kay Magistad, 'How Weibo is changing China', *Yale Global Online*, (9 August 2012), available at: http://yaleglobal.yale.edu/content/how-weibo-changing-china (accessed 1 January 2013).
31. Ibid.
32. 'More challenges to consular protection', *China Daily*, (16 January 2012), available at: http://www.chinadaily.com.cn/china/2012diplomats/2012-01/16/content_14456567.htm (accessed 21 December 2012).

Active Twitter Users in Select Countries, Q2 2012 (Million)

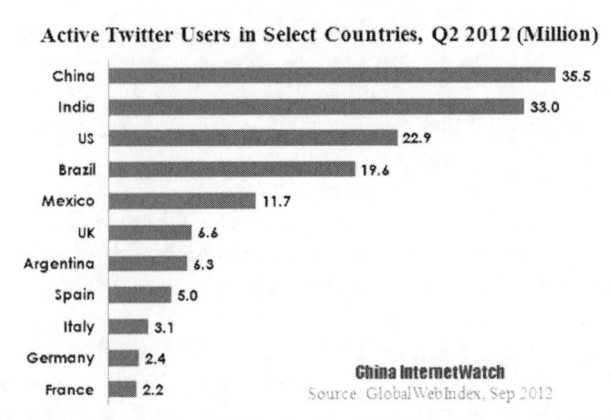

China	35.5
India	33.0
US	22.9
Brazil	19.6
Mexico	11.7
UK	6.6
Argentina	6.3
Spain	5.0
Italy	3.1
Germany	2.4
France	2.2

Source: GlobalWebIndex, Sep 2012

Figure 1. Active Twitter users. *Source*: 'China total active Twitter users exceed US', *China Internet Watch*, (September 2012), available at: http://www.chinainternetwatch.com/1707/china-total-twitter-users/ (accessed 21 December 2012).

She called the Chinese embassy for help, but was told the embassy was under no obligation to loan money to Chinese citizens abroad. Her story was soon forwarded thousands of times on Weibo. Many excoriated the Chinese embassy for being indifferent to its own citizens. While the Chinese embassy was unaccommodating, this tale has a happy ending when a local Chinese company provided assistance to this hapless, penniless girl stranded in a foreign land.[33] While these accounts seem trivial compared to kidnapping, ransacking or murder, they illustrate the capacity of Weibo to empower Chinese nationals to demand more consular protection overseas.

In the case of the Libya crisis, a person claiming to work for a Chinese state-owned railway company in Libya attracted widespread attention on the Internet after pleading for help on Weibo. The person, using the name Happy Xufeng, described an attack on the company's compound and posted photos of construction equipment and buildings in flames.[34] Although it's doubtful that Internet communications alone spurred Beijing into action, microbloggers put the spotlight on the precarious conditions on the ground as the crisis unfolded. Happy Xufeng and other microbloggers created drama, bringing intense media attention and public focus on the crisis amplifying the pressure for Beijing to act. Moreover, the political instability rocking Africa and the Middle East during the Arab Spring played into Beijing's decision making calculus in the handling of the Libya crisis. As Wang Yizhou, Professor of International Relations at Peking University, asserts, the Libya evacuation derived from 'social pressure due to the glare of the traditional media and Internet users forcing mainland leaders to respond in order to enhance their legitimacy and credibility'.[35]

---

33. Wang Wenwen, 'More Chinese seek consular help', *Global Times*, (18 June 2012), available at: http://www.globaltimes.cn/DesktopModules/DnnForge%20-%20NewsArticles/Print.aspx?tabid=99&tabmoduleid=94&articleId=715731&moduleId=405&PortalID=0 (accessed 21 December 2012).

34. Josh Chin, 'China vows to protect Chinese in Libya', *The Wall Street Journal* (online), (25 February 2011), available at: http://online.wsj.com/article/SB10001424052748703905404576164321645905718.html (accessed 21 December 2012).

35. Ed Zhang, 'China displays a bold face to world', *South China Morning Post*, (21 March 2011), available at: Lexisnexis.com (accessed 30 September 2011).

Features of China's Libya evacuation

As the situation in Libya deteriorated and Chinese nationals became entangled in the violence, Chinese President Hu Jintao and Premier Wen Jiabao issued an order on 22 February requiring all departments involved to try their best to guarantee the safety of Chinese nationals.[36] Over a 12-day period in February and March 2011, China evacuated more than 35,000 Chinese and other country nationals from civil war torn Libya.[37] Chinese official media called this the largest such operation China had mounted abroad since the Nationalists fled in 1949. The large-scale evacuation of Chinese nationals, including people from the mainland, Hong Kong, Macau and Taiwan and other foreign nationals, tested the crisis management capacity of the Chinese government.

The entire state was called upon to support the evacuation. Consular protection mechanisms and the ministerial-level joint conference system were activated. The State Council created an emergency headquarters, headed by Vice Premier Zhang Dejiang, to organize the evacuation. The Ministry of Foreign Affairs, the state-owned Assets Supervisions and Administration Commission of the State Council, the Ministry of Transport, the Ministry of Commerce, Civil Aviation Administration of China (CAAC) and relevant military departments jointly coordinated all possible conveyance to transport Chinese nationals out of Libya.[38] The Chinese Ministry of Transport issued an emergency notice calling on all transportation departments at home to cooperate with CAAC, to offer efficient service to the evacuees once they returned to China.[39]

During the massive operation, the Chinese government evacuated Chinese nationals via land, air and sea routes. Chinese evacuees took overland routes to escape Libya, by traveling to Tubruq, near the Libya–Egypt border, then entering Egypt through the Egyptian city of Salum. The Chinese Embassy in Egypt organized more than 100 buses to wait in Salum, Egypt to transport possible evacuees.[40] CAAC sent its first chartered plane to Libya's capital Tripoli to return evacuees to China on 23 February. CAAC dispatched flights to Egypt, Greece, Malta, Tunisia, Jordan, Sudan and Turkey, since the majority of evacuees had been transferred to these countries as an intermediate staging base.[41] Beginning on 1 March, CAAC dispatched 20 flights every day—chartering Air China, China Southern, China Eastern Airlines and Hainan Airlines (see Table 1). In addition, China's Foreign Ministry chartered planes from other countries to aid the withdrawal. The People's Liberation Army (PLA) Air Force also deployed four cargo planes to Libya to transfer evacuees to Sudan's capital Khartoum.[42] Most Chinese nationals,

---

36. Ding Ying, 'Out of Libya', p. 15.
37. Government of the People's Republic of China, 'Chinese President, Premier order "all-out efforts" to secure life, property of nationals in Libya', *gov.cn*, (Government of the People's Republic of China, 22 February 2011), available at: http://www.gov.cn/misc/2011-02/22/content_1808180.htm (accessed 21 December 2012).
38. Ding Ying, 'Out of Libya', p. 15.
39. *Ibid.*, p. 14.
40. *Ibid.*, p. 15.
41. An intermediate staging base (ISB) is a tailorable, temporary location used for staging forces, sustainment and/or extraction into and out of an operational area. Definition from United States Office of the Chairman of the Joint Chiefs of Staff, *Noncombatant Evacuation Operations, Joint Publication 3–68* (Washington, DC: CJCS, 23 December 2010), p. GL-7.
42. Ding Ying, 'Out of Libya', p. 14.

Table 1. China's Libya evacuation statistics

Civilian assets	Commercial aircraft chartered	
	Air China Airlines	27
	China Southern Airlines	22
	China Eastern Airlines	25
	Commercial vessels chartered	
	China Ocean Shipping Group passenger liner	4
	China Shipping Group passenger liner	3
	Greek commercial vessels	7
	Malta commercial vessels	Several
	Commercial vehicles chartered	
	Number of bus runs	100
Military assets	PLA Air Force	
	IL-76 transport aircraft	4
	PLA Navy	
	Xuzhou, Jiangkai-II class guided-missile frigate	1

Sources: Shi Jiangtao, 'Lessons to learn from Libyan evacuation; mission hailed as a patriotic success as the nation clocks up a number of firsts', *South China Morning Post*, (5 March 2011), available at: Lexisnexis.com (accessed 30 September 2011); Ding Ying, 'Out of Libya', *Beijing Review*, (10 March 2011), pp. 14–15; United States Office of the Secretary of Defense, *Annual Report to Congress: Military and Security Developments Involving the People's Republic of China 2011* (Washington, DC: US Department of Defense, May 2011), pp. 33 and 59, available at: http://www.defense.gov/pubs/pdfs/2011_cmpr_final.pdf (accessed 21 December 2012).

however, were evacuated from Libya by ship. After coordination with the Chinese Embassy in Greece, Greece sent seven chartered vessels to Benghazi. Evacuees were then transferred to Crete before they headed for China. Each Greek vessel transferred about 2,000 people each voyage. Malta also provided chartered vessels to transport Chinese citizens to its port city of Valletta. The China Ocean Shipping Company and the China Shipping Company directed their ships to nearby waters to participate in the evacuation as well. Finally, the PLA Navy sent a frigate from its antipiracy escort fleet in the Gulf of Aden to protect vessels.[43]

To comprehend the scale and complexity of this endeavor, China's Libya evacuation will be compared and contrasted to the United States' largest overseas evacuation of American citizens in recent history. During a three week period in July and August 2006, the United States evacuated 15,000 American citizens from Lebanon. On 12 July 2006, Hezbollah guerillas kidnapped two Israeli soldiers at Israel's border with Lebanon. The following day, Israel countered with a major military assault—bombing Lebanon's Beirut airport forcing its closure, bombing roads and bridges, and blockading Lebanon's ports.[44] Similar to the Chinese Libya evacuation, following the outbreak of hostilities the United States government determined it would need to implement a large-scale evacuation to ensure the

---

43. *Ibid.*, p. 15.
44. United States Government Accounting Office, *US Evacuation from Lebanon*, GAO-07893 (Washington, DC: GAO, 7 June 2007), p. 5.

protection of Americans. These operations are commonly referred to as noncombatant evacuation operations or NEOs by US military planners.[45]

In contrast to China's Libya operation, which was primarily implemented by civilian organizations and commercial transport, the American effort was predominately planned and executed by the United States military. Due to the size and scope, the US Ambassador in Beirut concluded that the State Department would not be able to safely evacuate thousands of Americans and requested the assistance of the Department of Defense (DoD). Moreover, because air and land evacuation routes were blocked or closed, the Lebanon NEO required military planning, capabilities and equipment. Additionally, sea routes required negotiation with Israel for safe navigation. Consequently, the US Military Central Command (CENTCOM) was designated the NEOs lead organization with support from European Command (EUCOM) and Transportation Command (TRANSCOM). Unlike the Chinese operation, the evacuation included an even mix of military and civilian air and sea transport (see Table 2). The US also employed a significant number of military personnel to provide security and safe passage for American citizens. Approximately 2,400 marines with the US Marine Corps' 24th Marine Expeditionary Unit (MEU), a marine unit specializing in evacuation operations, and approximately 3,000 sailors with the US Navy's Iwo Jima Expeditionary Strike Group (ESG) were deployed in the Lebanon NEO.

In comparing the American and Chinese operations, it is remarkable how the Chinese government was able to evacuate over twice the number of people in about the same timeframe using very few military resources. At the time of the Libya crisis, the Chinese government did not have the option to primarily use the military in planning and executing the evacuation because it lacked the capability and capacity to contribute more to a large overseas contingency. Nevertheless, the tremendously successful evacuation demonstrated the Chinese government's expanding capabilities to handle crises overseas. Without a doubt, this type of operation would not have been possible a decade ago.

The development of Chinese diplomatic capabilities

In the past decade, China has established institutional mechanisms to manage the overseas citizen protection system. The Libya crisis tested these institutions and revealed their growing maturity but also demonstrated their inadequacies. In 'putting people first', China upgraded emergency protection mechanisms at its embassies and consulates around the world. The Ministry of Foreign Affairs (MFA), Department of Consular Affairs set up a division for consular protection in May 2006, and in August 2007 it upgraded the division to the Center for Consular Assistance and Protection. The center is responsible for providing consular protection and assistance for Chinese citizens living abroad as well as issuing early warnings and relevant information. To

---

45. Noncombatant evacuation operations (NEOs) are operations directed by the Department of State or other appropriate authority, in conjunction with the Department of Defense, whereby noncombatants are evacuated from foreign countries when their lives are endangered by war, civil unrest or natural disaster to safe havens as designated by the Department of State. Definition from United States Office of the Chairman of the Joint Chiefs of Staff, *Noncombatant Evacuation Operations*, p. GL-8.

Table 2. United States' Lebanon NEO statistics

Civilian assets	Commercial aircraft chartered	Several
	Commercial vessels chartered	3
	Busses and other vehicles	Several
Military assets	US Air Force	
	Tanker aircraft	2
	Transport aircraft	2
	MH-53 helicopter	Several
	US Navy (approx. 3,000 sailors)	
	Amphibious ship	4
	Destroyer	2
	High speed vessel (catamaran)	1
	Command and control ship	1
	Oil tanker ship	1
	US Marine Corps (approx. 2,400 marines)	
	Marine Expeditionary Unit (MEU)	1
	CH-53 helicopter	6
	Interim Marine Corp Security Force	1

Sources: Iwo Jima Expeditionary Strike Group, 'Non-combatant evacuation Lebanon—July 2006', slide 4, (2006), available at: http://www.docstoc.com/docs/128011161/Non-Combatant-Evacuation-Lebanon (accessed 24 August 2013); 'Lebanon non-combatant evacuation operation (NEO) 2006', *GlobalSecurity.org*, available at: http://www.globalsecurity.org/military/ops/neo-lebanon-06.htm (accessed 24 August 2013).

deal with emerging challenges, Chinese diplomatic missions abroad have also established emergency response plans.[46] Although the fledgling consular protection system is progressing, it is hard-pressed to meet the growing demand. MFA employs 140 consular diplomats in Beijing and about 600 spread across more than 250 embassies, consulates and other representations abroad.[47] MFA's small representation overseas effectively guarantees Beijing's reliance on diplomatic channels to protect citizens abroad. Although the diplomatic contingency is small, Chinese embassies in the region effectively coordinated support for the Libya evacuation from Jordan, Turkey, Sudan, United Arab Emirates, Greece, Tunisia, Egypt and Malta.[48] Following the evacuation, Huang Ping, Director General of the Foreign Ministry's Department of Consular Affairs, boasted that the emergency mechanisms has become mature after being tested by tough tasks.[49]

Despite MFA's initiatives to reinvigorate consular protection, the Libya crisis revealed several shortcomings. One major shortcoming of the consular protection system was that the Chinese government did not know how many Chinese passport holders were in Libya—not to mention the overseas Chinese not holding passports.[50]

    46. Ding Ying, 'Out of Libya', p. 15.
    47. Duchâtel and Gill, *Overseas Citizen Protection*.
    48. Ding Ying, 'Out of Libya', p. 14.
    49. *Ibid.*, p. 14.
    50. Although statistics are incomplete and unreliable, there are an estimated 40 million overseas Chinese (华侨) living abroad with a presence in more or less every country of the world. See *Overseas Chinese Population Distribution* [Overseas Chinese Affairs Council, Republic of China (Taiwan)], available at: http://www.ocac.gov.tw/english/public/public.asp?selno=8889&no=8889&level=B (accessed 21 December 2012).

While it is difficult to accurately account for all citizens living abroad, an estimate is essential to planning an evacuation. In the case of the US Lebanon NEO, 50,000 Americans were estimated to be in Lebanon at the time.[51] US diplomatic posts overseas are required to maintain an F-77 report, an annual estimate of the number of American citizens in the country.[52] Recognizing the MFA deficiency during the Libya evacuation, in March 2012 a team of Chinese People's Political Consultative Conference (CPPCC) members proposed Chinese embassies and businesses work together to establish a database of Chinese staff abroad.[53]

Another deficiency that was uncovered was that the Chinese embassy had trouble providing early warning and security information to nationals in Libya. Again, in contrast, shortly after hostilities commenced in Lebanon, the US Embassy in Beirut issued warden massages[54] and press releases informing American citizens of the developing crisis and urging them to register with the embassy for possible evacuation.[55] To remedy the communication shortcomings during the Libya evacuation, beginning in November 2011, the MFA launched a website to provide updates and alerts. It also concluded an agreement with Chinese mobile phone operators to ensure that Chinese nationals traveling abroad receive a text message with basic security information, including the contact of the Chinese consulate and local police, upon arrival in a foreign country.[56]

During the Libya crisis, Beijing also activated the ministerial-level joint conference system for the protection of Chinese citizens and agencies abroad. Established as early as 2004, the Chinese government describes this system as a 'four-in-one' overseas security protection joint working mechanism which integrates the central government, local governments, Chinese embassies and consulates abroad.[57] This initiative enables various ministries and departments to cooperate and coordinate with each other. Despite these early initiatives to improve coordination, several ministries and departments involved in the planning and execution had difficulty synchronizing their overlapping functions contributing to the complexity of the Libya operation.

A major obstacle to effective crisis management and inter-ministerial efforts in China is the rank-conscious culture of China's political system which strictly

---

51. United States Government Accounting Office, *US Evacuation from Lebanon*, p. 3.

52. The report of potential evacuees, also known as the F-77 Report, identifies the numbers of potential evacuees at each embassy. Each embassy or consulate is required to submit to the DOS an annual report, on 15 December, of the estimated number of potential evacuees in its assigned area. These counts, however, are only yearly estimates. The accuracy of the estimate will vary with the speed and severity of the crisis. Definition from United States Office of the Chairman of the Joint Chiefs of Staff, *Noncombatant Evacuation Operations*, p. xiv.

53. Zhang Haizhou, 'Better protection of overseas citizens and assets proposed'.

54. Communications with potential evacuees may be via a warden system, which is a notification system used to communicate to the US population through wardens using telephones, faxes, emails and direct personal contact. A warden coordinator prepares lists of wardens and other contacts to cover areas of assigned responsibilities. The wardens prepare, update and maintain a list of phone numbers and addresses of US citizens residing in their assigned areas. Definition from United States Office of the Chairman of the Joint Chiefs of Staff, *Noncombatant Evacuation Operations*, p. xiv.

55. United States Government Accounting Office, *US Evacuation from Lebanon*, p. 25.

56. Duchâtel and Gill, *Overseas Citizen Protection*.

57. Ministry of Foreign Affairs of the People's Republic of China, 'Vice Foreign Minister Song Tao chairs the inter-ministerial meeting on safety protection of Chinese citizens and institutions abroad in Fuzhou', (27 May 2011), available at: http://www.fmprc.gov.cn/eng/zxxx/t826710.htm (accessed 21 December 2012).

observes 'rank that identify the relative importance of people, official agencies, public institutions, state-owned corporations, and geographic units'.[58] According to a Congressional Research Service (CRS) report,

> Among the rules that govern rank in China is that entities of equivalent rank cannot issue binding orders to each other. Often, they cannot even compel coordination, although Party entities and security agencies have more clout in that respect than other entities. An entity of lesser rank seeking to coordinate with an entity of higher rank faces a daunting challenge. Many analysts attribute the well documented communication problems between the PLA and the Foreign Ministry to the large gap in their respective ranks. The PLA's Central Military Commission is of equivalent rank to the State Council, China's cabinet, while the Foreign Ministry is a mere ministry under the State Council. For the Foreign Ministry to liaise with the PLA, it must report up to the State Council, which may have to report further up to the Politburo in order to secure PLA cooperation.[59]

In conjunction with the activation of the ministerial-level joint conference system, the State Council created an emergency headquarters, led by Vice Premier Zhang Dejiang, a Politburo member and at the time head of the State Council Production Safety Commission. The fact that the State Council delegated responsibility to the Vice Premier, a position of substantial rank, is significant. Given the rapidly deteriorating situation in Libya and the large number of Chinese nationals in danger, there was palpable pressure not to mishandle this mounting human calamity.

In comparison, during American NEOs bureaucratic rank normally does not complicate interagency operations. Interagency coordination is usually established through standardized procedures and memorandums of agreement (MOAs) and therefore is less reliant on the power of personality. According to US Joint Planning Doctrine, during a NEO, the in-country Ambassador is designated the responsible authority for the operation.[60] In the case of the Lebanon NEO, once the US Ambassador in Beirut requested military assistance, the military was able to take charge of the planning and execution of the operation according to established MOAs. The American operation did not require the Vice President or the Speaker of the House to take the lead in coordinating interagency operations. In contrast, the high level involvement of the Party during the Libya evacuation demonstrated how overseas crisis management lacked routinized and standard operating procedures. The *ad hoc* nature of the Libya evacuation and the apparent lack of procedures raise concerns regarding the capacity of the Chinese government to sustain its overseas crisis management system.

Following the evacuation of Chinese citizens from Libya, Chinese leaders reflected on lessons learned. NPC deputies and CPPCC members attending China's two most important annual meetings in March 2011, suggested that the government overhaul its emergency response system and strengthen risk assessments for Chinese nationals

---

58. Susan V. Lawrence and Michael F. Martine, *Understanding China's Political System* (Washington, DC: Congressional Research Service, 10 May 2012), p. 11, available at: http://www.fas.org/sgp/crs/row/R41007.pdf (accessed 1 January 2013).
59. *Ibid.*
60. United States Office of the Chairman of the Joint Chiefs of Staff, *Noncombatant Evacuation Operations*, p. III-1.

living and working abroad.[61] During the meeting, Li Chengbao, an NPC deputy from Qinghai Province, stated that China should carry out risk assessments on certain countries in areas of investment, security and economy to lower possible losses by Chinese enterprises.[62] A year later, a CPPCC team continued to work on proposals to better protect overseas citizens following two high profile kidnappings of more than 50 Chinese workers in Sudan and Egypt in early 2012.

Addressing the issue of moral hazard, the main component of the team's recommendations entailed Chinese businesses accepting more responsibility. The team suggested that Chinese firms evaluate host nations' political risks before making decisions on prospective projects.[63] Han Fangming, Deputy Director of the CPPCC's Foreign Affairs Committee, also advanced 'We don't need to go to some politically high-risk regions for economic interests only'.[64] Moreover, since 2011, the Chinese government has promoted workforce 'localization' as a best business practice.[65] Due to increasing international pressure, as well as the need to limit Chinese risk, Chinese authorities are encouraging SOEs to change existing business practices overseas and create jobs for locals. However, implementation of 'localization' is nonlinear and varies from one SOE to another and from one business sector to another.[66] Given uneven workforce 'localization', the Chinese population in Africa and high risk regions continues to rise. Notwithstanding these initiatives, there is growing recognition among Chinese officials that China will have to confront more challenges in protecting its interests abroad.

The development of Chinese military capabilities

The Libya evacuation marked the PLA's first overseas deployment to evacuate Chinese citizens. Operationally, the evacuation marked some very significant milestones in China's military outlook. For the first time, the PLA worked with civilian authorities to implement a large-scale operation. In another first, the PLA Navy deployed a JIANGKAI-II class frigate, *Xuzhou*, which had been operating in the Gulf of Aden, to provide security escort for chartered vessels,[67] while the PLA Air Force deployed four IL-76 long-haul transport aircraft to assist with evacuating Chinese citizens.[68] Speaking on the sidelines of the annual gathering of the NPC in March 2011, retired Major General Luo Yuan said, 'It is a good start. The rare overseas mission will help our military get more experience and specifically, the use

---

61. 'China's evacuation of nationals from Libya hailed by attendees of top annual meetings', *Xinhua*, (3 March 2011), available at: http://news.xinhuanet.com/english2010/china/2011-03/03/c_13758215.htm (accessed 21 December 2012).
62. *Ibid.*
63. Zhang Haizhou, 'Better protection of overseas citizens and assets proposed'.
64. *Ibid.*
65. Antoine Kernen and Katy Nganting Lam, 'Workforce localization among Chinese state-owned enterprises (SOEs) in Ghana', *Journal of Contemporary China* 23(90), (2014), doi: 10.1080/10670564.2014.898894.
66. *Ibid.*, doi: 10.1080/10670564.2014.898894.
67. United States Office of the Secretary of Defense, *Annual Report to Congress: Military and Security Developments Involving the People's Republic of China 2011* (Washington, DC: US Department of Defense, May 2011), p. 59, available at: http://www.defense.gov/pubs/pdfs/2011_cmpr_final.pdf (accessed 21 December 2012).
68. *Ibid.*, p. 33.

of navy ships and aircraft means our long-range capability has seen a remarkable improvement'.[69]

Although largely symbolic, this deployment enabled the PLA to demonstrate a commitment to the protection of Chinese citizens living and working overseas. However, the PLA's sealift and airlift capacity is still insufficient to sustain a large overseas contingency. The Navy's contribution was due in part to the proximity of its ongoing antipiracy missions in the Gulf of Aden. The *Xuzhou* was easily diverted to escort commercial vessels in the Libya evacuation. Unlike the Lebanon NEO, in which the US employed four amphibious ships with a carrying capacity of between 1,000 and 1,200 people, the Chinese Navy did not employ any amphibious assets capable of transporting evacuees. In total, US Navy ships and Marine Corps helicopters evacuated over 7,000 American citizens during the Lebanon NEO.[70] During the Libya evacuation, the PLA's only lift capacity was showcased by the Air Force. The Chinese Air Force's participation came at the tail end of the operation and was limited to four transport aircraft evacuating little over 1,700 people.[71] Beijing-based defense expert Song Xiaojun said the evacuation exposed many flaws, especially an inability to project military power. 'Our missile frigate is not big enough to carry many people and the aircraft are large old Russian-made transport jets'.[72] Despite the PLA's modest role in the evacuation, its participation portends more coordination and involvement in protecting citizens overseas. If the PLA's global mission is expected to expand, then its shortcomings in the Libya evacuation undoubtedly justify increased defense spending in long-range lift capabilities.

In addition to long-range lift capabilities, military operations, such as a NEO, require basing rights for logistical support and port access to refuel and resupply. During the Libya evacuation, China's military attachés from Europe and the Middle East coordinated logistical support for the operation. For instance, China's military attachés prearranged the use of Khartoum, Sudan airport to refuel IL-76 transport aircraft headed to and from Libya and the use of Oman seaport to replenish the *Xuzhou*.[73] The Libya operation clearly showed the PLA's expanding access to ports due in part to their ongoing deployments to the Gulf of Aden and participation in peacekeeping operations in Africa. Moreover, there are indications that the PLA may be contemplating more basing agreements. On 2 December 2011, the Seychelles invited the PLA to set-up a military base on the archipelago to strengthen the fight against piracy.[74] To expand its global capability, inevitably, the PLA will need to

---

69. Shi Jiangtao, 'Lessons to learn from Libyan evacuation; mission hailed as a patriotic success as the nation clocks up a number of firsts', *South China Morning Post*, (5 March 2011), available at: Lexisnexis.com (accessed 30 September 2011).
70. Iwo Jima Expeditionary Strike Group, 'Non-combatant evacuation Lebanon—July 2006', slide 3, (2006), available at: http://www.docstoc.com/docs/128011161/Non-Combatant-Evacuation-Lebanon (accessed 24 August 2013).
71. Gabe Collins and Andrew S. Erickson, 'Implications of China's military evacuation of citizens from Libya', *The Jamestown Foundation*, (10 March 2011), available at: http://www.jamestown.org/programs/chinabrief/single/?tx_ttnews%5Btt_news%5D=37633&cHash=7278cfd21e6fb19afe8a823c5cf88f07 (accessed 2 January 2013).
72. Shi Jiangtao, 'Lessons to learn from Libyan evacuation'.
73. United States Office of the Secretary of Defense, *Annual Report to Congress*, p. 59.
74. 'Seychelles invites China to set up anti-piracy base', *Agence France-Presse*, (2 December 2011).

consider future basing agreements with host nations to ensure landing, embarkation and transit rights are in place to support any long-range operation.[75]

When it comes to operational planning and execution, geography is an important component. Geography worked in China's favor in the Libya evacuation. Since Libya is situated along the Mediterranean Sea, the Chinese were able to utilize sealift assets from nearby European countries and use surrounding countries as intermediate staging bases. Functional overland transportation systems also permitted the use of intermediate staging bases outside Libya. Beijing was also able to charter commercial assets to augment the limited PLA air and sealift. If similar circumstances had occurred in a landlocked nation in Africa, evacuation would have been extremely difficult since sealift would be out of the question. Additionally, commercial air carriers may not be able to participate in a hostile environment.[76] Moreover, remote locations in Africa may not have suitable intermediate staging bases readily accessible. Therefore the only option available may be military transport. Given the limited PLA lift capacity, the Chinese government could not perform an evacuation on the scale of the Libya crisis from a remote African nation.

The Libya operation represents the evolving mandate of the PLA due in part to China's expanding global interests. On Christmas Eve 2004, newly minted Chairman of the Central Military Commission, Hu Jintao signaled the PLA's expanding mission while addressing the armed forces. He outlined a set of 'new historic missions' for the PLA, calling on the armed forces to protect China's expanding national interests and adopt a larger role in promoting international peace and security.[77] According to James Mulvenon, these 'new historic missions' stem from traditional concerns such as political stability and territorial integrity, but also new requirements and challenges 'created by China's increasingly global interests and entanglements'.[78] Mulvenon concludes that Hu Jintao's address justified continued resource priority for PLA modernization beyond the Taiwan issue. It provided the conceptual framework for broader global requirements and prompted the PLA's growing involvement in international peacekeeping efforts, antipiracy operations, humanitarian assistance and disaster relief, and the evacuation of Chinese citizens from overseas trouble spots.[79]

The implications of China's new diplomatic imperative

Since initiating the 'Go Global' economic policy, Beijing has upheld the maxim 'keeping a low profile' as the basis of its global strategic posture. In the early 1990s, Deng Xiaoping formulated 'keeping a low profile' or 'lie low and hide your talent'

---

75. United States Office of the Chairman of the Joint Chiefs of Staff, *Noncombatant Evacuation Operations*, p. B-4.
76. Hostile environment is an operational environment in which hostile forces have control as well as the intent and capability to effectively oppose or react to the operations a unit intends to conduct. Definition from United States Office of the Chairman of the Joint Chiefs of Staff, *Noncombatant Evacuation Operations*, p. GL-7.
77. United States Office of the Secretary of Defense, *Annual Report to Congress*, pp. 3–4.
78. James Mulvenon, 'Chairman Hu and the PLA's "new historic missions"', *China Leadership Monitor* no. 27, (9 January 2009), p. 2, available at: http://media.hoover.org/sites/default/files/documents/CLM27JM.pdf (accessed 21 December 2012).
79. United States Office of the Secretary of Defense, *Annual Report to Congress*, p. I.

(韬光养晦) as a means to give China breathing room to slowly build into a global power.[80] According to Zhu Weilie, 'keeping a low profile' proffers guidelines for Beijing's approach to foreign affairs:

> China should not seek a leadership role in the international arena, but take a low profile and concentrate on China's own affairs in the first place while observing and coping with international affairs calmly in the dramatically changing world and making full use of advantages while avoiding disadvantages.[81]

Another underpinning of China's global strategic posture is the principle of noninterference in the internal affairs of others. Since the early 1950s, China has adhered to the Five Principles of Peaceful Co-existence: mutual respect for sovereignty and territorial integrity; mutual nonaggression; noninterference in each other's internal affairs; equality and mutual benefit; and peaceful coexistence. In particular, the policy of noninterference affirms China does not interfere in other countries' internal affairs and respects 'the right of the people of other countries to independently choose their own social system and path of development'.[82] Noninterference is a key principle in China–Africa cooperation and is 'regarded as one of the strong selling points to African elites, as it is a distinguishing feature from "Western" aid'.[83] Promoting itself as an alternative model to liberal democracies, China's advocacy of noninterference has helped China win the support of weak African state allies and access to the regions immense resources such as oil, minerals and the like. Patently, the noninterference principle contributes to Chinese firms' investment in states with questionable human rights practices, rampant corruption and overall poor governance.

While these traditional diplomatic guidelines were effective in the past, they are proving insufficient in protecting Chinese interests abroad. Arguably, in the past, the Chinese government used the maxim 'keeping a low profile' and the noninterference principle to mask its impotence abroad. Since China is growing in capability and confidence, will it need to maintain this unassuming diplomatic posture to justify its inaction? The effect of China's rise and the expansion of China's global interests, make it increasingly difficult to maintain this strategic posture in practice as the Libya evacuation showcased. With increased global activity, Sven Grimm avers that the noninterference principle is facing challenges—'a strict "hands off" attitude becomes more and more difficult'.[84] If China is to continue the 'Go Global' economic strategy, it will need to be more proactive in protecting its interests, investments and Chinese citizens overseas. In order to meet these challenges, Beijing has three options or a

---

80. Deng Xiaoping's (韬光养晦) statement: 'To keep low profile, be good at pretending and hiding, never take the lead, act out selectively, preserve ourselves, and gradually and quietly expand and develop', from Jiang Zemin, *Jiang Zemin Wenxuan* [*Selected Works of Jiang Zemin*] (Beijing: Renmin Press, 2006), p. 202.
81. Zhu Weilie, 'On diplomatic strategy of "keeping a low profile and taking a proactive role when feasible"', *Global Review* 6(3), (May/June 2010), p. 4.
82. People's Republic of China White Paper, *China's Peaceful Development* (Information Office of the State Council the People's Republic of China, 6 September 2011), available at: http://www.gov.cn/english/official/2011-09/06/content_1941354_2.htm (accessed 21 December 2012).
83. Sven Grimm, 'China–Africa cooperation: promises, practice and prospects', *Journal of Contemporary China* 23(90), (2014), doi: 10.1080/10670564.2014.898886.
84. *Ibid*.

combination thereof: (1) depend on host countries to provide security; (2) encourage Chinese businesses to source their own security; and (3) utilize the Chinese state security apparatus such as the PLA or MPS (Ministry of Public Security) in protective services.

Since entering the global economy, like most countries operating overseas, China relies heavily on option one. The Chinese government works primarily with host governments and their security personnel and armed forces to protect its citizens abroad. In a state with strong institutions, relying on host nation's security is usually not a problem; however, it is problematic in weak states that lack functioning government and security organizations. However, Chinese firms often operate in countries that struggle to provide basic security for their own citizens, let alone provide security for foreign businesses and nationals. According to Hess and Aidoo, Chinese firms have distinguished themselves as being less 'risk-averse' than their Western competitors operating in regions others' avoid.[85] If China relies exclusively on host nations for security, it will need to expect better governance from host countries. Paradoxically, expecting quality host nation security and services could be viewed as interfering in the internal affairs of others.

Next, Beijing could require Chinese businesses to take more ownership of their security. Beijing could mandate that Chinese SOEs include security as a notable line item in their annual budgets to limit moral hazard associated with high risk ventures overseas. Beijing could also contract Chinese-owned security companies to go abroad to protect Chinese firms and workers in dangerous regions. These security companies, unfortunately, would not have any legal authority in a host nation limiting their effectiveness. Again, this option could appear to be heavy handed and at odds with the strategy of 'keeping a low profile'. Moreover, Chinese businesses have not shown the propensity to take ownership of their security. Currently, many Chinese firms lack security procedures to prevent incidents and handle exigencies. Although some firms have risk assessment units, most have not yet established a chief security officer position.[86]

Finally, Beijing could employ the PLA or other state security apparatus to protect citizens abroad. Option three is a proactive and less reactive approach to international security further obliging Beijing to reconsider its global strategy. This option is gradually gaining traction in Beijing when it comes to dealing with overseas exigencies. The PLA's antipiracy mission in the Gulf of Aden and the Libya evacuation are quintessential illustrations of Beijing employing the PLA to protect Chinese interests and citizens overseas. Also demonstrating a gradual predisposition to utilize Chinese security organizations overseas, in October 2011, the MPS intervened following the murder of 13 Chinese merchant sailors along the Mekong in northern Thailand. After a multinational investigation, six foreigners suspected of murdering the Chinese sailors were extradited to China to stand trial. This was a rare instance, where foreigners who commit crimes against Chinese nationals outside the country were brought to justice before a Chinese court. The MPS also directly

---

85. Steve Hess and Richard Aidoo, 'Beyond the rhetoric: noninterference in China's Africa policy', *Africa and Asian Studies* 9, (2010), p. 366.
86. Duchâtel and Gill, *Overseas Citizen Protection*.

negotiated with Myanmar, Thailand and Laos to reach an agreement on joint river patrols.[87] These cases clearly represent a trend towards more intervention and assertiveness abroad to protect Chinese nationals.

Recognizing the limitation of past policies, Chinese scholars have begun to voice and promote a shift in Chinese foreign policy. In a March 2012 *Beijing Review* interview, Professor Wang Yizhou, Vice Dean of the School of International Studies of Peking University, explains a new diplomatic concept he coined, 'creative involvement' (创造性介入), which calls on China to actively play a bigger role and voluntarily get involved in international affairs.[88] He believes China is entering a new phase in its diplomacy since it has become a leader in world affairs and has expanding overseas interests. He also enumerates various examples of 'creative involvement' already demonstrated by Beijing—China's mediation in Sudan and South Sudan in 2007, China's initiatives in the six-party talks on the Korean Peninsula nuclear issue, China's antipiracy mission off the coast of Somalia, and finally China's evacuation of citizens from Libya in 2011.[89] When asked if 'creative involvement' is a betrayal of China's principle of noninterference, Professor Wang dismisses the incongruity in the two concepts,

> Although advocating active participation in international affairs, 'creative involvement' has essential differences from interventionism. It calls for active contact and involvement instead of intervention by force. It definitely is not an opposition to our traditional diplomatic principles, but an enrichment of these principles.[90]

Whatever path Beijing chooses, the Chinese will have to take a more proactive posture and modify its foreign policy principles in order to protect its interests and citizens overseas. Officially, the Chinese government will continue to maintain the noninterference principle and the humble maxim 'keeping a low profile' because they still serve China's diplomatic interests. However, the Libya evacuation, antipiracy missions in the Gulf of Aden and joint Mekong River patrols, demonstrate a more capable, assertive and confident China. As the Libya evacuation revealed, China is willing, or being pushed, to be more assertive to protect its interests abroad. The confidence Beijing displayed in the large-scale evacuation of Chinese citizens from Libya, suggests a pending shift in China's strategic posture and foreign policy outlook—in other words, China is coming out of its shell.

Conclusion

China's leaders are beginning to experience the daunting responsibilities of a first world power. No longer is Beijing acting in a world with limited expectations of its behavior. Because of its meteoric economic growth and precipitous rise in power, its

---

87. Cai Kai, 'Mekong murder trial a model of regional judicial cooperation', *Xinhua*, (20 September 2012), available at: http://news.xinhuanet.com/english/indepth/2012-09/20/c_131863289.htm (accessed 21 December 2012).
88. 'New direction for China's diplomacy', *Beijing Review*, (8 March 2012), available at: http://www.bjreview.com.cn/world/txt/2012-03/05/content_439626.htm (accessed 21 December 2012).
89. *Ibid.*
90. *Ibid.*

citizens' expectations have also grown. Increasingly, the Chinese public expects the government to manage crises at home and abroad. Specifically, the safety of Chinese citizens overseas has become a new diplomatic imperative for several reasons. First, many Chinese businesses are investing in projects overseas—many in high risk regions such as Africa and Southwest Asia. Chinese businesses opt for these precarious locations because of the potential high pay-off. Second, further complicating these ventures, instead of employing locals, Chinese businesses bring their own skilled workers from China to work on major projects. Third, because of China's growing prestige and wealth, domestic pressure is increasing to protect citizens overseas. China's growing middle class and internationally astute citizens expect Beijing to act as a first world power and protect its citizens abroad. Furthermore, because of the Internet and near instantaneous communication, the maltreatment of Chinese friends and family is quickly disseminated.

China spent at least US$152 million (1 billion yuan or 28,000 yuan per evacuee) on the mass evacuation of its citizens from Libya.[91] The large-scale Libya evacuation served as a wake-up call for Beijing. There is growing realization among Chinese officials that conducting business in high risk countries has many drawbacks, and Beijing needs to be prepared to manage and mitigate emergency situations. In the past decade, China has established institutional mechanisms to manage the overseas citizen protection system. Additionally, the PLA has begun to expand its capabilities and missions to support this system. The Libya crisis tested these institutions and demonstrated its maturity but also revealed their inadequacies. Moreover, the *ad hoc* nature of the Libya evacuation operation and the apparent lack of standard operating procedures raise concerns about the capacity of the Chinese government to sustain its overseas crisis management system. The Libya operation also highlighted the evolving mandate of the PLA and the justification for increased defense spending in long-range capabilities due in part to China's expanding global interests. To deal with emerging global challenges, China will therefore continue to develop and expand its diplomatic and military capabilities.

Since initiating the 'Go Global' policy, the combination of the noninterference principle and 'keeping a low profile' maxim has been the cornerstone of China's global strategic posture. However, with Chinese vital national interests tethered to the stability of weak states, its current foreign policy approach is untenable. Overseas crisis management, especially the protection of Chinese abroad, is increasingly incompatible with Beijing's current foreign policy construct. As the Libya evacuation case revealed, the implications of China's rise and growing global interests make it increasingly difficult to maintain this strategic posture in practice. If China is to continue the 'Go Global' strategy, it must be proactive and assertive in protecting its interests, investments, and Chinese citizens overseas. Thus, China's expanding global interests will compel Beijing to reassess its global strategic posture and foreign policy principles in order to meet future challenges.

---

91. 'Libyan case proves China's emergency capability', *China Daily*, (13 March 2011), available at: Lexisnexis.com (accessed 30 September 2011).

China's Exceptionalism and the Challenges of Delivering Difference in Africa

CHRIS ALDEN and DANIEL LARGE

This article explores the notion of 'China's exceptionalism' in Africa, a prominent feature in Beijing's current continental and bilateral engagement. 'China's exceptionalism' is understood as a normative modality of engagement that seeks to structure relations such that, though they may remain asymmetrical in economic content they are nonetheless characterised as equal in terms of recognition of economic gains and political standing (mutual respect and political equality). This article considers the burden that the central Chinese government has assumed through its self-construction and mobilisation of a position of exceptionalism and, concurrently, the imperatives that flow from such rhetorical claims of distinctiveness in terms of demonstrating and delivering difference as a means to sustain the unity and coherence of these rhetorical commitments.

Beijing operates a distinctive normative mode in conducting its foreign policy relations with Africa: resting on an historically-informed framework, this is reinforced by contemporary rhetoric emphasising political equality, mutual benefit, sovereignty, non-interference and 'win–win cooperation'.[1] This position has meant that, to date, Chinese diplomacy has officially operated on a different basis from the more overtly hierarchical, prescriptive power relations of established external powers. China's official approach can be condensed into the notion of 'exceptionalism', understood as a term distilling a normative modality of engagement. It is geared toward ensuring 'mutual benefit' and 'win–win' outcomes at continental and bilateral levels. These principles are not unique as concepts informing the rhetoric of different

---

*Chris Alden is a Reader in International Relations at the London School of Economics and Political Science. Daniel Large is research director of the Africa Asia Centre, Royal African Society at the School of Oriental and African Studies. The authors would particularly like to thank Jonathan Holslag and the EU–China Academic Network annual conference, Brussels, December 2008 at which this paper was first presented. For providing feedback on the first draft, we would also like to thank Ricardo Soares de Oliveira, Gavin Williams and participants at an Oxford seminar on 19 January 2009; Lucy Corkin and participants at a SOAS seminar on 16 February 2009; Mario Esteban, participants and especially Yoon Park at a University of Johannesburg conference on 29 August 2009, and two anonymous reviewers for this journal. We naturally remain responsible for any errors.

1. See Barry Sautman, *Friends and Interests: China's Distinctive Links with Africa*, Center on China's Transnational Relations Working Paper No. 12 (Hong Kong: The Hong Kong University of Science and Technology, 2006).

external actors in Africa. China's Africa policy, however, does take particular form in so far as it accentuates a basic but fundamental difference in its relationship with the continent as compared to other actors—notably in a shared history of colonialism and experience as a developing country—while at the same time promoting political principles based on a stronger conception of state sovereignty, non-interference, territorial integrity and political equality. The position of exceptionalism, overall, seeks to structure relations such that though they may remain asymmetrical in economic content, they should remain equal in terms of recognition of economic gains and political standing (mutual respect and equality). This has been the authoritative source of much of China's engagement with Africa and, concurrently, a bone of contention with various African constituencies, Western governments, donors and NGOs alike, and is contributing to a recasting of Africa's relationship with the outside world.

Yet in projecting 'exceptionalism' as a basis for its successful Africa policy, the Chinese government has introduced a set of challenges arising from this foreign policy of difference, which imposes a considerable burden on Beijing to sustain the unity and coherence of its rhetorical commitments. What follows begins to explore some of the issues that China's own principles give rise to and to suggest possible ways in which a different perspective on China's mobilised principles might productively open an alternative interpretative window onto China's contemporary African relations. Given that China's African engagement has been too often treated with a problematic combination of assumed linearity—the apparent inexorable expansion of China in Africa—and economic reductionism—ascribing motivations exclusively to resource-diplomacy or material motivation, we aim to present a more conceptually oriented analysis in a manner geared toward stimulating discussion of broader themes. The first section in what follows briefly contextualises current China–Africa relations and the role of ideas in China's relations with Africa. The notion of 'exceptionalism' and the challenges this poses for China is then elaborated upon, followed by an assessment of possible means by which reconciliation might be achieved. Finally we conclude by positing that an alternative variation of 'neo-modernisation with Chinese characteristics' is at play in Africa.

Four caveats should be noted at the outset. First, this article does not claim to present a narrowly or in any way rigorous theoretical treatment. However, being conceptually attuned, it seeks to assert the importance of a more serious treatment of the ideational dimensions of China's foreign policy. It aims, therefore, to introduce another element to the debate on Chinese foreign policy in Africa and not to replace or downplay important material factors.[2] Second, while scholarly analysis of China–Africa has become more sophisticated and progressed beyond what was formerly a narrow and frequently binary subject, this paper is not concerned with rehearsing a critique of China; rather, it points to aspects which might be considered as arising from these self-generated tensions as well as seeking to rescue China's rhetoric as received in Africa from the dilemmas and problems of its own making. Third, with respect to the literature on 'Chinese exceptionalism' (and, for that matter, of other states): rather

---

2. See Giles Mohan and Marcus Power, 'New African choices? The politics of Chinese engagement', *Review of African Political Economy* 115, (2008), pp. 23–42, for a recent, theoretically oriented account.

than broaching issues of cultural or racial essentialism, the term here is used only in so far as it delineates a Chinese position characterised by its own proponents as unique to the point of being exceptional both in itself, and in comparison to other external actors in Africa.[3] In this light, a subject considered in different ways elsewhere is taken up in the Africa context where key aspects of China's self-identity and claims made in its ideology—notably China's position as a great nation, absolute respect for state sovereignty and a foreign policy based on immutable universal principles such as justice and equity rather than expediency—can be said to 'constitute a claim to entitlement by virtue of China's ontological status rather than its behavioural characteristics. In effect, they are a demand that others recognize and respect China's exceptional qualities'.[4] Finally, we suggest that prevailing Western approaches to key areas of debate on contemporary Chinese foreign policy in Africa might be reconsidered in so far as these impose prior narratives and sets of interpretative criteria. Rather than advancing any particular relativist argument, the intent here is to argue for greater interpretative latitude in approaching the official Chinese position that blends fresh energy of the present with ideas and approaches that reprise previous phases of externally-directed development, and multiple attendant tensions, in Africa.

Expansive relations amidst economic uncertainty

After decades of obscurity, China's relations with Africa have become a regular feature of media attention, research and policy engagement.[5] This was catalysed by the third Forum for China Africa Cooperation (FOCAC) in November 2006, which proclaimed 'the establishment of a new type of strategic partnership between China and Africa featuring political equality and mutual trust, economic win–win cooperation'.[6] Three aspects of what still remains an emergent relationship are worth noting by way of context. The first is the striking dynamism of China's relations with Africa.[7] This was seen most obviously at a general level in rising China–Africa trade: in 2007 this reached US$73 billion, and in 2008 total trade reached nearly US$107 billion, meaning that the trade target of US$100 billion set at FOCAC III was achieved two years early and China was acclaimed as Africa's second leading trading partner after the United States, having surpassed France as one of the continent's key investors. While trade has been affected by the recent global economic downturn, the expansion of a multifaceted Chinese engagement has continued. This is evident in

---

3. There is a longstanding debate concerning China and International Relations theory, as well as approaches to China in other disciplines as part of what has been called 'the tiresome apartheid of Chinese exceptionalism—promoted by scholars in and out of Chinese Studies, as well as political leaders around the world'. See Louise Edwards, 'Review article: The "problem of China" and Chinese exceptionalism', *Journal of Contemporary History* 43(1), (2008), pp. 155–164. Our approach is different, however, in engaging the Chinese government's *self-proclaimed* distinctiveness that can be considered exceptional.

4. Stephen Levine, 'Perception and ideology in Chinese foreign policy', in Thomas W. Robinson and David Shambaugh, eds, *Chinese Foreign Policy: Theory and Practice* (Oxford: Clarendon Press, 1995), p. 44.

5. See Daniel Large, 'Beyond "dragon in the bush": the study of China–Africa relations', *African Affairs* 107(426), (2008), pp. 45–61; Robert Rotberg, ed., *China into Africa: Trade, Aid and Influence* (Washington, DC: Brookings Institution Press, 2008); Arthur Waldron, ed., *China in Africa* (Washington, DC: The Jamestown Foundation, 2009).

6. *Declaration of the Beijing Summit of the Forum on China–Africa Cooperation* (draft), (16 November 2006), available at: http://english.focacsummit.org/2006-11/16/content_6586.htm.

7. See, for example, recent issues of *Africa–Asia Confidential*.

the expansion of educational and cultural exchanges, emergence of new business areas like financial services (seen in the Industrial and Commercial Bank of China's US$5.6 billion purchase of a 20% stake in Standard Bank in late 2007), the further development of agricultural cooperation and recalibrated long-term supply agreements for infrastructure, inclusive of other social offsets not seen in earlier Chinese deals with African states. The progress of the US$5 billion Africa Development Fund, overseen by the China Development Bank, has been impeded by such factors as capacity constraints and uncertainty from smaller Chinese businesses about its operations, but is nonetheless set to underwrite substantial investment by Chinese companies in Africa. The rolling out of China's 'special economic zones' pledged in 2006 continues, the first such zones being established in Kitwe (Zambia), Mauritius and Luanda (Angola), with competition to secure the others, even as other China-backed free trade and export processing zones are established independently (including three in Nigeria). Second, the diversity of Chinese actors is increasingly important in shaping official relations by the experiences of ground actors. An overlooked driver of Chinese engagement is that of small and medium sized enterprises engaged in a variety of business activities across Africa. Attention has mostly been directed toward the larger state-owned enterprises, particularly the likes of CNPCC, Sinopec, CNOOC or others engaged in high-profile energy security foreign policy imperatives, and with good reason. However, Africa remains a destination and operating space for a diverse array of Chinese entrepreneurial endeavours, framed as it has been in terms of frontier imagery as the last open continent of strong economic opportunity and potential.[8] Third, adaptation on the part of African governments and Beijing has continued. With notable exceptions, officially harmonious state–state relations are nevertheless not immune to political and economic gravity. Investment protection amidst burgeoning economic activities has become more important to Beijing and Chinese companies as part of a changing engagement profile that, broadly speaking, has been in the process of transitioning from an entry/start-up phase into more routine, normalised business relations on the back of these committed investments. In places, notably those resource-rich African states with more developed Chinese investment, this expansion and consolidation of Chinese interests has been accompanied by a process and logic of deepening political involvement. China has thus far put its faith in elite ties and African-led multilateralism to handle issues rising from its deepening engagement but the cross-cutting interests of central state, provincial and private actors makes this increasingly difficult to manage, and harder to effectively control.

China remains the most important rising force in Africa and its position of enhanced significance in African affairs appears set to have a sustained impact beyond the short–medium term.[9] However, the exuberant optimism that has accompanied—and in part been engendered by—the apparently inexorable Chinese rise throughout the continent has been tempered in light of global economic change. As China's economy enters

8. See Jing Gu, 'China's private enterprises in Africa and the implications for African development', *European Journal of Development Research* 21(4), (2009), pp. 570–587.

9. See Chris Alden, Daniel Large and Ricardo Soares de Oliveira, eds, *China Returns to Africa: A Rising Power and a Continent Embrace* (London: Hurst, 2008); and Ian Taylor, *China's New Role in Africa* (Boulder, CO: Lynne Rienner, 2009).

uncharted waters against the background of wider global economic turbulence, its relations with Africa also face the same reality. The contingencies that have accompanied the thickening of relations ought to be recognised: there is nothing inevitable or preordained about the easy, linear expansion of China's own domestic development and its relations with Africa likewise. As Wu Bangguo, chairman of the standing committee of the National People's Congress, reminded the China–Africa Business Cooperation Conference in Cairo in May 2007, China faces its own challenges in sustaining its development.[10] This illustrates how the rhetoric of the Chinese government can not only generate raised expectations about China's relations with developing countries but additionally underlines, in the face of domestic economic and political imperatives, the constrained ability of China to contribute to the development of other developing countries. The pressure on China to meet high expectations within Africa as a benevolent development partner, an 'all weather friend' that can deliver irrespective of turmoil in international markets, continues.

China's constitutive foreign policy rhetoric

Different accounts of China's relations with Africa have commonly ascribed Chinese engagement to resource supply needs bolstered by geopolitical imperatives. These are important factors underpinning China's interest in Africa.[11] However, the quotidian tendency to reduce Chinese motivation exclusively to economic need, a dynamic particularly evident in mono-causal realist explanations foregrounding explanations of interest-driven resource demand and power politics, should be resisted.[12] Resource supply imperatives for China as a globally-connected manufacturing centre are undoubtedly a critical aspect of its Africa relations but in and entirely of themselves do not provide a full explanation for the form and trajectory of China's Africa policy. Crude economic reductionism, sometimes framed in terms of a misleadingly atavistic, inaccurate and politically charged terminology of 'imperialism' or 'neo-colonialism', is limited on its own as a means of appraising current relations wherein political factors, including those—most notably China's relations with Japan—that have been overlooked in Western commentary are increasingly important in directing and shaping Beijing's foreign policy.

Though commonly dismissed as masking ulterior motives or cloaking hard interests, the official rhetoric of the central Chinese government's relations with Africa is significant in ways that go beyond any simple, misleading and binary material/discursive opposition ('reality'/'rhetoric', 'practice/discourse'). The principles of China's Africa relations matter: these structure and influence foreign policy formation, decision-making and implementation both in terms of China's pan-Africa relations as well as bilaterally between China and currently 49 very different African states. China's foreign policy rhetoric, in other words, should be taken as importantly

---

10. See 'Full text of Wu Bangguo's speech at the China–Africa Business Cooperation Conference in Cairo 21 May 2007', *Xinhua*, (21 May 2007).

11. For a recent account, see Wenran Jiang, 'Fuelling the dragon: China's rise and its energy and resources extraction in Africa', *The China Quarterly* 199, (2009), pp. 585–609.

12. See, for a recent example, Peter Hitchens, 'How China has created a new slave empire in Africa', *Daily Mail*, (28 September 2008).

constitutive of its relations with Africa in exercising a structuring influence to date in official state–state relations and having real-world impacts. Foreign policy principles, like those of other actors engaged in business or the business of development, will invariably diverge from stated motivations, action and experience. Liberalism's language of human rights or humanitarian intervention, for instance, can be utilised to express genuine motivation and also be more instrumentally appropriated: it will always be messy and hard to disentangle. The principles of Chinese foreign policy are likewise important to take seriously as historically-informed contemporary articulations of political viewpoints, even if at times these can blur or intermesh with other political or economic dynamics,[13] or be the object of disagreement and active critique.

The question of China's exceptionalism raises a fault-line likely to remain important in influencing the trajectory of how expanding relations in Africa develop over time. This concerns the way in which China's official foreign policy rhetoric as a discursively enacted normative ideal relates to and coheres with discourse as a descriptive feature of increasingly multifaceted, complex Chinese experience throughout the African continent. While an established, officially sanctioned narrative of China's ancient historical connections continues to be propagated, and has remained remarkably constant,[14] an alternative historical interpretation of China's Africa relations might be offered. China's episodic post-colonial relations with Africa were characterised by shifting periods of expanding and contracting relations that broadly synchronised with the PRC's domestic politics and foreign policy imperatives vis-à-vis the West, the Soviet Union and Taiwan. In this regard, two general, contrasting trajectories of historical engagement, which themselves exhibit variation, might be advanced as a means to conceive of today's relations. During the revolutionary Maoist period, China's 'thick' rhetoric of socialist solidarity and revolution was mobilised across what was otherwise mostly a 'thin' content of actual relations with Africa.[15] By way of contrast, what has become more apparent with the development of relations in the post-1989 period (and particularly since 1996 following President Jiang Zemin's Africa tour and the first FOCAC in Beijing in 2000), is that the 'thickening' of multidimensional Chinese relations with Africa has meant that official Chinese principles are spread in a 'thinner' manner over an increasingly substantive and diverse content of ties with the continent. One result is that the Chinese government's foreign policy rhetoric does not always reflect or comfortably accommodate the trajectory of change in today's comparatively new historical phrase of relations. The broad contrast suggested here between the 'thick' rhetoric and 'thin' content of the Maoist era on the one hand and the 'thickening'

---

13. See Michael H. Hunt, *The Genesis of Chinese Communist Foreign Policy* (New York: Columbia University Press, 2006); Chris Alden and Ana Cristina Alves, 'History and identity in the construction of China's Africa policy', *Review of African Political Economy* 115, (2008), pp. 43–58.

14. See Julia C. Strauss, 'The past in the present: historical and rhetorical lineages in China's relations with Africa', *China Quarterly* 199, (2009), pp. 777–795.

15. To loosely adapt Michael Walzer, *Thick and Thin: Moral Argument at Home and Abroad* (Notre Dame, IN: University of Notre Dame Press, 1994). See also Philip Snow, *The Star Raft: China's Encounter with Africa* (London: Weidenfeld and Nicolson, 1988). By this, of course, we put China's present engagement in comparative perspective to that of the past, on which see also Deborah Brautigam, *Chinese Aid and African Development: Exporting Green Revolution* (London: Macmillan Press Ltd, 1998); and Jamie Monson, *Africa's Freedom Railway: How a Chinese Development Project Changed Lives and Livelihoods in Tanzania* (Bloomington, IN: Indiana University Press, 2009).

content and 'thinner' rhetoric of the current moment on the other might suggest that the importance of Chinese rhetoric in informing the actual content of relations should not be overemphasised.

Given this experience, China's foreign policy rhetoric can serve to both constrain and enable: it constrains by imposing restrictions and storing up policy challenges, but it also enables by continuing to offer a tremendously successful discourse of a particular form of intervention into Africa to date. Here, for African governments in particular but also more widely, the virtues and political benefits of China's unchanging political principles can be favourably contrasted with periodic Western impulses to alter their Africa policy, which produces a contrast between the image of chameleon-like Western powers constantly changing their political colours and China as a constant all-weather friend. Nonetheless, questions remain over the efficacy of what have proved a short-term tactical advantage that has accrued to China in the presentation of its Africa policy codified under the rather different historical circumstances of post-1949 Chinese foreign relations. In particular, the official contemporary Chinese rhetoric on Africa has been presented as one of a paradoxical involved detachment: it is at one and the same time a language of common aims and activity exhorting an expansion of ties for 'common development', but it also appears to be predicated upon a basic non-involvement of China in African affairs defined in terms of 'internal politics' of states. As such, and given that it can be far more easily operationalised as a rhetoric of intention rather than real-world effects, the principles of China's approach appear to be more comfortable on paper with the absence of a substantive, more politicised involvement. The assorted tensions in China's Africa relations today might therefore be said to come as little surprise.

Exceptionalism and the moral basis of Chinese engagement

The decline of formal ideology expressed in Marxism–Leninism–Mao Zedong Thought observed during the 1990s has prompted the more recent observation that: 'For practical purposes, it's the end of [Marxist] ideology in China'.[16] This has been superseded by more recent attempts to articulate a coherent conceptual basis for Chinese foreign policy, notably in the abortive 'peaceful rise' line and President Hu Jintao's 'harmonious society' in China and its global concomitant, a 'harmonious world'.[17]

The moral basis of China's power in Africa is in itself a form of 'exceptionalism' given expression and codified in the language, framing and conduct of official state relations. It is informed by a discourse of difference and similitude: China is a power represented as being categorically different to all previous external powers engaged in Africa, a difference based fundamentally on its seminal experience as a developing country with a similar experience of (semi-)colonial subjugation. As such, China's

---

16. Daniel Bell, *China's New Confucianism* (Princeton, NJ: Princeton University Press, 2008), p. 8.
17. See Suisheng Zhao, ed., *Chinese Foreign Policy: Pragmatism and Strategic Behavior* (Armonk, NY: M.E. Sharpe, 2004); Bonnie S. Glaser and Evan S. Medeiros, 'The changing ecology of foreign policy-making in China: the ascension and demise of the theory of "peaceful rise"', *China Quarterly* 190, (2007), pp. 291–310; Chih-Yu Shih, *China's Just World: The Morality of Chinese Foreign Policy* (Boulder, CO: Lynne Rienner, 1993).

exceptionalism is a modality of engagement that structures relations such that they may remain asymmetrical in economic content but equal in terms of recognition of economic gains and political standing. The Chinese government's distinctive mode of conducting its Africa relations is founded in a historically-informed framework defined by equality, mutual respect and benefit, sovereignty and non-interference, as well as the competitive practices that have contributed to its business expansion.

China's exceptionalism is rooted in the Chinese government's interpretation of the famous 1955 Bandung conference's principles which effectively laid out a framework for post-colonial states' responses to the challenges of nation building in the international system; China's related, Five Principles of Peaceful Coexistence, its key principles on foreign aid (particularly the Eight Principles for China's Aid to Foreign Countries from 1964) and its own assessment of its unblemished record of application of these principles in its foreign policy towards Africa.[18] Drawing from Bandung has the advantage that its principles have been influential in shaping the construction of Africa's regional sub-system and the foreign policy of most African states as well. First, sovereignty, equality and non-interference were seen to be a bulwark for weakly legitimated regimes coming to power in the wake of colonialism and the competing claims to recognition both within and without the territorial boundaries of newly formed states. Secondly, mutual benefit—a developmental principle of Bandung that allows conformity to a variety of practices from socialist-era barter trade to complex project financing—breaks decidedly with the donor–recipient model of development dominant amongst OECD countries. The modality of negotiation, reinforced by China's own experience as a developing country which involved a doctrine of experimentation, ranging from the ideological extremes of the early revolutionary period to the market gradualism of the 'opening and reform' policy, puts a premium on achieving practical outcomes recognised as such by all parties. It is an interest-based approach that is designed to ensure 'ownership' of the development process. Third, official Chinese interpretation of their experience in Africa, which casts (and in places white-washes) its uneven record of engagement with revolutionary parties and regimes as one of seamless constancy, and suggests that China's relationship with Africa has already demonstrated conformity to the continent's key aims, namely the political project of independence and the developmental project of economic welfare and nation-building. A final dimension concerns the Chinese government's non-conditional approach to its political relations with African states (albeit with the exception of the One China principle). Taking the form of a commitment against public prescription and an aversion to imposing its preferred political direction upon a state partner, this is in part an operationalised extension of political equality but extends into a number of areas of 'internal politics' from state policy to development strategy. It does much to distinguish Beijing's approach from the more intrusive policies of Western states, IFIs or the development sector more generally that codify and promote change agendas from without. Somewhat in contrast to the Maoist impulse to export world revolution, China today claims to accept political diversity from a baseline of not overtly dictating to others.

---

18. Xu Jian, 'The theory and practice of the Five Principles of Peaceful Coexistence', *International Studies* 1, (January 2005), pp. 22–42. See Kweku Ampiah, *The Political and Moral Imperatives of the Bandung Conference of 1955: The Reactions of US, UK and Japan* (London: Global Oriental, 2007).

The flipside is support for such regimes as the National Congress party in Sudan or ZANU in Zimbabwe that has done much to damage China's international reputation.

These four constitutive features of exceptionalism underscore China's preference for order and stability in the international system, and its desire to work with fellow developing countries on a basis of political equality and mutual economic gain. As articulated in the Chinese government's Africa Policy of January 2006, it claims to be a durable interest-based framework that will hold fast to the nation building needs of African states and, as such, will not indulge in activities that undermine these central goals.[19] This position provides the moral basis of its expressions of power in Africa by asserting that Chinese foreign policy will, unlike the West, be guided by principles that ensure that the relationship will not devolve into one of exploitation. At the same time, it is important not to artificially frame China's relations with Africa in a vacuum: in comparative regional context, there are interesting parallels to be made with other regional dimensions of contemporary Chinese foreign relations. Despite the uneven application and meaning of the principles of Chinese foreign policy in the wider scheme of China's different regional relations, there are also common themes that unite it. In the case of China–Latin America relations, for example, a similar rhetorical basis underpins official relations.[20]

Challenges of delivering difference

China's approach has been more than sufficient to confer legitimacy and enable it to foster flourishing relations with many African governments. Part of China's actual and perceived success can be attributed as a judgement on the record and conduct of Western powers in Africa, but much can also be said to be a positive recognition of the actual and potential benefits of serious Chinese engagement. Nevertheless, this position is being tested in various ways by the growing complexity of ties. In this continuing process, the consolidation and sustainability of Chinese relations with Africa poses challenges to China's exceptionalism. This can be seen both from above, in terms of how the core principles of Chinese foreign policy in Africa square with the nature of growing ties, and also from below in terms of how the increasingly multifaceted Chinese experience is putting accepted rhetoric under strain and is already—and may further—contribute toward a revision of official rhetoric through myriad quotidian ground encounters.

Just as it contains noted advantages and benefits, Beijing imposes a number of notable challenges on itself by continuing this rhetoric of difference. The first, perhaps most visible and controversial area where this is manifest, is the tension between the principle of non-interference and its actual and experienced meanings when related to a more involved Chinese role. Commentators have been quick to cite episodes in Zambia and Sudan as evidence of a shift beyond non-interference, but this idea is problematic and too easily simplified in view of the case-specific tactical

---

19. See *China's Africa Policy*, (January 2006).
20. This can be seen in China's policy on its relations with Latin America, where there is a similar emphasis on 'equality, mutual benefit and common development' or 'independence, full equality, mutual respect and non-interference in each other's internal affairs'. See *China's Policy Paper on Latin America and the Caribbean*, (5 November 2008).

manoeuvring on display in both instances and the prevailing centrality, amidst policy discussion and pragmatic practical adaptation, of the principle in Chinese foreign policy.[21] Non-interference as a policy principle opposed to holding an active, public position in regard to the internal politics of a given state may be maintained by Beijing as a distinctive marker of difference from European states or the United States. As it transpires in terms of the content of political relations, this in practice sits uncomfortably with the Chinese government's support for a variety of regimes from Chad to Equatorial Guinea.[22] Furthermore, in places the principle has emerged as an apparent constraint on securing China's established interests and given rise to policy debate about its advantages and disadvantages and possible revised meanings.[23] China's approach in relation to Sudan has been usefully defined as insisting on 'influence without interference' and in many ways this captures a policy position that is both coherent to an extent and also ambiguous.[24] Nonetheless, the kidnapping and killing of five Chinese workers in Sudan's region of Southern Kordofan in October 2008 illustrated in tragic and dramatic manner the political complexities of a key engagement built up since the mid-1990s and in turn the need by China to ensure its workers and investments can be protected, a challenge that invariably must entail a more involved political engagement that deals with divergent interests between China and the central Sudan government.[25]

A second area is the strain imposed by the distance between the principle of sovereign equality and actual asymmetry of power evident in China's relations with an immense variety of African states. Julius Nyerere, the former President of Tanzania, in describing his country's relations with the PRC, summed up a wider pattern of African state relations with China in describing these as the 'most unequal of equal relationships'.[26] Simultaneously capturing co-existent political equality and asymmetrical power relations, this description remains appropriate today when China's principle of political equality remains a live normative commitment informing diplomatic practices and relations with African states; yet at the same time it is hard to square with broader economic and political relations, whether these be with the minority of resource-rich key Chinese trade partners—Angola, South Africa, Nigeria, Sudan—or other less strategic partners.

A third area concerns the underlying structure of China's trade relations with Africa. Chinese trade and investment is concentrated in a comparatively small number of resource-rich states, which produce a majority of Africa's trade with

---

21. The controversy during the Zambian presidential elections in late 2006 focused on the Chinese ambassador's response to opposition candidate Michael Sata's playing the Taiwan card, thus arguably in part broaching an issue in 'domestic' Chinese politics; China's position on Sudan/Darfur undoubtedly evolved but has maintained non-interference as a basic principle. See Jonathan Holslag, 'China's diplomatic maneuvering on the question of Darfur', *Journal of Contemporary China* 17(54), (2008), pp. 71–84.

22. Mario Esteban, 'The Chinese *amigo*: implications for the development of Equatorial Guinea', *The China Quarterly* 199, (2009), pp. 667–685.

23. Linda Jakobson, 'The burden of "non-interference"', *China Economic Quarterly* 11(2), (2007), pp. 14–18.

24. See Li Anshan, 'China and Africa: policy and challenges', *China Security* 3(3), (2007), p. 77.

25. See Daniel Large, 'China's Sudan engagement: changing Northern and Southern political trajectories in peace and war', *The China Quarterly* 199, (2009), pp. 610–626.

26. To paraphrase Nyerere's description of China–Tanzania relations. Quoted in George T. Yu, *China and Tanzania: A Study in Cooperative Interaction* (Berkeley, CA: Center for Chinese Studies, University of California, 1970), p. 97.

China. Thus in 2008, 79% of all African exports came from just five countries and 93% of exports came from ten countries; five countries accounted for some 55% and ten countries for 75% of all exports from China to Africa.[27] The basic character of trade flows—Africa exporting raw materials and China exporting manufactured products—replicates Africa's economic relations with other external powers. Amidst Chinese competition for Africa's nascent manufacturing sector—initiatives such as voluntary export ban with South Africa have been tried—these have also fallen foul of forces of economic globalisation. There, optimistically speaking, appears to be a certain potential in terms of the relocation of Chinese manufacturing into Africa as an effort to correct to a degree basic underlying economic issues. The Chinese government's insistence on its exceptional condition—that it will remain a benevolent partner for Africa despite the hard facts about the nature of its commodity based trading relationship—places a significant burden on China's foreign policy to ensure that it is conducted in a manner that retains this moral high ground against the backdrop of its evolving commercial and attendant interests on the continent.

The question of China's principal–agent dilemma is, fourthly, in many respects the heart of the problem facing the consolidation of sustainable Chinese ties with Africa. The proliferation of sub-state actors, ranging from provincially-backed firms to smaller, more independent businesses, and their behaviour seems increasingly to suggest their own autonomy from Beijing. Chinese diplomats will often assert, when confronted by evidence of malfeasance by a Chinese company, the lack of knowledge of the firm in question and the absence of means to control such conduct. This situation is a deliberate outgrowth of decisions taken as far back as the mid-1980s when the Chinese government began authorising provinces and the three largest municipalities to conduct trade policy through local offices.[28] By the 1990s, the richer of these, like Guangdong and Zhejiang, were able to use their financial reserves to launch trade initiatives and fund delegations to explore the economic opportunities to be found in Africa. More recently, in a clear expression of the gap between the official declaratory policy towards Africa and the emerging reality on the ground, Beijing has turned to these same sub-state actors to help fulfil the ambitious targets announced at FOCAC III in 2006. The package of incentives then put forward by the Chinese government and administered by the China Development Bank, which included export credits and insurance for firms investing or relocating in Africa, has served as a spur for greater expansion of commerce into the continent. In responding to such challenges as grievances concerning business practices, increased security threats, or demands for protection by Chinese nationals in Africa, the central Chinese government is attempting to direct a diverse array of actors over which its ability to exercise control is not as great as widely assumed.[29]

The consolidation of China's newly acquired standing in the continent presents challenges that are at once the ordinary stuff of preserving these interests but also, and

---

27. Mark George, *China Africa Two-Way Trade—Recent Developments*, DFID Policy Briefing Paper, (30 January 2009).
28. Chen Zhimin and Jian Jinbo, *Chinese Provinces as Foreign Policy Actors in Africa*, South African Institute for International Affairs (SAIIA) China in Africa Policy Report (Braamfontein: SAIIA, 2009), pp. 2–3.
29. Bates Gill and James Reilly, 'The tenuous hold of China Inc. in Africa', *The Washington Quarterly* 30(3), (2007), pp. 37–52.

more profoundly, reassessing its position within the international system.[30] The claim that China's ties with Africa are distinctly different from those of the West, rooted as has been argued in historical interpretation and reinforced by the official contemporary rhetoric, is more significant than it first appears in that it cuts at the very identity of China and the nature of its relations with Africa. Chinese diplomacy on the continent has successfully operated on a different basis from the more overtly hierarchically structured power relations of established powers or international development agencies. China today is still a developing country but one whose very success is transforming its economy and that of the globe's, a process which projects the vagaries of China's changing status upon its relations with Africa and the rest of the world.

African states and governing elites today consider linking their economic and political direction to a Chinese future precisely because of its accomplishments and the presumed economic path that it is proceeding along. This has historical precedents: during socialist China's relations with Tanzania in the 1960s, 'the primary appeal of the Chinese model to the Tanzanian elites', as Yu observed, lay 'in its perceived technical competency and rapidity toward modernity'.[31] Today, however, China's confirmed economic development and global position presents less a perception of a route toward modernity than a confirmed example of its partial achievement and the prospect of yet further development. At the same time, China's own demonstrable developmental success poses an obvious contradiction that—while historically correct—effectively undermines the oft-stated diplomatic mantra of Beijing that China is the largest developing country relating to Africa as the largest developing continent. This is testament to China's enduring 'dual-identity syndrome of great power versus poor country', in seeking recognition as a great power but also being a developing country.[32] It is hard, particularly when engaging many African actors, for Beijing to convincingly maintain that China is a developing country with lingering and pervasive poverty if African delegations are taken only to such places as the glimmering vision of modernity that is contemporary Shanghai. There are naturally multiple images of and associations with China in highly diverse African contexts, including that of rural poverty, but in many African contexts, the image of China as a developed country is also upheld and is one that can support its developmental rhetoric by reference to concrete achievements in China.[33]

The limits of reconciliation through rhetoric

FOCAC III was an opportunity to celebrate the achievements in Chinese–African relations on the fiftieth anniversary of diplomatic relations with Egypt, and chart the way forward through declared intent and a set of targets. FOCAC IV, held in Egypt in November 2009, contrasted with Beijing in November 2006 in terms

---

30. 'Conclusion: all over Africa', in Alden *et al.*, eds, *China Returns to Africa*, pp. 371–376.
31. George T. Yu, *China and Tanzania: A Study in Cooperative Interaction* (Berkeley, CA: Center for Chinese Studies, University of California, 1970), p. 33.
32. X. Wu, 'Four contradictions constraining China's foreign policy behavior', in Suisheng Zhao, ed., *Chinese Foreign Policy*, pp. 58–65.
33. As the FOCAC III Declaration asserts, 'The African countries are greatly inspired by China's rapid economic development'.

of timing and tone. With an emphasis partly on fulfilling aims set out at the Beijing summit of 2006, this also marked a further set of commitments to expand and deepen relations in a variety of areas from climate change to health, agriculture or science and technology, as well as measures to provide concessional financing support to African countries and businesses. Overall, this position suggests that the emerging contours of relations are becoming normalised. Greater involvement has already exposed different Chinese interests to African politics and brought unanticipated challenges. Chinese actors are becoming a more established part of the economic and socio-political landscape in many parts of Africa through expanding business relations, experience and the progressive deepening of a wider range of links. This process whereby a range of Chinese interests become a more ordinary part of life is one that, in parts, has already and is likely to further erode the popular reputation for exceptionalism China has enjoyed from below. While bringing benefits and holding strong potential, China's re-engagement also renews familiar questions of power, states and constraints on development, including Africa's unfavourable position in the world economy. Identifying a possible 'divergence of interests' between China and Africa over the longer-term, given the emerging tensions between the principles and nature of Chinese engagement, Snow's observation remains pertinent:

> a frank quest for profits by both China and its African partners might well, in the end, prove a more solid basis for their future relationship than the continuing attempt to sustain a rhetorical unity which has sometimes disguised the pursuit of profoundly different goals.[34]

The Chinese government's established practice of characterising controversy in ambiguous language—embracing market capitalism but insisting that there is no break with high principle ('socialism with Chinese characteristics')—is one way that this division between proclamations and reality is being managed. These sorts of rhetorical contortions, while possibly satisfying to a Chinese audience, have very limited appeal to African audiences long accustomed to mythologies as different as the discourse of racial superiority articulated under colonialism to the tarnished hopes of nationalism and fables of development under a corrupted independence. In fact, the proximity between these classic discourses of power which maintain notional equality and development as principles while conducting relations on a wholly different basis and Chinese rhetoric of mutual benefit in Africa appear, through the prism of African historical experience, closer than officials in Beijing may realise.

At the same time, the challenges posed by the consolidation of China's ties with Africa are introducing new dynamics that could potentially upset even these modest efforts at achieving rhetorical reconciliation. China's comparative advantage—articulated through the rhetoric of exceptionalism—in both being different and behaving differently from the West stands to be diluted if it changes its rhetoric (and all the more so its actions on the continent). However, as China seeks to adapt its policies to the pressures and necessities of protection of its new found interests, it will increasingly feel the need to adopt policy positions that reflect these changing circumstances. For instance the failure of governments in Ethiopia, Nigeria, South

---

34. Philip Snow, 'China and Africa: consensus and camouflage', in Robinson and Shambaugh, eds, *Chinese Foreign Policy*, pp. 283–321.

Africa and Sudan to protect Chinese citizens from targeted violence has brought a sharp, albeit conducted behind closed doors, response from Beijing.[35] The openness of global information sources has meant that the death of Chinese citizens in Africa is a feature of Chinese media and Internet interest, raising the stakes for a Chinese government that is becoming increasingly sensitive to accusations of neglect of its citizens' interests at home and abroad.[36] So too the problems faced by firms operating in the African context, where experiences of corruption, labour and legal disputes (whether self-imposed or otherwise) are affecting their business prospects. The result of all of this is to introduce stresses on Chinese foreign policy towards African states that must inevitably find their way in the content of China's African policy. While the rhetoric of exceptionalism and the framework of political engagement it operates through may in fact endure, much as the mantra of 'socialism' is clung to despite the changes to the economy, the strain of credibility for African audiences could come at a political cost.

In this regard, an important question for the most ardent proponents of China's exceptionalism has to be, even under the most benign conditions of Chinese engagement with the continent, whether Africans would acknowledge that win–win/mutual benefit was in fact taking place (beyond elite levels) and indeed possible in any longstanding relationship with an outside power. In other words, if one assumes Chinese intentions are fully honourable, the perceptual barrier on the part of Africans who incline to see in any external actor's displays of power a sign of nascent colonial intentions, would find a way of rendering these actions in a negative light. After all, for Africans the lessons of history suggest that the actions of foreign powers ultimately devolved into some form of self-serving exploitation of the continent, its people and resources. It is an attitude that is further conditioned by centuries of practices conducted by African elites who seek to employ external power in the service of their personal and regime interests against domestic opponents and, as often, societal interests.[37] This powerful, if problematic and simplistic, perspective is effectively 'coded' into African political discourse and contributes to the endurance of that cardinal principle of African international relations, that of solidarity. Indeed, there are parallels with the search for modernity and the perpetuation of a notion of 'victimhood' as a source of identity within China. Chinese officialdom's constant invocation of historical figures like Admiral Zheng He is a recognition of this seminal impulse within Africa and the necessity of combating a deeply held set of beliefs.

Finally while China may wish to claim exceptionalism on the basis of actions, in the end, what may truly distinguish China's engagement with Africa from that of the traditional Western powers is tied to the spectacle of Chinese migration.[38] Unlike Europe, which foreswore large-scale migration to Africa in the aftermath of the continent's political independence, the steady stream of Chinese immigrants to the continent is changing the social and economic landscape of many African countries.

---

35. Interview with senior Chinese official, September 2008.

36. Simon Shen, 'A constructed (un)reality on China's re-entry into Africa: the Chinese online community perception of Africa (2006–2008)', *The Journal of Modern African Studies* 47(3), (2009), pp. 425–448.

37. Jean Francois Bayart, 'Africa in the world: a history of extroversion', *African Affairs* 99(395), (2000), pp. 217–267.

38. Yoon Park, *Chinese Migration to Africa* (Braamfontein: SAIIA, 2008); E.M. Mung, 'Chinese migration and China's foreign policy in Africa', *Journal of Chinese Overseas* 4(1), (2008), pp. 91–109.

That they are arriving in cities and towns across the continent without necessarily being encouraged to do so by the Chinese central government does not take away from the African perception of a new and sometimes troubling displacement of the traditional order of things. This factor, arguably more than any other, is shaping the reaction of African society to China's engagement and is one which is largely outside of Beijing's control. And while personal relations between Chinese and Africans can be good, there is clear evidence of disquiet if not hostility on the part of some segments of African society. No amount of official rhetorical posturing will offset the perceptions founded in localised, personal experiences.

Towards African development with Chinese characteristics?

If the possibilities of reconciliation through rhetoric are limited, then what avenues are open to China? Far from relying on finely crafted words alone, Chinese officials put much stock in the practical application and accompanying demonstration effect of African experiences with China in constructing its Africa policy. This faith in the power of practical experience of development to 'win hearts and minds' of sceptical Africans, all the while openly pursuing an interest-based policy on the continent, is remarkable in its surety and ambition. At one level, it harbours an abiding conviction in the cultural fastness of Chinese society which, despite the emerging pluralism inherent in the contemporary Chinese engagement with Africa, will nonetheless convey and produce a singular, positive experience amongst Africans. According to this reading, the Chinese retail store owner will be as capable and interested in facilitating a positive relationship with local Africans as the Chinese oil company manager or diplomat. At another level, this assumes a radical re-orientation of the development process in Africa away from the Western-dominated approach to one which draws from Chinese perceptions, practices and experiences to help re-produce in Africa the development achievements that made contemporary China what it is today. It is the latter in particular, the promulgation of what might be termed African development with Chinese characteristics, that is causing consternation within traditional Western development circles.[39]

The conundrum faced by Western analysts and policy makers attempting to form an interpretation of Chinese strategies of development co-operation in Africa is misplaced. Rather than force the square pegs of OECD–DAC categories into the round holes of China's policies, Chinese actions in this sphere might best be understood as introducing a form of African development with Chinese characteristics, one which folds together classic assertions of modernisation theory, mercantilist self-interest and actual development experience within the open-ended rhetoric of 'South–South co-operation'. Informing this is a visceral understanding of the challenges of fomenting economic development in post-conflict context alongside the imperatives of nation-building under guidance of the governing party, the very conditions which the Chinese Communist Party (CCP) faced upon taking power in October 1949. A drive towards modernisation, echoing the classic discourse of the

---

39. See, for example, Moises Naim, 'Rogue aid', *Foreign Policy* 159, (March/April 2007), pp. 94–95; for a more considered, authoritative new account, see Deborah Brautigam, *The Dragon's Gift: The Real Story of China in Africa* (Oxford: Oxford University Press, 2009).

1960s rather than the problematic concept(s) of development, is a more accurate source of understanding of the Chinese approach in Africa. At a basic level, classic modernisation theory posits that political stability goes hand-in-hand with the effort to re-structure traditional agriculture along mechanised lines and, concurrently, a shift over into industrialised forms of production.[40] Indeed, the CCP's primary task has remained—unashamedly so—to break the back of feudalism and modernise China in its economic and political forms along lines that allow the country to assume its putative rightful place amongst the top table of nations.[41] This is contrasted with the more tortured positions adopted by Western development advocates who, true to liberal impulses, envisage a form of utopian intervention that 'does no harm' and absolves itself of acting as anything other than in what it defines as the interests of the target population.[42] The difference that China brings to its contemporary form of modernisation is a deep suspicion of overarching blueprints borne of their own experience of command-style economics, a bottom-up form of *praxis* whose very eclecticism allows it to adapt to changing circumstances and whose ultimate measure of success is to be found in the outcome, rather than the process itself. It is a script for economic development written not in the windy halls of the World Bank's H Street building in Washington nor the universities of the West but in the farming communities of Sichuan and on the shop room floor of hundreds of new industries built on the marshland of Shenzhen.

Practically speaking, the Chinese approach to facilitating development in Africa translates to a conscious recognition that the material requirements of economic improvement are dependent upon the pre-requisites of hard infrastructure—provisions for transport, communication—to mobilise the factors of production—capital, land and labour—in Africa. This adaptive approach to development flows from Beijing's own experience as it shifted from command-style modernisation to a more decentralised, market-based approach and the rapid economic gains that followed. Where once these initiatives in Africa would have been directly done by the Chinese state (as in the case of the Tazara railroad), this has evolved into a partnership in which the state retains a strategic role in agenda setting and financial resource provisions but devolves responsibilities for implementation to a mix of state, provincial and private interests.

The parallel in political terms is the Chinese recognition of the necessity of ensuring that the pre-requisites for nation-building are in place and the most significant of these is a defence of sovereignty. The formative requirements of modern nationalism are political stability with which the possibilities and prospects for long term, sustainable development are intertwined—indeed are crucially dependent upon this feature—and these are most simply supported through a defence of Westphalian principles. True to their adaptive impulse, the Chinese in Africa have come to distinguish between what they view as an acceptable face of external involvement—that which is approved by the

---

40. For an overview of the literature on political development, see Myron Weiner and Samuel Huntington, eds, *Understanding Political Development* (Boston, MA: Little Brown, 1987).

41. See Mao Zedong's famous 'New China' speech on 1 October 1949, echoed by subsequent leaders and policies.

42. To use the oft-quoted mantra, originally in the literature on conflict and development, sparked by Mary B Anderson, *Do No Harm: How Aid Can Support Peace—or War* (Boulder, CO: Lynne Rienner, 1999).

host and authorised by multilateral institutions—from 'interference' which damages the overarching imperatives for nation building and political control by the ruling party. Indeed, as scholars have noted, the Chinese attitude and support for forms of multilateral intervention has expanded in scope since the 1990s as Beijing has come to view peacekeeping as a necessary accompaniment to establishing 'peace and stability' in certain countries.[43]

The cumulative result of these features is a road map for African development which deliberately ignores some of the key shibboleths of the contemporary Western development 'canon'. It is an approach grounded not in the development discourse as formulated by economists and an ensemble of development industry experts but rather one constructed in line with practical experience. This practice-based methodology is mirrored by an outcomes based approach: impact is ultimately the only worthwhile focus for assessment, relegating the standard concerns of process to a secondary status. Projects, rather than budget support and other forms of financial transfers, are preferred for their tangible end products as well as the ability of the Chinese to exercise a large measure of control over issues like expenditure and management of quality control. The use of Chinese implementing agents, while controversial in some quarters, conforms to the logic of an interest-based approach to co-operation as well as the need to exercise project oversight. And yet, at the same time, the push for local inclusiveness does exist—both in terms of agenda setting for particular projects and involvement of the local community—and is most evident in the joint government-to-government negotiations at the project development phase. The importance of recipient African governments and their desire to press for greater inclusion of local actors (sub-contractors, suppliers etc.) as well as the robust regulatory and enforcement environment all play a part. Technology and skills transfer is standardly not privileged in any Chinese sponsored projects unless negotiations require their inclusion, as Beijing's extensive technical training and exchange programmes which run in parallel with particular projects are seen to be the preferred route. In the end, China's ability to point to its own development successes as concrete examples of the application of this approach is a powerful expression against which technical and abstract criticism will find difficulties countering.

Conclusion

The Chinese government faces the challenge of meeting high expectations in Africa as one consequence of the optimism accompanying China's prominent ascendancy throughout the continent. These are in part produced by its official rhetoric that puts Beijing in the position of needing to demonstrate that it is different to other external powers not merely in its declared principles but also in the substantive nature of its relations with Africa, not just in the comparative short term of emergent relations but more importantly over the medium–long term. By taking the ideational dimension of China's foreign policy toward Africa more seriously, the challenges of reconciling ideal aspirations with policy prescriptions and, ultimately the myriad of experiences of Chinese and Africa actors *in situ*, become more readily apparent.

---

43. Bates Gill and James Reilly, 'Sovereignty, intervention and peacekeeping: the view from Beijing', *Survival* 42, (2000), p. 44; Alan Carlson, 'Helping to keep the peace (albeit reluctantly): China's recent stance on sovereignty and multilateral intervention', *Pacific Affairs* 77, (2004), p. 13.

In endeavouring to assert the need to consider and engage the ideational dimensions of China's foreign policy in Africa, this article has sought to contribute toward better positioning approaches to China's engagement that has been widely seen as a largely material encounter. By exploring the notion of the normative modality of engagement condensed in the idea of China's exceptionalism, we have also suggested that the Chinese government self-generates a number of issues and challenges facing its expansive role in Africa that, in the changing circumstances of a thickening, more embedded role in the continent, are becoming more prominent in the face of greater political exposure. China's adaptive, pragmatic approach to its engagement with Africa, which has seen flexible shifts in key positions without abandonment of the rhetorical forms of engagement, is only now beginning to be tested against the hard realities of the continent.

One current trajectory of developing relations producing multiplying challenges that Beijing must confront is the meaning of its core constitutive rhetoric of exceptionalism beyond its state-centric focus. How can 'win–win' or non-interference be extended beyond the state level to encompass and engage political pluralism in a highly differentiated landscape of Africa's politics? How can the principle of treating African governments on the basis of equality be reconciled with underlying asymmetries of power? Can the Chinese modalities of development induce a process of economic change which will lead to tangible development gains for Africa and, concurrently, stable economic ties for China? Finding solutions to these pressing concerns will set the pattern for Chinese–African ties and the shape that their future will take.

Bashing 'the Chinese': contextualizing Zambia's Collum Coal Mine shooting

BARRY SAUTMAN and YAN HAIRONG

The 2010 shooting of 13 miners at Zambia's small, privately-owned 'Chinese' Collum Coal Mine (CCM) has been represented by Western and Zambian politicians and media as exemplifying the 'neo-colonial' and 'amoral' practices of 'China' and 'the Chinese' in Africa. CCM has been used to provide a sharp contrast to the supposed ways of the Western firms that own most of Zambia's mines. Embedded in racial hierarchy and notions of strategic competition between the West and China, the discourse of the CCM shootings further shapes conceptions of global China and Chinese overseas. While examining all the oppressive conditions that have given rise to protest at the mine, we contextualize the shooting and subsequent conflicts. In analyzing CCM's marginal and troubled development, we discuss aspects of the 2010 shooting incident known to miners and union leaders, but ignored by politicians and media. We look at the shooting's political fallout, focus also on the epilogue that was the 2012 CCM riot—in which one Chinese person was murdered and several others seriously injured—and trace the sometimes violent discontent manifested at other foreign-owned mines in Zambia since their privatization in the late 1990s. The empirical data for this detailed study derive from hundreds of documentary sources and interviews with union leaders, workers, officials and others in Zambia from 2011 to 2013.

> The Chinese don't care about human life and do just what they want. (Leonard Kapwizi, father of a miner wounded in the 2010 CCM shooting)[1]

Collum Coal Mine (CCM) lies down a rutted, unpaved road, scores of kilometers from the highway into Southern Province, five hour's drive from Zambia's capital Lusaka, in a hilly area with no Internet access. A 1966 article averred that 'Perhaps there's a more remote or hotter spot in Zambia, though it's hard to imagine'.[2] CCM is small and highly exploitative, resembling private mines in China that its government has closed. Conditions for Zambian and Chinese workers are much worse than at Zambia's medium-sized, Chinese state-owned copper mines. Yet, since 2010, when supervisors Xiao Lishan and Wu Jiuhua wounded 13 protesting Zambian miners, CCM has been central to global controversies of China-in-Africa.

---

*Barry Sautman is a political scientist and lawyer at Hong Kong University of Science & Technology. Yan Hairong is an anthropologist at Hong Kong Polytechnic University. A much more elaborate and fully-footnoted paper is available from the authors.

1. Bivan Saluseki, 'The dark side of coal', *The Post*, Zambia, (15 November 2010).
2. Jon Miller, 'Nkandabwe Coal Mine', *Zambia Magazine*, (April 1966), pp. 24–26.

Global media depict the CCM shooting as a crime of 'China' or 'the Chinese'. A US government broadcast put it that '[T]he miners staged a spontaneous but non-violent protest. Instead of negotiating with their staff, Chinese managers fired on them with shotguns'.[3] Such articles neglect the fact that CCM's founder is not 'China', but an Australian with a 'Chinese face'; that the shooters had reason to fear for their lives; that only two of those shot were hospitalized; and that several Chinese were hurt. At the insistence of the Chinese Embassy, CCM compensated the victims, deplorable aspects of work were improved (although oppressive conditions remained) and CCM had to apologize to its Zambian workers and to Zambia's Chinese-owned firms.

Large-scale violence occurs in many parts of Africa and beyond; yet the CCM incident where no-one died has often featured in discourse, both in anti-Chinese incitement by the Patriotic Front (PF)—Zambia's main opposition party until 2011 and now its ruling party—and the Western-generated narrative of China-in-Africa. It jibes with longstanding Western tropes of Chinese cruelty and disregard for human life and with implications that Western investors practice corporate social responsibility (CSR) and have a sense of decency that obviates reputational risks, while venal, 'mercantilist' Chinese spurn such standards. These notions mesh with assertions by politicians such as Hillary Clinton and David Cameron that the Chinese state's authoritarian capitalism fosters a neo-colonialism that subjugates Africans and negates Western democracy promotion and philanthropy.[4]

Most recountings of the CCM incident are de-contextualized, Manichean morality tales of powerful Chinese attacking defenseless Africans. We contextualize the shooting by viewing the development of this marginal mine as part of neoliberal Zambia and examine aspects of the incident known to miners, union heads and Chinese community leaders, but ignored by politicians and media, including the discontent at *most* mines in Zambia in the decade between their privatization and the CCM shooting. Our alternative interpretation of the shooting contends that singling out Chinese deflects attention away from neoliberal structural ills and says more about racial and ideological pre-conceptions that politicians and media bring to bear in bashing 'the Chinese' than about the actual Chinese presence in Africa.

Zambia's neoliberal reform and CCM's troubled marginality

Neoliberal reform created CCM as a small private mine in Zambia. Under World Bank and IMF pressure from the late 1990s to early 2000s, the Government of the Republic of Zambia (GRZ) privatized through a corrupt process 280 parastatals that accounted for 85% of Zambia's formal economy, with 29% sold to foreign firms, especially to those from Britain, Zambia's former colonial master. Outcomes for Zambians were highly negative. For example, privatized mining firm workers were divided into casuals, contract workers and permanents, and conditions of service

---

3. 'Chinese, African leaders gather in Beijing for talks', *Voice of America*, (18 July 2012).
4. Suisheng Zhao, 'A neo-colonialist predator or development partner? China's engagement and rebalance in Africa', *Journal of Contemporary China* 23(90), (2014), doi: 10.1080/10670564.2014.898893.

deteriorated, even as the larger multinationals made huge profits when copper prices doubled and tripled.

CCM is far different: an abandoned mine made into a struggling one that uses crude technology and has unreliable customers, it is a private, Zambian-registered firm, with no parent in China, limiting the Chinese state's ability to influence it.[5] The GRZ and private firms created the mine as an open-pit in 1965,[6] but it closed in 1968 due to flooding and better quality coal from nearby Maamba Collieries. It was a reservoir to irrigate small farms until 2002, when sole bidder and naturalized Australian Xu Jianxue created CCM *ex nihilo* as an underground mine. Xu, from Leping, Jiangxi, came as a Chinese construction aid team translator to Zambia in 1991, when there were only 300 Chinese there.[7] He later founded Yangts Jiang Enterprise, which has done many public construction projects. His four brothers came later, bringing money and becoming CCM managers.

CCM has low quality coal and large mines dropped it as they replaced coal-fired smelter furnaces with electric ones, leaving a customer base of irregularly-operating breweries and cement factories that can also buy from Zimbabwean mines.[8] A Zambian scholar reported in 2007 that CCM 'operate[s] with only rudimentary equipment, the shaft was dug with picks and shovels, and ore was brought to the surface in buckets. Workers lacked safety equipment and clothes'.[9] Before 2011, CCM kept going only due to strikes at Maamba Collieries.[10]

Chinese journalists who have reported on CCM have observed that

> Unlike major state-owned enterprises, private Chinese companies are often forced to scrape bones for meat. They operate in areas with weak profitability and low added value. They often have a hard time avoiding risk and their operational problems can revolve around disputes over wages, workplace conditions and benefits.[11]

CCM has had poorly-paid casual workers and shuts down often. Worker numbers have fluctuated wildly, with many miners working only when demand is high: in October 2010, there were 600 Zambians and 70 Chinese; the next month, 855, including 62 Chinese; in June 2012, 489 locals and 49 Chinese, etc. Recruited locally, they are aged 16–30 and mainly low-skilled.[12] Most have no formal education or an education that lasted for only a few years. Many speak only Tonga, a local language,

---

5. Interview, Pan Wenxiu, Chinese Chamber of Commerce in Zambia head, Lusaka, 11 August 2012.
6. Chen Zhu and Zhang Boling, 'Zanbiya Kelan Meikuang qiangji shijian' ['The Zambia Collum Coal Mine shooting incident'], *Xin Shiji Zhoukan* no. 42, (27 October 2010).
7. *Ibid.*; Interview, Ambassador Zhou Yuxiao, Lusaka, 9 August 2011.
8. Interview, Mooya Lumamba, Director, Mine Safety Department, Kitwe, 19 August 2011.
9. Neo Simutanyi, *Copper Mining in Zambia: the Development of Privatization*, ISS Paper no. 165 [Institute for Security Studies (S. Africa), July 2008], p. 9, available at: www.iss.co.za/uploads/PAPER165.PDF.
10. Interview, Charles Mukuka, President Mineworkers Union of Zambia (MUZ), Lusaka, 15 August 2011; Saluseki, 'The dark side of coal'.
11. Zhang Boling and He Xin, 'The killing of a miner in Zambia', *Caixin*, (20 August 2012), available at: http://english.caixin.com/2012-08-20/100426275.html.
12. See also 'ZCCM IH to appoint evaluator for Collum Coal Mine', *Zambian Mining Magazine*, (15 January 2014) ('500 workers'), available at: http://www.miningnewszambia.com/zccm-ih-to-appoint-evaluator-over-collum-coal-mine/. The approximately 90 percent share of locals in CCM's workforce indicates that it is not only in Chinese state-owned enterprises in Africa, but even in some family-owned ones, that there is a modicum of localization. Antoine Kernen and Katy Nganting Lam, 'Workforce localization among Chinese state-owned enterprises (SOEs) in Ghana', *Journal of Contemporary China* 23(90), (2014), doi: 10.1080/10670564.2014.898894.

and most cannot read or write. Their production quota varies day to day, depending on how much coal is needed and how many miners show up for work. They have a fixed daily, non-production based wage, because Chinese supervisors say Zambian workers do not know how to calculate.[13] Like Xu, Chinese staff are mostly from Leping, on three-year contracts, paid US$1,000–1,500 a month, and live in a mine site residential compound. Being expensive to bring over, they are not laid off mid-contract. Mining teams contain 20–30 Zambians and one Chinese, who uses simple Tonga mixed with English to communicate with the Zambians. Many Chinese supervisors have low education levels; the 2010 shooters had seven years.[14]

Beatings have occurred at CCM, as elsewhere. The Gemstone & Allied Workers Union of Zambia (GAWUZ) has complained, but victims take money to withdraw their charges.[15] On such beatings, a CCM union activist related,

> These happened when Chinese tell Zambians to do certain things, but maybe Zambians didn't understand or maybe they didn't want to do it. So Chinese beat them. This tended to happen with new guys (Zambians). Some Zambians only speak Tonga and there are times when Chinese workers and Zambian workers thought that each was insulting the other.[16]

CCM efforts to get officials to suppress illegal mining on its property have led to conflicts with local people. Safety issues have been aggravated by lack of first aid and failure to pay into workers' compensation funds. Officials shut the mine in 2005 for not issuing miners with protective clothing. Southern Province Minister Alice Simango, visiting in 2006, wept on seeing most workers without shoes or shirts. The GRZ declared CCM unsafe and shut it until it supplied protective clothing and hired Zambian foremen and safety personnel. CCM later 'issue[d] workers protective clothing, but workers often sell it for beer, so managers allow them to go underground in casual clothes'.[17]

An outbreak of cholera hit the mine in 2009 and villagers complained in 2010 that coal effluents had polluted streams and a lake. A visiting official described unsound sanitary conditions at CCM and nearby areas, including drinking water unfit for consumption. A CCM miner stated in 2010 that because workers had no masks, they cough, 'spit black' and have chest pains. Housing for miners is highly deficient. GAWUZ's president noted that

> Among the Chinese workers, some don't have toilets in their quarters. Among the black workers, they have even less. They generally have six or seven people living in a room, with cooking done outside .... It's as if they live in a village.[18]

In the run-up to the shooting in 2010, CCM was closed again for not providing workers with structures for changing clothes and underground toilets and 22 miners were injured when an oxygen machine burst underground, although that equipment

---

13. Interview, Chinese CCM workers, Nkandabwe, 23 August 2011.
14. Interview, President Nyumbu, Lusaka, 12 July 2013.
15. Interview, Sifuniso Nyumbu, President, Gemstone & Allied Workers Union of Zambia, Lusaka, 22 August 2011.
16. Interview, CCM worker and GAWUZ branch secretary Moddy Chigonke, Nkandabwe, 23 August 2011.
17. Lumamba interview.
18. Interview, GAWUZ President Sifuniso Nyumbu, Lusaka, 23 August 2011.

was not under CCM control. Three more miners were injured at CCM in a 2010 accident.

Not surprisingly, there have been strikes at CCM. In 2006, workers received only K150,000 (US$35) a month, but that did not result in notable strikes. The first major walkout was a five-day strike in 2008, after an accident in which a miner was killed and three others injured. Wages were also a factor, the lowest was K10,040 (less than US$3) a day and the highest K15,000. CCM pledged to raise the monthly salaries to K600,000, but failed to do so. When workers struck, police held three for the duration. GAWUZ said that police threatened to shoot strikers and helped CCM evict those fired for 'mentioning the collective agreement should be honored'.[19] Managers stated that strikers 'threw stones, broke the windows of vehicles and threatened violence', forcing Chinese to hide in buildings.[20] GAWUZ's president affirmed that stoning occurred after a Chinese shot into the air. The strike failed, although the Mine Safety Department (MSD) did close the shaft where the worker died.[21]

In another strike over pay and safety in 2009, workers 'threw missiles at Chinese managers'. Shaft manager Xu Jianrui said '300 miners opened the doors and entered the Chinese premises to fight them ... about 10 people were injured and their aim was to kill Chinese managers ...'.[22] Workers chased managers into the hills. CCM negotiated with GAWUZ and workers got raises, protective gear and a promise of supervisorial courtesy. A half-year later however, when the 22 miners were injured, workers tried to lynch a Chinese CCM employee arranging for them to get medical care. A group of Chinese workers told us

> Once *lao hei* [blacks] think they have money, they don't come to work; when they think they have no money, they come to work. Thus, if you raise their wages, you can't count on them coming to work. When they have money, they spend it on beer; they don't save money. [Workers] *nao yi ci, you yi ci'* [each time they make noise, they get something, i. e. a raise] ... *Lao hei* are afraid of the British, but not of Chinese ... when *lao hei naoqilai* [blacks come to make noise, i.e. protest], they use stones, spades and other tools and that it is terrifying. They will attack the walls and roofs. There's no safety here; you don't know whether you can keep your life.[23]

CCM was, in short, seen as almost irretrievably troubled. Soon after the shooting, the Home Affairs Minister formed a committee to investigate the plight of miners and consider setting up a fast-track court to deal with it, although a Labor Ministry official recalled that this was done 'to protect the company's interest'.[24]

Chinese journalists termed CCM management culture '*jiang hu*'—marked by informality, irregularity and crudeness regarding the use of force and ethical questions and based on low profit, low value-added, low problem-solving capacities as a small private firm. Unlike Chinese SOEs, CCM is unregulated by China's

---

19. Letter from Nyumbu to Inspector General of Police, 4 March 2008; letter from Nyumbu to managing director, CCM, re 'Post-mortem of labor withdrawal which took place 28 February to 4 March, 2008 and unlawful detention of employees', 13 March 2008 (letters in GAWUZ office).
20. 'Chinese mine operates with no safety measures', *Lusaka Times*, (28 February 2008).
21. Chen and Zhang, 'Zanbiya Kelan Meikuang qiangji shijian'; Nyumbu interview, 22 August 2011.
22. 'Chinese coal mine employees down tools', *Lusaka Times*, (2 December 2009).
23. Chinese CCM workers interview.
24. Interview with Assistant Commission of Labor Venus Seti, Ministry of Labor, Lusaka, 12 August 2011.

Ministry of Commerce and can only be exhorted to adopt CSR principles. The mine, the journalists argued, continues due to the need for local jobs and bribery.[25] Doubtless too, CCM is one of those firms a Chinese newspaper said does 'not quite understand [African] culture or customs, particularly the legal system'.[26]

After the shooting, grave problems remained, with no MSD inspectors within 100 km, inspections only every three–four months, no qualified managers, engineers or Zambian managers. In 2012, a miner was killed and two injured by a rock fall after blasting, allegedly due to a lack of qualified explosives personnel. The Minister of Mines ordered CCM to implement safety measures or lose its licenses. Miners, who said they earned K20,000 (US$4) a day, threatened to strike unless safety and pay improved. When the Engineering Institute of Zambia (EIZ) found no registered engineers at CCM in 2012, it fined the firm and sought an injunction against the mine's operation. A few months later, the GRZ also threatened to close it. Brian Kashimu, a new mine manager and 2011 election PF parliamentary candidate, came to CCM in August 2012, labeled most underground managers as incompetent, urged a GRZ investigation, and called on China's ambassador for assistance. The Minister of Mines stated he would bring the CCM issue to the Cabinet. The four Xu brother mine managers then fired Kashimu and, a few weeks later, another miner died accidently. Miners reportedly 'blam[ed] their Chinese bosses' for his death and demanded Kashimu's reinstatement.

In February 2013 the GRZ, citing safety conditions and CCM's refusal to declare its production and pay royalties, finally cancelled CCM's mining licenses. The Chinese ambassador and Zambian parties, unions and NGOs endorsed the takeover and shutdown. The GRZ said there was no nationalization and the Minister of Mines claimed 'We have received a lot of companies, both local and foreign firms, interested to run the mine'.[27] Yet, in quick succession, three shafts were flooded, the mine warehouse looted, and two caretaker workers severely burned in another shaft's fire that may have been set and raged for weeks. Eldest brother Xu Jianxue sued to recover the mine and China's ambassador averred that he might succeed, given that at least this Xu was well-liked and well-connected and the GRZ needed the mine to reopen. A quarrel over the value of the mine's assets prevented their transfer to the GRZ.[28] Thus, even after being shut down, CCM remained troubled.

Bashing the Chinese, shooting the Zambians: violence at the mines

CCM miners did not see the increase in their October 2010 pay they expected from a CCM/GAWUZ agreement. Many Shaft 3 miners accepted that, but Shaft 2 miners did not and marched to Shaft 3, where the three shifts totaled 150 workers, only 17–18 of them Chinese.[29] Due to poor English, managers were unable to communicate with the

---

25. Chen and Zhang, 'Zanbiya Kelan Meikuang qiangji shijian'.
26. 'Local people are key to firms' success in Africa', *China Daily*, (9 December 2011).
27. 'Investors keen to run Collum Coal Mine', *Daily Mail*, Zambia, (2 April 2013).
28. 'State suspects sabotage in Collum Mine fire', *Daily Mail*, Zambia, (22 July 2013); Interview with Ambassador Zhou Yuxiao, Lusaka, 10 July 2013; 'State to reassess Collum Mine', *Times of Zambia*, (4 December 2013).
29. Interview with CCM Shaft 3 foreman, Nkandabwe, 23 August 2011.

200–400 protestors. Home Affairs Minister Mhkondo Lungu said miners 'advanced as a mob' toward the managers[30] and 'After protracted arguments, the managers fled and the Zambians gave chase'.[31] Described as shouting and cursing, they pushed toward the shaft area entry gate, behind which stood Chinese employees. A manager claimed 'The miners attacked the employees [who] had no other choice but to shoot in the air'.[32] A Zambian eyewitness reported that day that protestors stoned Chinese supervisors. That morning, workers had rocks in their hands as they gathered. They 'said to the Chinese, "We'll beat you if you don't increase the salary"'. Two supervisors first shot into the air and then

> Zambians got fear and started to throw stones. Then the Chinese started to shoot in the crowd and people started to run. The Chinese were also very scared and didn't intend to shoot at first. But that's what happened. The Chinese had the fear that these blacks might kill us.[33]

Inspector General of Police Francis Kabonde said miners went to the management office and threatened to 'manhandle' and 'beat up' managers. Police commanders stated supervisors feared for their lives.[34] It was unclear whether they shot at the crowd or near it.

A wounded CCM miner later said that 'We weren't going to hurt them, but maybe the Chinese didn't understand that'.[35] That is because violence is not uncommon in Zambian miners' strikes and protests.[36] From 2005 to mid-2008, Swiss-based Enya's Chambishi Metals had three strikes and Swiss-based Mopani Copper Mines (MCM) had two.[37] Canada-based First Quantum's Kansanshi mine had one, Enya's Luanshya Copper Mine (LCM) had two and UK-based Konkola Copper Mine (KCM) had one. From mid-2008 to mid-2011, besides two strikes at the Chinese Chambishi Mine and one at the small Chinese–Australian Albidon nickel mine, workers at (white) South African-owned Chibuluma mine also struck, as did miners at Australian-owned Lumwana mine. There was a strike at KCM and two at MCM. Workers at Maamba and GRZ-owned Ndola Lime demanded managers be fired.

Strike violence is never against whites and Chinese industry managers view Zambian miners as deferential to white, but not Chinese bosses, with violent protests against Chinese and Indians, the groups the PF has attacked.[38] CCM supervisors would have known about violence in strikes. Thousands of redundant miners who

---

30. Daily Parliamentary Debates, Zambia, 20 October 2011.
31. 'Opposition politicians lambast the Lusaka government's timidity after Chinese managers shoot local mine workers', *Africa–Asia Confidential*, (25 October 2010).
32. 'Zambians riot after miners are shot', *Wall Street Journal*, (18 October 2010).
33. Chigonke interview. A Chinese source stated that wounded miner Bowas Syapwaya said supervisors shot *after* miners threw stones. 'Zambia police to charge Chinese mine managers with attempted murder', *Global Times*, (18 October 2010).
34. 'Chinese mine managers face arrest', *Daily Mail*, Zambia, (18 October 2010); '11 Miners injured at Chinese run Collum Mine in Southern Zambia', *Steel Guru*, (18 October 2010).
35. Barry Bearak, 'Zambia Uneasily Balances Chinese Investment and Workers' Resentment', *New York Times*, (20 November, 2010).
36. There is also a larger context of violence against Chinese in Africa, of which most Chinese there are aware. Shaio H. Zerba, 'China's Libya evacuation operation: a new diplomatic imperative—overseas citizen protection', *Journal of Contemporary China* 23(90), (2014), doi: 10.1080/10670564.2014.898900.
37. Interview with Rayford Mbulu, President, MUZ, 26 August 2008.
38. Interview, Mr Wang, Shandong Zhengyuan Geology & Resource Exploration Co., Lusaka, 1 August 2012.

worked for the Indian firm Binani, Luanshya Mine's owner from 1997 to 2000, rioted in 2002, firebombed government offices and beat local officials. That riot came after Zambian police killed several Luanshya miners during protests in 1998, 1999 and 2002. In 2005, KCM strikers planted explosives that damaged a mine with miners in it, smashed the cars of managers and stoned one, trashed a school and looted offices. In 2006, rumors of delayed wages triggered a riot. Workers attacked the residences of the Chinese managers and destroyed cars and other property. Several miners were shot, either by police, private guards or a Chinese construction contractor's manager. After the incident, rioters attacked Chinese and their businesses in Lusaka.

In 2008, 500 workers building Chinese-owned Chambishi Copper Smelter struck. They chanted anti-Chinese slogans, blocked roads, set ablaze a hostel for Chinese workers, held Chinese managers hostage and threw stones at them, leaving one toothless. In a 2009 riot at KCM, miners stoned and looted India Villa, where Indian employees live, attacked students at a KCM-run high school, burnt property belonging to Yangts Jiang, which was building houses for KCM miners, torched the firm's housing for its Zambian workers, looted food from a miners' union storehouse, tried to burn KCM vehicles, blocked roads, broke windows at a court house and allegedly attacked Chinese and Indians. CCM supervisors would also have heard that in January 2010, at the nearby GRZ-owned Maamba Colleries

> police saved [Managing Director Stephen] Mutambo from being killed by workers, who had metals and stones ready to harm him ... [P]olice guarded Mutambo and whisked him out of the mine area from the workers who were ready to lynch him.[39]

There was also Pythias Chinene's murder of his Chinese supervisor Zhong Tinghui in February 2010 at CCM's 'mine farm' at Chongwe, Lusaka province. Chinene was sentenced to death and Chinese at CCM would have known about this. China's ambassador has stated that another, unreported killing of a Chinese supervisor happened at CCM itself: 'Two weeks before the [shooting] incident, CCM workers had called away a Chinese and said they wanted to talk to him about something. He was beaten to death with hoes'.[40]

A Chinese community leader stated that on the day of the CCM shooting, 300 protestors pushed down a wall separating them from supervisors and were charging at them when the latter, fearing for their lives, shot.[41] After the shooting, miners vandalized the site. GAWUZ President Nyumbu noted that 'the idea [was] to kill anything Chinese', that CCM lost a lot of property, and that people from villages where miners live joined in while Chinese at CCM 'went underground and hid'.[42] Protestors destroyed the roof of a mine shaft, took away water pumps, generator batteries and electrical appliances, and pushed coal buckets into the shaft. Several Chinese sustained head and thigh injuries, apparently from stonings.

---

39. 'Maamba Collieries director survives lynching from miners', *The Post*, Zambia, 22 January 2010.
40. Interview with Ambassador Zhang Yuxiao, Lusaka, 9 August 2011. There had been at least one other murder of a Chinese by a Zambian fellow-employee, in 2009; both worked at a Lusaka hotel construction project. 'China expresses serious concerns about safety of its nationals', *Lusaka Times*, (6 August 2009).
41. Interview with Li Weixiang, head, China/Africa Chamber of Commerce, Lusaka, 8 August 2011.
42. Nyumbu interview, 22 August 2011.

The CCM shooting made simplistic

The shooters were charged with attempted murder and CCM was given two weeks to clear up labor problems or face government action. Two 'ring leaders' of the assault on the Chinese were also arrested, but not prosecuted. The two supervisors were released on bailed and the trial was set to begin in January 2011. The accused failed to appear at that point, but days later they showed up and were arrested, confounding Zambian diaspora bloggers who claimed that the GRZ had smuggled them out of Zambia to placate 'the Chinese'. A court later declared the re-arrests unlawful, so the accused were again bailed and a trial set for March 2011.

The Chinese Embassy urged the GRZ to punish both the shooters and the 'inciters of the riot'. It advised CCM to pay medical bills for the wounded, compensate them and meet workers' pay demands. The ambassador stated that 'our people need to learn the labor laws of your country to avoid situations like what happened [and] learn your culture'.[43] An Embassy official said that 'After the incident, we invited all Chinese companies in Zambia to come in for meetings and rethink what can be improved at their operations'.[44] The Chinese embassy had learned that Xu Jianxue was an Australian citizen, but as media presented him as only Chinese, China's ambassador said 'The lesson to be drawn is that we should pay attention to labor relations ... No matter whether he's Chinese Chinese or Chinese Australian, all people who bear a Chinese face should do things reasonably'.[45] A Chinese community leader described the public self-criticism that Xu was made to give:

> After the incident the Embassy had a meeting with 200 enterprises and demanded that Xu Jianxue make a public admission of wrongdoing and that he should do it both at the meeting and with the media. The Ambassador said at the meeting, 'CCM, your gunshot had wiped out the good relationship that we have won through hard work in those years'.[46]

CCM owners also apologized to workers. One recalled that 'there was a meeting. All the workers came. A manager came out to apologize to the workers. Officials were present, big officials from Lusaka and the Chinese ambassador'.[47] Shaft 3 manager Xu Jianrui apologized to Zambians at a meeting with the Southern Province minister. CCM pledged to compensate victims. It met with the Mineworkers Union of Zambia (MUZ), agreed to recognize it, provide safety gear and underground toilets, and hire a human resources manager and interpreter. The lowest monthly basic wage was doubled to K450,000—the highest wage before the shooting—with monthly housing, transport, meal and underground allowances raising the total lowest pay to K970,000. Miners received back pay for days not worked.[48] At a meeting the Labor Commissioner called, CCM and injured miners agreed to compensation of K375m, about US$80,000. The seriously injured worker got K45m, plus school fees for five

---

43. 'Chinese investors should respect Zambia's laws', *The Post*, Zambia, (28 November 2010).
44. Simon Mundy, 'Zambian workers alienated by cultural and linguistic divide', *Financial Times*, (21 January 2011).
45. Ambassador Zhou interview, 9 August 2011.
46. Li Weixiang interview.
47. Chigonke interview.
48. Interview, Shaft 3 miner, 23 August 2011.

years; the others got K20m–35m and all agreed that the supervisors should not be prosecuted.

While the controversy about the shooting raged in early 2011, Mary Musyalike, a heroin trafficker under death sentence in China, was returned to Zambia. Critics of 'the Chinese' claimed that was a quid-pro-quo for the release of the CCM shooters and launched an online torrent of racial invective against Chinese and Musyalike, an ethnic Lozi. Mercy Agness Mwale, another Zambian heroin smuggler sentenced to death in China, was also returned. These actions likely responded to GRZ requests, but belied a claim by the MUZ's president that if a Zambian were to shoot a Chinese in China, he would be instantly killed. The released Zambians had committed a crime for which drug smugglers of diverse nationalities have been executed in China, but the Zambians were not 'instantly killed'.

When the trial of the two Chinese supervisors was about to begin, the GRZ's Director of Public Prosecutions (DPP) entered a *nolle prosequi* ('unwilling to pursue'), due to the lack of witnesses. As the wounded miners had agreed that the shooters should not be prosecuted, it would have been anomalous to testify against them, while at least one eye witness had said miners stoned Chinese just before the shooting, several Chinese were injured and great damage was done to CCM property. These factors, together with earlier stonings and other violence at CCM and elsewhere, plus the murder of one or more Chinese supervisors at CCM facilities, arguably influenced the shooters' state of mind. Their use of shotguns, lethal only at short range, was also relevant. Arguments that the supervisors lacked intent to murder and acted in self-defense thus would have made it hard to prove a case of attempted murder beyond a reasonable doubt.

In fact, before the *nolle prosequi*, the charges had already been reduced to 'attempt to cause grievous bodily harm with intent to maim'.[49] If the shooters first fired in the air and then not directly at miners, that would negate intent to inflict grievous bodily harm.[50] Even if the shots were direct, self-defense applied, based on the shooters' fearful state; indeed, the then-Minister of Labor said the supervisors 'used guns to defend themselves'.[51] After a 2012 riot at CCM (see below), Ambassador Zhou stated that 'The Chinese employees [said] that, if we did not do something previously [in 2010] in self-defense, we could have been killed [in the 2012 riot]'.[52] What is legally allowed is not necessarily what is politically or morally warranted and thus the Chinese Embassy regarded CCM as culpable for the shooting,[53] but it is likely that given a fair trial the shooters would have been acquitted no matter who testified.

The Chinese in Zambia hold views on the CCM incident that contrast with those of Zambian and Western elites. The Chinese ambassador perceived it to be a labor

---

49. Barry Bearak, 'Zambia drops case of shooting by Chinese mine bosses', *New York Times*, (5 April 2011).
50. See M. K. Magistad, 'Chinese investment at the cost of respect in Zambia?', *PRI's The World*, (4 October 2011), available at: www.thworld.org/2011/10/chinese-investment-respect-zambia-collum/.
51. 'Zambia orders Chinese Collum Mine managers to compensate the victims', *Steel Guru*, (2 November 2011).
52. 'Chinese diplomat hopes Zambian labor violence limited', *Voice of America*, (10 August 2012), available at: www.voanews.com/content/zambia-chinese.../1483697.html.
53. The culpability attributed by the Chinese Embassy in this instance resembles that attributed by Chinese officials in Ghana after clashes there between Chinese miners and police. See Fei-Ling Wang and Esi A. Elliot, 'China in Africa: presence, perceptions and prospects', *Journal of Contemporary China* 23(90), (2014), doi: 10.1080/10670564.2014.898888.

relations problem. A Chinese community leader in Zambia saw CCM's practices in a business context:

> Those Chinese who deal with poor Zambians don't have much themselves. They are not intentionally harming people, but are not generous, as all their money is earned bit by bit through hard labor: [such a Chinese] wants to win the world with his own bare hands (*chi shou kong quan da tian xia*). He's not like a foreigner who already has some kind of accumulation: cars, houses, etc. and who can give you more benefits. But the Chinese who employ poor Zambians had themselves just been '*dagong de*' (working for a boss).[54]

Managers of large Chinese enterprises in Zambia have felt that their differences from smaller private firms are so obvious they did not need to be concerned about the CCM incident's impact.[55] They are 'water from the well that does not mix with water from the river' (*jing shui bu fan he shui*), as Chinese say. PF leader Michael Sata, now GRZ President, did not make such distinctions, however. He stated 'You see the Chinese are above the law'[56] and Zambians are 'shedding more innocent blood at the hands of these merciless so-called investors'.[57] Sata claimed 'the Chinese' would be treated lightly, as they had corrupted the GRZ electoral process, but offered no evidence; in fact, the CCM shooters were at the time still scheduled to be tried for attempted murder. Opposition party youth went to the Chinese Embassy to demand an apology for the shootings, which the PF and the United Party for National Development (UPND) related to the general Chinese presence in Zambia and upcoming elections. Sata decried the GRZ stance that those shot should be compensated, arguing that it was intended to avert criminal prosecution. PF MP Yamfwa Mukanga—who became Minister of Mines in 2012—said that 'the Chinese will continue to shoot people'.[58] The Zambia Federation of Employers (ZFE), which has resisted minimum wage increases and the demands of Zambian unions, stated, with no hint of irony, that 'it is disgraceful to see investors mistreating local workers who fight for their rights'.[59] After Sata won in 2011, the CCM non-prosecution was counted as a reason for his success.[60]

Zambian media stress ethnicity where Chinese are accused of crimes. Soon after the CCM incident, one Bo Khan was arrested for allegedly threatening to shoot his maid and gardener because they wanted to visit a store. The media focused on his assumed Chinese ethnicity. Zambian media identify the ethnicity of Asians arrested or even investigated, while whites are typically only termed 'expatriates'. Chinese convicted

---

54. Interview, Li Weixiang, President, Association of Chinese Corporations in Zambia, Lusaka, 10 August 2012.
55. Interview with Wang Chunlai, CEO of NFCA, 15 August 2011.
56. 'Chinese investors are above the law: Sata', *The Post*, Zambia, (17 October 2010).
57. 'Rupiah's defence of Sinazongwe Mine crimes is scandalous: Sata', *The Post*, Zambia, (29 October 2010).
58. Zambian Parliament, Debate, 25 November 2010, available at: www.parliament.gov.zm/index.php?option =com_content&task = view&id = 1316&Itemid = 86&limit = 1&limitstart = 3.
59. 'Zambia Federation of Employers disappointed with labor minister over minimum wage announcement', *Lusaka Times*, (12 July 2012); 'Federation of Employers speaks out on Chinese gunmen', *Zambian Watchdog*, (19 October 2010), available at: www.zambianwatchdog.com/ > p = 9492.
60. Sata's remarks should be seen in the light of his wide range of attacks on the Asian, particularly Chinese, presence in Zambia, which were obvious from the beginning of his run for the presidency in 2006 and continued until after he captured that office in 2011. See Barry Sautman, 'The Chinese defilement case: racial profiling in an African "model of democracy"', *Rutgers Race and the Law Review* 14(1), (2013), pp. 87–134. In that regard, Sata spurned the idea of South–South cooperation that is often promoted in Africa/China relations. See Sven Grimm, 'China–Africa cooperation: promises, practice and prospects', *Journal of Contemporary China* 23(90), (2014), doi: 10.1080/10670564.2014.898886.

of crimes receive especially long sentences. In 2012, a Chinese engineer and Chinese driver who had worked building Ndola stadium were arrested for possessing a half kilo of ivory bangles. The defendants argued that they did not know such possession was illegal and bought the items as gifts, but got five and seven years hard labor, respectively. There is much online racist abuse and calls for harsh retaliation against Chinese by Zambians, who use anti-Chinese terms such as 'choncholis', 'ching chongs', 'squinty eyes', 'yellow savages', 'Chinese piglets' and 'chinks'. Some avow hatred for Chinese and call for mass deportations or 'xenophobic attacks'.[61] CCM's private owners are conflated with the Chinese state, although its general manager at the time of the shooting was not an official, but a 24-year-old who, before his recent hiring, worked busing tables at a Lusaka Chinese restaurant.

Diaspora bloggers accused the GRZ of entering the *nolle prosequi* to please 'the Chinese' or as a trade for Zambian drug smugglers in China. MUZ President (now Deputy Minister of Labor) Rayford Mbulu said that 'government was putting its relationship with China ... above its duty to protect workers'.[62] The International Trade Union Confederation (ITUC) conflated privately-owned CCM with 'China', blamed the shooting on CCM managers' language inabilities and, without regard to the legal basis of the *nolle prosequi*, said that 'the decision ... casts doubt on the independence of the Zambian judiciary when private foreign investors are involved'.[63] PF deputy leader Guy Scott and UPND head Hakainde Hichelma both held that the case involved miners being gratuitously shot. Further afield, the tale became taller. US Congressman Donald Payne (D-NJ) said of the CCM shooting that 'they have opened fire on workers who protested about poor working conditions in Zambia, Chinese soldiers just fired on them ...'.[64] A US journalism professor claimed 'Zambians want to know ... whether bribes were paid in order to obtain the exoneration of the Chinese managers', even though no Zambian media claimed bribery.[65] A Reuters story warned that 'it remained to be seen what the reaction on the street would be'.[66] No overt street reaction occurred, although many Zambians did fault the Chinese for the incident[67] and Western media cast it as representative of conduct by 'China' and 'the Chinese' in Africa.

Bashing the Chinese, literally: the 2012 CCM riot

The 2010 CCM incident's aftermath was the 2012 CCM riot, in which Chinese were literally bashed, one to death. It came a month after the GRZ's 'surprise' gazetting of

---

61. The US State Department has also fixed exclusively on Chinese firms and individuals in discussing human rights violations related to working conditions in Zambia. US State Department, *2010 Human Rights Report: Zambia*, (11 April 2011), available at: www.state.gov/g/drl/rls/hrrpt/2010/af/154376.htm.
62. 'Charges dropped against Chinese in Zambia', *Associated Press*, (5 April 2011). Indian investment in Zambia is also about US$3b: 'India's FDIs increase', *Times of Zambia*, (1 November 2012).
63. 'Zambia: charges dropped against two Chinese supervisors', *States News Service*, (6 April 2011).
64. House Comittee on Foreign Affairs, 'Briefing on China and US interests', (19 January 2011), Financial Markets Regulatory Wire, (19 January 2011).
65. G. Pascal Zachary, *Africa Works*, (10 April 2011), available at: http://africaworksgpz.com/2011/04/10/in-zambia-china-rules/.
66. 'Zambia drops mine shooting charges against Chinese', *Reuters*, (4 April 2011).
67. See Gerard van Bracht, 'A survey of Zambian views on Chinese people and their involvement in Zambia', *African-East Asian Affairs* no. 1, (2012), pp. 54–97 at (pp. 63–64).

a new minimum wage. Employer outcries and requests for delay made it initially unclear whether the new minimum was in force. GRZ officials and union leaders noted also that the new minimum only affected workers not covered by contracts.[68] The Labor Minister added that 'employers, trade union officials and general workers did not understand the new minimum wage implementation'.[69] Some unionized workers did not know the new minimum was legally irrelevant to them; others saw it as creating a moral entitlement to a raise, even if they already earned above it. That was so at CCM. A February 2012 collective bargaining agreement (CBA) provided that most CCM workers earn more than the new minimum wage: 30,000 a day or K780,000 (with allowances, K1.2–1.3m).[70]

Responding to the new minimum, MUZ met with two Xu brothers on 2 August. A 25% wage hike was agreed,[71] the first such agreement with a union since the new minimum decree. MUZ's General Secretary noted that 'the amounts agreed upon by the union and management were far above the minimum wage recently announced and gazetted by the government'.[72] According to MUZ officials, after the 2 August wage hike agreement, CCM's Zambian human resources manager was tasked with briefing workers on it, but he waited.[73] Pay day at CCM was 4 August. That morning, workers learned their July pay checks would not include a raise. Managers knew the new minimum did not affect CCM workers, that under the agreement the raise was to start in August, and that July salary checks were already prepared using the old rate. They nevertheless brought a bag of cash to hand out the wage increase along with pay checks.[74] Workers did not know that and did not come to collect their pay, going instead to the local police to seek a permit to protest. The police demurred and called the CCM managers to come, but they were afraid and refused. Four hours of rioting ensued, after '[T]he miners ... went back to the mine and started beating and throwing stones at anyone they found working, shaft by shaft'.[75]

Starting between noon and 1 pm, some 300 people attacked Shaft 2, injuring two Chinese employees and several Zambian security guards and miners. By 2 pm, miners and villagers had moved on to Shafts 3 and 6, injured two more Chinese and stole their belongings. At about 3 pm, the crowd arrived at Shaft 5. Five Chinese surveyors and constructors working there saw the crowd coming and retreated. Rioters robbed the Chinese residences, looted the mine offices of computers and other

---

68. Minimum wages are legally relevant to unions only when a CBA expires. Interview, Ministry of Labor Principle Labor Officer Khadija Sakala, Lusaka, 8 August 2012.
69. 'Govt to explain new minimum wage', *Times of Zambia*, (6 August 2012).
70. Interviews, Sifuniso Nyumbu, Lusaka, 3, 5 and 12 August 2012; 'Collective agreement between Collum Coal Mines and Mine Workers Union of Zambia', (20 February 2012).
71. Nyumbu interview, 12 August 2012; Interview, Wang Dong, Chinese Embassy, Economic and Commercial Counsellor's Office, 8 August 2012.
72. 'Collum Mine saga: MUZ condemnation timely', *Times of Zambia*, (7 August 2012). The CCM miners thus did not demand 'salary arrears following the revised minimum wage'. Zhao, 'A neo-colonialist predator or development partner?'. Rather, miners were apparently unaware that the new minimum wage did not apply to them, both because they had a union bargained-for contract and because they already made more than the minimum wage.
73. Interview, Charles Muchimba, director of research, MUZ and Webby Mushota, director of occupational health and safety, MUZ, Kitwe, 16 August 2012.
74. Interview, Ambassador Zhou Yuxiao, Lusaka, 7 August 2012.
75. 'Wu cremated as business continues at Collum Coal Mine', *Daily Mail*, Zambia, (19 August 2012). See also 'Minimum wage to be harmonized', *Tumfweko*, (6 August 2012) ('irate miners mobilized themselves and started beating up fellow miners found working').

valuables, and destroyed facilities. Chinese fled down the 240 meter Shaft 5's 40 degree slope. Rioters did not enter the shaft, but threw stones and bricks and pushed a trolley down the shaft. Four Chinese got out of the trolley's way by hiding in an emergency escape path. Wu Shengzai, who was surveying and had been in Zambia since 2009, was already injured and not fast enough. He was run over by the trolley and died on the spot, his corpse grossly disfigured. Rioters later spread to the furthest shaft, no. 1, where they stoned the Chinese employees' residential area and tried to enter. At 5 pm, the police arrived and rioters dispersed. The Chinese Embassy stated that 'the Chinese were extremely restrained to avoid intense confrontations during the incident':[76] due to CCM orders after the 2010 shooting, none of the Chinese were armed during the 2012 riot.

Besides the one dead, four middle-aged Chinese were hospitalized with serious wounds; four others had lesser wounds. The 30-plus Chinese at CCM barricaded themselves indoors. When Labor Minister Fackson Shamenda visited, Chinese at CCM 'mobbed' him, stating 'their lives were in danger', 'they were gripped with fear' and needed protection. They said 'the mine was experiencing disturbances every three or four months [and] when such occur, a government official is sent to the mine but problems persist'.[77] PF leader Shamenda stated 'I don't know why there is always tension between Chinese investors and workers at Collum'.[78]

Chinese in Zambia were highly agitated. Some wanted to organize a protest, an action that the Embassy opposed and did not take place. Ambassador Zhou stated that 'All Chinese nationals are now frightened'.[79] Because 'many Chinese [were] fearful to leave their homes' and 'Chinese managers were scared that Zambians would harass them again if they resumed operations',[80] CCM shut down. When Shafts 3 and 6 were reopened a week after the riot, the GRZ ordered them closed again.

Police soon arrested 12 alleged rioters including miners and villagers who 'are the people who just went to terrorize the Chinese'.[81] Police continued to look for rioters, including three suspected of killing Wu. A CCM manager asserted that Wu was killed by villagers, likely because initial reports quoted Minister Shamenda as saying Wu's killers were 'ordinary thugs and not miners' and 'just a bunch of criminals who took advantage of the disturbance'.[82] Alex Sindebuka, the first to be charged with murder, turned out to be a CCM miner. Ten weeks later, the 'mastermind', 27-year-old Slyvester Siyanchebani, was arrested. 'Locally known as "Savage"', he was a Shaft 3 miner.

An analyst stated the violence was not directly related to wages, but 'instigated by a vocal minority that insisted on carrying out retaliatory attacks against the Chinese management'. He added that 'Many of these people congregate near mining sites.

---

76. 'Ambassador visits injured Chinese in Zambia', *China Daily*, (7 August 2012).
77. 'Chinese at Collum Mine call for protection as Shamenda blames MMD', *Lusaka Times*, (7 August 2012).
78. 'Rioting Zambian coal miners at Chinese owned mine kill Chinese manager', *Agence France Presse*, (5 August 2012).
79. 'China's Ambassador to Zambia warns of deteriorating relations', *Zambia Reports*, (7 August 2012); 'Fear grips Chinese', *The Post*, Zambia, (7 August 2012).
80. 'Chinese community in Zambia demands justice following murder of mine manager', *Zambia Reports*, (10 August 2012).
81. 'Fear grips Chinese', *The Post*.
82. 'Chinese mine boss killed', *Daily Mail*, Zambia, (6 August 2012).

Some have been fired from mines. Their motivation is to exploit situations like this for their own gain to loot when they can'.[83] Support for that idea appeared when Zambian workers at Chinese mine construction firm 15 MCC threatened to burn the Chinese SOE-owned Mulyashi Mine they had just built in Luanshya, Copperbelt, if they were not offered jobs there. Two Chinese were also severely beaten in Luanshya by 15 unemployed youths who reportedly resented that Chinese were employed in the mine.

Deputy Minister of Home Affairs Stephen Kampyongo blamed the 2012 CCM riot on rising tempers after the MMD government did not convict the 2010 shooters. A Zambian Oxford University Ph.D. student and *Post* columnist said the riot was 'a culmination of years of workers' frustrations and sea of discontent against the Chinese over alleged low wages and exploitative working conditions'. He implied that the violence was based on accurate perceptions that Chinese are worse employers than other foreigners and Zambians, as well as on Zambians' disappointment that the PF government had not chased away 'the Chinese'.[84]

Such views ignore the role of PF incitement in inflaming miners' violence. A prominent NGO leader pointed to just that as the riot's main cause. Sam Mulafulafu, Zambia Executive Director of the international Catholic charity Caritas stated

> [T]he cold relationship between the Chinese and the Zambians was a problem that had been building up from last year's electoral campaign ... Immediately [after] PF came into power, there was an outbreak of conflict between Chinese-owned firms and the workers ... The PF government must admit that they are solely responsible for this problem of Chinese–Zambian workers' [conflict].[85]

Several GRZ ministries reportedly 'consult[ed] various stakeholders to address issues that sparked off the protest'. Police were stationed at the mine, but also ordered 'to hunt down and interrogate [GAWUZ] officials, who may [have been] involved in [the] protest',[86] although no evidence of any such involvement appeared. Expectedly, the government ignored the role of anti-Chinese incitement and a local MP complained, when the mine reopened again six weeks after the riot, that its root cause had not been addressed.

The 2012 riot was presented globally through Western media, which mostly conflated CCM with Chinese firms or Chinese in Zambia/Africa. A *New York Times* article asserted that 'Chinese companies in Zambia have long been accused of mistreating and underpaying their workers'.[87] Nothing was said about contestations of the accuracy of the accusations or who the accusers were and what their political agendas might be.

Western media assumed the murdered Wu Shengzai was an authority figure who lorded it over Zambians, i.e. a manager, supervisor, 'mine boss' or even 'Chinese official', although he was, in reality, surveying to prepare Shaft 5 for operation.

---

83. Quoted in Zhang and He, 'The killing of a miner in Zambia'.
84. Sishuwa Sishuwa, 'Understanding reactions to Chinese investment in Zambia', *The Post*, Zambia, (14 August 2012).
85. 'Fear grips Chinese', *The Post*.
86. 86. 'Chief Sinazongwe sorry', *Daily Mail*, Zambia, (11 August 2012).
87. Lydia Polgreen, 'Zambia: Chinese supervisor dies during protest at mine', *New York Times*, (6 August 2012).

Reuters wire service claimed 'animosity towards [Chinese] is growing as Zambian workers accuse firms of abuses and underpaying' and that 'critics' warn Chinese firms 'are importing their poor track record on workers' rights'.[88] Yet, no evidence indicates that the riot was part of a trend. It had been three years since Zambians last used violence at a Chinese-owned mine other than CCM, in part because these mines have adapted to Zambia's labor relations regimen.[89]

An official German radio broadcaster quoted Deputy Labor Minister Rayford Mbulu as stating that the riot was 'a lesson to employers to respect workers'. It added that in the 2010 CCM incident 'Chinese managers shot and wounded Zambian workers who were protesting over poor wages'.[90] The report did not indicate how murder might lead to respect and made no mention of the violent context of the 2010 shooting.

A *Wall Street Journal* article asserted that the GRZ was 'struggling to contain mounting anger among miners, who accuse China of exploiting Zambia's resources and taking advantage of its workers'. Zambian miners, however, seek to *expand* production and affirm that every foreign-owned mining firm they strike against takes advantage of them. The article however portrayed Chinese firms as uniquely and harshly exploitative, with the only miner quoted reportedly stating 'The Chinese are just here to make a profit, to make their country rich ... We are slaves in our country'.[91]

An article in the *Los Angeles Times* claimed that 'Zambia frequently sees confrontations between Chinese mine management and workers' and recounted every incident at Chinese-owned mines from 2005 to 2012, but none at other mines, while quoting extensively from a highly flawed 2011 Human Rights Watch report on Chinese copper mining in Zambia.[92]

An article by *The Telegraph* (UK) Chief Foreign Correspondent David Blair about the 2012 riot focused on the 2010 shooting. It claimed 'Beijing' successfully pressured the GRZ to not indict the shooters, although they had been indicted, and said CCM pays employees less than the minimum wage for shop workers, although most were paid more. Blair quoted Michael Sata's claim that British colonialism was better than the Chinese presence.[93]

The Chinese are worst

On Chinese mining in Zambia, 'the debate is clearly informed by racist assumptions',[94] embedded in racial hierarchy, exemplified by a remark of (the famously white) Guy Scott, PF deputy leader and, since 2011, GRZ Vice President.

---

88. 'Zambian miners kill Chinese supervisor over pay', *Reuters*, (5 August 2012).
89. See Yan Hairong and Barry Sautman, 'Beginning of a world empire? Contesting the discourse of Chinese copper mining in Zambia', *Modern China* 39(2), (2013), pp. 131–164.
90. 'Arrests over deadly clashes at Chinese-owned Zambian mine', *Deutsche Welle*, (6 August 2012).
91. Peter Wonacott, 'China investment brings jobs, conflict to Zambia mines', *Wall Street Journal*, (5 September 2012).
92. Robyn Dixon, 'Enraged Zambian miners kill Chinese manager', *Los Angeles Times*, (6 August 2012).
93. David Blair, 'Zambian miners crush a Chinese manager to death', *Telegraph*, (6 August 2012).
94. John Lungu and Alastair Fraser, *For Whom the Windfalls: Winners and Losers in the Privatization of Zambia's Copper Mines* (Lusaka: Civil Society Trade Network of Zambia, 2007), p. 53.

He proclaimed in 2007 that 'People are saying: "We've had bad people before. The whites were bad, the Indians were worse but the Chinese are worst of all"'.[95] Western media also depict Chinese as the worst employers in Zambian mining, even though commonalities of work in Chinese-owned copper mines and the larger mines of Western-based firms far outweigh differences. The latter are much more profitable than the Chinese-owned mines, but still pay miners only subsistence wages. Non-Chinese mines' safety records are also no better than those of the main Chinese-owned mines.[96] Yet PF-fostered incitement ensured that 'the Chinese' are subject to racial animus and pilloried in global media, while Western firm managers escape race-based opprobrium.

The discourse of the CCM shooting is about how distinctively bad it is to work for 'the Chinese', who would rather shoot workers than improve their lot. Yet, a coal mine long GRZ-owned and a nearby white-owned stone production facility have also had deplorable conditions and worker discontent, but unlike at CCM, the owners' ethnicities play no role, as Zambian workers are not incited against co-nationals or whites.

Maamba Collieries Ltd (MCL), with one of southern Africa's largest coal deposits, is Zambia's main coal mine. Despite state support, it was failing in the mid-2000s and had many labor disputes. A strike at MCL in 2006 became a riot, in which 800 miners demanding payment of wage arrears and removal of managers stoned management housing and battled police. In 2009, the GRZ contracted MCL to Oriental Quarries, a Zambian Indian company, and Singapore-based NAVA Bharat. When these firms also did not pay workers, they struck. Another outbreak occurred at MCL in 2010 and included attempts to stone and lynch its Zambian Managing Director.

There is also a mining operation that GAWUZ President Nyumbu has called 'much worse than CCM'. Zambezi Natural Stone Enterprise (ZNS), owned by white Zimbabweans, creates on-demand tiles for flooring, paving, etc. It has factory workers and miners, but the owners did not recognize the miners as employees. ZNS workers, who are migrant villagers from afar, live in grass huts, receive no protective clothing and work from 06:00 to as late as 18:30. There are frequent injuries, but no medical treatment or workers' compensation. Miners are only paid if they find valuable stone and the owners determine how much they pay for it. Although they are union members, ZNS factory workers were paid the minimum wage, which applies only to non-union members. The owners were unwilling to negotiate, let alone enter into a CBA, with GAWUZ.[97]

CCM, MCL and ZNS have not had greatly different workers' conditions, despite the latter two having advantages CCM does not enjoy: MCL has been state-backed; ZNS has a ready market. These three nearby enterprises share levels of exploitation and worker discontent, but have ethnically-disparate managements, which puts paid to the idea that CCM's poor labor relations are because 'the Chinese are the worst'.

---

95. Chris McGreal, 'Thanks China, now go home', *Guardian*, UK, (7 February 2007).
96. Barry Sautman and Yan Hairong, *The Chinese are the Worst?: Human Rights and Labor Practices in Zambian Mining*, Maryland Series in Contemporary Asian Studies (Baltimore, MD: University of Maryland, 2013).
97. Nyumbu interview, 22 August 2011.

Racial neoliberalism and the discourse of the CCM shooting

The CCM shooting is depicted as gratuitous cruelty, not a response to reciprocal violence in the context of a crude small enterprise management, and impliedly as an act that only 'the Chinese' would commit. The incident at this remote mine hit a nerve in Zambia not because the shooting was unprecedented in labor conflicts, but because PF had already created an anti-Chinese discourse. There is scant political difference among parties in Zambia. A PF government minister has stated that 'Zambian politicians do not differ on principle'[98] and virtually all support neoliberalism. The PF distinguished itself mainly by an anti-Sinicism that made China 'a new subject ... in the spotlight ... a political football [with] a political undertow in Zambia', as a leading Zambian analyst has noted.[99]

PF channeled discontent with neoliberal reforms into its anti-Chinese campaign, fixed in part on the wage gap between Chinese and other mines, a gap that reflects differences in size and profitability and is narrowing. The impact of Chinese investment and migration on Zambia has been exaggerated and made a source of Zambians' woes, especially for small traders and miners, who experience the worst of privatization, deregulation and free trade. Sata claimed 80,000 Chinese were given work permits in Zambia, but the GRZ reported 2,340. Chinese-owned mines produced only 5% of foreign-owned mines' copper in Zambia in 2010,[100] but Sata singled out 'the Chinese' as 'invaders' and 'infestors' and said the problem is 'not only [in] Zambia—it is all Cape to Cairo where the Chinaman is'.[101]

Racism is so embedded in neoliberalism that a concept of racial neoliberalism has been adumbrated.[102] The US discourse of the 'anti-market' behavior of 'welfare queens' and black ghettos for example was central in creating popular support for welfare cuts. Neoliberalism makes immigrants to Canada economic contributors, but also 'effectively demonized as deviant, criminalized and tarnishing the supposed Canadian way of life'.[103] Neoliberalism in 'post-racial' societies largely removes, through privatization, race from the public to the *private* sphere. In contrast, 'the Chinese' are staged as problematic in Zambia's *public* sphere, allowing neoliberalism to escape scrutiny. Contradictorily viewed as 'good investors' but 'bad employers',[104] Chinese will likely be singled out for as long as any 'Chinese face' is associated with exploitation and oppression at enterprises like CCM.

In the larger world, tropes of gratuitous cruelty and disregard for human life are axioms of anti-Chinese racism, expressed in contexts of 'strategic rivalry', a view spread out from the West since the nineteenth century. The villain of early twentieth century British novelist Sax Rohmer's Fu Manchu tales and later films was portrayed

---

98. 'Zambians haven't learnt to differ on principle: Luo', *Daily Mail*, Zambia, (17 July 2013).
99. Interview, Professor Oliver Saasa, Lusaka, 18 July 2007.
100. Interview, Gao Xiang, Deputy CEO, CNMC Luanshya Mine (CLM), Luanshya, 17 August 2011; 'China pledges increased investment under PF', *The Post*, Zambia, (1 October 2011).
101. Barry Sautman and Yan Hairong, 'African perspectives on China–Africa links', *China Quarterly* 199, (2009), pp. 729–760 (at p. 752).
102. David T. Goldberg, *The Threat of Race: Reflections on Racial Neoliberalism* (Hoboken, NJ: John Wiley, 2011).
103. David Roberts and Minelle Mahtani, 'Neoliberalizing race, racing neoliberalism: placing "race" in neoliberal discourse', *Antipode* 42(2), (2010), pp. 248–257 (at p. 252).
104. See Sautman and Yan, *The Chinese are the Worst?*.

as challenging the white race for world domination and described as a Chinese person 'whose very genius was inspired by the cool, calculated cruelty of his race'.[105] These notions have remained commonplace. 'Human wave tactics' that Chinese forces used during the Korean War were ascribed to an Asian contempt for life and William Westmoreland, US commander in Vietnam, opined that 'The Oriental does not put the same high price on life as does a Westerner'.[106] Jeanne Kirkpatrick, former US ambassador to the United Nations, said in 1999 'the Chinese do not value human life and might be willing to suffer retaliatory consequence for the psychological benefit of striking American soil with a missile'.[107] In 2010, an ex-aide to Ronald Reagan wrote that 'China' has 'a much lower valuation on human life than it should have', citing the CCM shooting,[108] while British singer/songwriter Morrissey said the cruelty of Chinese made them a 'subspecies',[109] yet continued to be widely popular in the West. Such ideas, applied to China and the Chinese, proliferate on the Internet.

Notions of Chinese cruelty and disregard of human life have also been around in Africa, initially among white settlers, for more than a century. In 1905, Chinese miners who escaped from indentured labor into South Africa's veldt were said by a local newspaper to have 'an Asiatic contempt for life in their blood and Chinese cruelty and callousness in their hearts'.[110] These ideas are still present in Western writings on Chinese in Africa. An analyst of Chinese firms in conflict zones in Africa stated that they, compared to Western firms, 'appear to be more tolerant to physical losses' and 'ready to accept human losses'.[111] A UK journalist's report on Chinese activity in Zambia said of Chinese attitudes toward the loss of lives in a mining accident, that 'To them 50 people are nothing'.[112] A white South African journalist specializing in the Chinese presence in Africa said it entails 'an apparent disrespect for human life', as shown by the CCM shooting, which he claimed involved only 'miners shot while presenting a list of grievance to Chinese managers'.[113]

Conceptions of Chinese as numerous, widespread and inured to pain have been linked to notions of Chinese cruelty and disregard for human life. The idea that the Chinese do not value human life because of their large population is now found among Africans as well. Workers at Zambia's Chinese-owned Chambishi Mine told a researcher in 2008 that 'the Chinese do not value life because there are so many of them in China'.[114] When unsubstantiated claims were made that Chinese engineers

---

105. Sax Rohmer, *Fu Manchu* (Sheffield: PJM Publishing 2008 [1910]), p. 327.
106. Quoted in Derrick Jackson, 'The Westmoreland mind-set', *Boston Globe*, (20 July 2005).
107. Cal Thomas, 'China's espionage coup', *Baltimore Sun*, (31 May 1999).
108. Peter Hannaford, 'A tale of two mines', *American Spectator*, (22 October 2010).
109. 'Morrissey reignites racism row by calling Chinese a "subspecies"', *Guardian*, UK, 3 September 2010.
110. Gary Kynoch, '"Your petitioners are in mortal terror": the violent world of Chinese mineworkers in South Africa: 1904–1910', *Journal of Southern African Studies* 31(3), (2005), pp. 531–546.
111. Egbert Wesselink, 'Who should engage and how: governments, business, civil society', in IKV Pax Christi, ed., *Chinese State-owned Enterprises and Stability in Africa* (Clingendael Institute, 30 May 2008), pp. 12–13, available at: www.clingendael.nl/.../20080825_asia_report_expert.
112. Peter Hitchens, 'In China, 5,000 people die and there is nothing. In Zambia, 50 people die and everyone is weeping', *Mail on Sunday*, London, (28 September 2008).
113. Kevin Bloom, 'Does Africa need China?', *Daily Maverick*, (21 October 2011), available at: http://dailymaverick.co.za/article/2010-10-21-analysis-part-i-does-africa-need-china.
114. Jamie Whitlock, *Digging for Prosperity: Mining and Labor Practices in Chambishi, Zambia*, unpublished M.Sc. thesis, University of Oxford, 2008, p. 9.

working in rural Zimbabwe cruelly killed dogs for food, an opposition newspaper thundered that 'This kind of cruelty cannot be allowed to be extended to the workers at the construction sites, mines, restaurants and retail shops where the Chinese claim protection from well-placed government officials'.[115] Such assessments reflect a narrative of Chinese cruelty and disregard of human life that now takes the CCM shooting as a prime example, a narrative likely to expand as Chinese become more prominent globally and thus impinge on Westerners' long-running leading role.

The CCM incident shows that China's government is not indifferent to malfeasance abroad, because 'Anything a Chinese firm does in Africa will be seen as representative of China as a whole. Anything a Chinese immigrant does in Africa will be seen as representative of all Chinese people',[116] a standard not applied to Western firms or individuals. Chinese state involvement however is a form of crisis management—urging Chinese firms to respect the rights of local communities, workers and the environment—while its ability to regulate enterprises is undermined by its own neoliberal reforms. Chinese diplomatic outposts also regard it as 'inconvenient' and often unavailing to influence non-SOE firms, like CCM, that are state of incorporation citizens with no parent company in China.

Even if the Chinese state mitigates practices of miscreants with 'Chinese faces', the discourse will remain unbalanced, as China/Africa links now have an outsized significance within a perceived strategic competition between the West and China. Narratives of Chinese activities in Africa will continue to be incorporated into the larger China Threat discourse, which has the cat's proverbial nine lives in the West. The narrative of the CCM shooting will feed the contention that Chinese menace not only the West, but even vulnerable grassroots Africans. If, however, the narrative of the CCM shooting is itself problematized, the discourse of China/Africa links may not remain the complete binary it is now presented to be.

---

115. 'It's not for party-less Mutambara to tutor us about unfair Chinese practices', *News Day*, (20 June 2011).
116. Wang Xiaojuan, 'Thorns in the African dream (2)', *Chinadialogue*, (31 January 2012), available at: www.chinadialogue/net/article/show/single/en/4749.

Workforce Localization among Chinese State-Owned Enterprises (SOEs) in Ghana

ANTOINE KERNEN and KATY N. LAM

Chinese state-owned enterprises (SOEs) have gradually localized their workforce since they began operating in Ghana in the 1980s. Examining their workforce localization patterns, the Chinese SOEs in Ghana appear to be diverse in their business practices and highly autonomous from the Chinese state. Our hypothesis on the substantial autonomy of Chinese SOE overseas subsidiaries, which is consequent to the lack of management control from the Chinese central authority since the Chinese economic reform, contrasts the dominant assumption in the China–Africa debate, in which Chinese SOEs are depicted as closely linked to the Chinese state and/or as the arms of the new Chinese policy in Africa. The workforce localization process of Chinese SOEs in Ghana is largely determined by factors like profit maximization objective, market competition and political pressure. The localization experience is similar to those of Western companies in Africa where complete workforce localization takes a long time to achieve.

Introduction

During our fieldwork in Ghana, country directors of Chinese state-owned enterprises (SOEs) expressed their intention to hire as many locals as possible. Rather than just a discourse, they consider that workforce localization (属地化/本土化) is necessary for different reasons ranging from reducing labour costs to adapting to local political contexts. Nevertheless, the human resources localization pattern—degree, motivations and strategies—is non-linear and varies from one SOE to the other. Each of the Chinese firms seems to have its own localization formula influenced by its specific internationalization trajectory and experience in Ghana.

Viewing the workforce localization practice from a bottom-up perspective, it appears that Chinese SOEs in Ghana enjoy a high degree of managerial autonomy in their business operation. This observation is certainly not in line with the prevailing view in which Chinese SOEs are usually depicted as a state apparatus and an executor of the new Chinese policy in Africa.

---

* Antoine Kernen is a Senior Lecturer and Researcher in Social Sciences of the University of Lausanne, Switzerland. Katy N. Lam is a doctoral candidate in Social Sciences at the University of Lausanne, Switzerland. The authors are thankful for the financial support from the Swiss National Fund to conduct the current research.

Our hypothesis on the substantial autonomy of Chinese SOE overseas subsidiaries does, however, correspond with the literature on Chinese SOEs in China, which, in general, highlights the lack of management control from the state authority since the beginning of the reform in 1978. This loose control continues to be a subject of debate. Different Chinese newspapers published several articles before the 18th Chinese Communist Party Congress (November 2012) calling for strengthening control of SOE overseas businesses.[1] State ownership does not necessarily signify an effective state control of the SOEs. Nevertheless, studies of China and Africa still pay scant attention to this dimension.

Focusing on workforce localization, the objective of this article is to provide a disaggregated perspective of how the internationalization process of Chinese SOEs is taking place in Ghana.[2] The Chinese state alone is far from the only actor that shapes the process. The difference of localization practices suggests that the SOE subsidiaries themselves are among the drivers framing the process in which the 'vision' and personal experience of the country directors (who are Chinese) are also determinants.

In addition, the local (national)[3] context certainly influences the business practices of Chinese SOEs in Ghana. Although China is a powerful actor in the African continent, scholars recently pointed out the importance of the local agency in the China–Africa encounter.[4] We will see later how localization strategies of Chinese SOEs are adapted to respond to forces from local politics and Ghanaian employees.

This article is based on our ethnographical research in Ghana, using semi-directive interviews and participant observation along with secondary resources. On top of two long field studies in Ghana, our research experience on Chinese SOEs in China and shorter visits in other West African countries provide empirical reference for comparison. The outcomes of several years of research allow us to propose this bottom-up vision of the Chinese presence in Africa.

The article will first give a brief summary of Chinese SOE reform and SOE internationalization in Ghana. It will then explain diverse localization patterns in several sectors in which Chinese SOEs are active, especially the construction and telecommunication sectors, and for each type of employee as well as the current limits for SOEs to implement workforce localization in Ghana.

China reform and internationalization of Chinese SOEs in Ghana

SOE reform: changing relationship with the Chinese state

State-owned enterprises have been a major target in China's economic reform since 1978. During the first phase of the reform between 1978 and 1993, the SOEs were

---

1. Bao Chang, 'SOE urged to manage overseas businesses', *China Daily*, (17 October 2012).
2. This article will include Huawei Technologies, which is in fact not a state-owned enterprise but a wholly employee-owned company though many question if it has a close link with the Chinese government; see http://english.caixin.com/2010-08-11/100169742.html.
3. In this article, we use 'local' to refer to Ghana and Ghanaian instead of 'national' to avoid causing any confusion.
4. Giles Mohan and Ben Lampert, 'Negotiating China: reinserting African agency into China–Africa relations', *African Affairs* 112(446), (2013), pp. 92–110.

granted managerial autonomy in developing new market-oriented business, though they continued to play significant social roles in urban China. Until 1993, a state worker could not be laid off and SOEs not only had to secure existing posts but also provide new jobs to the urban population. Before the social security system reform, the SOEs provided a retirement pension and access to medical care to a large part of the urban population. Continuing to bear social costs, SOEs contributed to a smooth transition during the first decade of the reform.[5]

Nonetheless, the managerial autonomy of SOEs encouraged rent-seeking behaviour as managers were rewarded for their success, but were not penalized for their failure in meeting targets. Most financially profitable businesses have been absorbed by newly created subsidiaries of SOEs. Social burdens (retired workers' pension, redundant workers, etc.) and business debt were kept in the original SOE and the most profitable business was transferred to subsidiaries that progressively took up their autonomy. In such a way, many SOEs have been 'eaten' by their subsidiaries that led to an informal privatization.[6]

Weak control and the continuous social burden of the Chinese SOEs in the first phase of the reform led to the financially poor performance of many state enterprises. In 1996, the Chinese government adopted the policy 'keep the large and let the small go' (抓大放小), in order to reinforce and concentrate control on enterprises in strategic sectors.

To 'keep the large', the largest Chinese SOEs at the national level were merged into groups. This merging process still continues and has resulted in 113 groups today.[7] These groups, which are also called 'central enterprises' (中央企業), are now under the supervision of the SASAC (State Assets Supervision and Administration Commission). These groups can often take advantage of a monopoly position and still benefit from preferential access to credit.

The rest of the SOEs were 'let go'—sold or leased—and have undergone diverse corporatization processes, modelled from Western-style corporations.[8] Many SOEs were also transferred to be under the supervision of provincial governments.[9] At the same time, the provincial governments also tried (but with lesser funds) to build up smaller groups by merging enterprises in the same sector. Therefore, apart from the national industrial groups under the supervision of the Chinese central government (SASAC), many smaller groups are supervised directly by provincial governments. In order to compete with the centrally controlled (and protected) groups, the provincial groups have quickly developed their international business. Nowadays,

---

5. For example: Barry Naughton, *The Chinese Economy: Transition and Growth* (London: MIT Press, 2007).

6. Antoine Kernen, *La Chine vers l'économie de marché. Les privatisations à Shenyang* (Paris: Karthala, 2004).

7. For a list of the Chinese Central SOEs, see http://www.sasac.gov.cn/n2963340/n2971121/n4956567/4956583.html.

8. Yi-min Lin and Tian Zhu, 'Ownership restructuring in Chinese state industry: an analysis of evidence on initial organizational changes', *The China Quarterly* 166, (2001), pp. 305–341.

9. Antoine Kernen, 'Shenyang: privatisation in the vanguard of Chinese socialism', in Beatrice Hibou, ed., *Privatising the State* (London: Hurst, 2004); Jean C. Oi, 'The role of the local state in China's transitional economy', *The China Quarterly* 144, (1995), pp. 1132–1149; Jean C. Oi, 'Patterns of corporate restructuring in China: political constraints on privatization', *The China Journal* no. 53, (January 2005), pp. 115–136; Ross Garnaut et al., *China's Ownership Transformation: Process, Outcomes, Prospects* (Washington, DC: World Bank, 2005).

some Chinese provincial groups are often more internationalized than the big centrally-controlled groups.[10]

The outcomes of this second phase reform were not only establishing industrial groups at the central (national) and provincial levels, but also selling and closing thousands of small SOEs. This restructuring process resulted in massive lay-offs—around 35–40 million people lost their job during this process.[11] The Chinese state enterprises as a whole have become a much more economic-oriented entity whereas their previous social responsibility has largely diminished.[12]

Even though Chinese SOEs are still 'state-owned', the role of the 'state' has largely evolved in terms of 'ownership' and 'control'. The multi-shareholders ownership structure of Chinese SOEs has transformed the relationship between the central state and the Chinese SOEs. Even for the central SOEs which are under the direct supervision of the central state through SASAC, how the co-existence of enterprise autonomy and central control can function is also questioned.[13] The provincial SOEs under the supervision of provincial governments maintain an even more distant relationship with the central state as the provincial governments have been granted a greater autonomy in managing provincial economic sectors.

First SOEs in Ghana—construction sector as pioneer in the 1980s

The oldest SOE remaining in Ghana is the China State Hualong Construction Engineering Co. Ltd, which entered the Ghanaian market for the construction of the Chinese Embassy. Its parent enterprises are a provincial one from Gansu province and the China State Construction Engineering Co. Ltd. The latter is one of the four Chinese SOEs which received authorization in November 1978 to operate overseas. At the beginning of the reform, the Chinese government adopted a policy for internationalization and outward investment.

> In November 1978, the Central Committee of the CCP and the State Council jointly approved the establishment of the China State Construction Engineering Co. Ltd which specialized in overseas engineering and construction works and labour services. In 1979, three other such companies were sanctioned by the State Council. They were the China Civil Engineering and Construction Corporation, the China Road and Bridge Engineering Co. Ltd. and the China Complete Set Equipment Import Export Co Ltd.[14]

The previous international experience gained through implementing Chinese aid projects in the Mao era and the international project bidding opportunities made the construction sector an ideal sector for selection to develop business abroad since

---

10. As per the Engineering News-Record (ENR), numerous Chinese provincial groups are among the top 225 International Contractors; see http://enr.construction.com/toplists/Top-International-Contractors/001-100.asp.
11. Li Peilin, 'The professional reintegration of the Xiagang: a survey in Liaoning province underscores the importance of vocational training courses', *China Perspectives* no. 52, (April 2004).
12. Yingyi Qian and Jinglian Wu, 'China's transition to a market economy: how far across the river?', in Nicolas. C. Hope, Dennis Tao Yang and Mu Yang Li, eds, *How Far Across the River? Chinese Policy Reform at the Millennium* (Stanford, CA: Stanford University Press, 2003).
13. Kjeld Erik Brødsgaard, 'Politics and business group formation in China: the party in control?', *The China Quarterly* 211, (2012), pp. 624-648.
14. Eunsuk Hong and Laixiang Sun, 'Dynamics of internationalization and outward investment: Chinese corporations' strategies', *The China Quarterly* 187, (2006), pp. 610–634.

the first day of the reform.[15] These four pioneer enterprises have undergone restructuration and merging since then, but they still remain important international contractors in the construction sector.

Since the 1980s, Chinese construction SOEs have been using international cooperation projects as a stepping-stone to enter the international market. Though first projects can be Chinese cooperation projects, it is not rare that a SOE enters into a country market through projects financed by other governments and institutions. The China Water & Electric Corporation (CWE) moved to Ghana by winning a Japanese-supported assistance project in the early 1990s.[16] Afterwards, it obtained a water-well digging project financed by Germany and Denmark, and more recently projects of highway construction and rural electrification supported by the Ghanaian and Qatari governments. Likewise, some Chinese companies have obtained projects from other international organizations like the World Bank.

In Africa, most of the Chinese SOEs are in fact provincial SOEs. This is especially the case in the construction sector—in Ghana, 11 out of 20 Chinese companies in the sector are provincial, many from interior and relatively poor provinces such as Gansu (甘肃), Shaanxi (陕西) and Jiangxi (江西). Examining closely the business profile of these non-coastal provincial enterprises, the African market represents an important market for their overseas business and even as part of their core business. However, these provincial enterprises usually get lesser support from the Chinese central government. In fact, the only concrete 'help' the provincial groups can receive from the Chinese government is to obtain a Chinese cooperation project as an internationalization stepping-stone. In the construction sector in Ghana, this is the case for seven provincial groups and for all (two) centrally controlled groups.

After the initial cooperation project, the SOEs' subsequent projects are usually non-Chinese financed projects. They have to obtain them through open bidding, in which price and quality are the primary factors rather than diplomacy. During the three decades, China State Hualong Construction Engineering Co. Ltd, for instance, has been involved in numerous projects financed by the Ghanaian government, other donors, international institutions and private companies. The Chinese SOE overseas subsidiaries have been in direct competition with foreign competitors in the continent since the 1980s.

Fishery and telecommunication sectors. Apart from the construction sector, the rest of the Chinese SOEs in Ghana are active in the fishery and telecommunication sectors. The Shandong Fishery and the China National Fisheries Corporation (CNFC)[17] expanded their business in Ghana by establishing a joint venture company with Ghanaian fishery companies under the Chinese technical cooperation programme.[18] In Ghana, there are a few state-owned and private Chinese fishing companies. All of them operate in partnership with local companies. In Ghana, the

---

15. Jean-Paul Larçon, ed., *Chinese Multinationals* (New Jersey: World Scientific, 2008).
16. Interview with the Director of the China Water & Electric Corporation Ghana, the Ghanaian branch of the China Water International, 4 February 2010.
17. The CNFC set up business in West Africa starting from 1985. See Zhu Xiaolei, 'CNFC: 25 years' fishing and shrimping in Africa', *Africa Magazine*, (31 May 2011).
18. Interview with the manager of a Chinese state-owned fishery enterprise in February 2010.

fishing business is reserved for Ghanaians; foreign companies can participate through partnerships with local fishing companies in the 1990s.[19] Therefore, in terms of ownership, the Chinese SOE fishery subsidiaries in Ghana are not purely Chinese.

The two main Chinese telecommunication companies entered the Ghanaian market around the mid-2000s, but are mainly involved in projects with private and multinational telecommunication operators. The largest Chinese telecommunication company in Ghana is Huawei Technologies, which is an entirely employee-owned company, though some suggest that the company has close ties with the Chinese government.[20] The second largest telecom company is the state-owned ZhongXing Telecommunication Equipment Corporation (ZTE), but about 70% of its shares are traded publicly in the Shenzhen and Hong Kong stock exchange markets.[21]

The impact of the going out strategy policy

The 'going out' strategy (走出去) was officially proposed for implementation in the report of the Tenth Five-Year Plan for National Economic and Social Development in 2001[22] as an outcome of China's entry into the WTO. The 'going out' policy encourages Chinese companies to look for business opportunities overseas, by simplifying the procedure and softening the control on overseas investments. The Chinese enterprises responded quickly because international opportunities are more attractive than those available in China. Chinese outward investments grew rapidly from 2.5 billion in 2002 to 84.2 billion in 2012.[23] Simplifying outward investment procedures is one more step towards the marginalization of the role of the Chinese state on its economy. The Export Credit Agency was created at the same time as the policy adoption, however it is nothing unique, as most Western countries also possess similar institutions.

Accompanying the 'going out' policy is also a significant increase of Chinese cooperation projects in Africa, following the establishment of the Forum on China–Africa Cooperation (FOCAC) in 2000. In Ghana, more than ten Chinese SOE arrived after 2000 through winning Chinese cooperation projects. Unlike some Chinese SOEs that came earlier for a cooperation project but left immediately upon termination, Chinese SOEs now had a clearer intention to stay and to prospect for business opportunities following their first contract in Ghana.

The first Chinese state-owned enterprises arrived in Ghana in the early 1980s, two decades before the 'going out' policy was formalized in 2001. Around 25 Chinese state-owned enterprises are currently operating in Ghana; six of them arrived before the liberalization of the overseas Chinese investments. Therefore, one of the real impacts from the introduction of the 'going out' policy is the increasing competition

---

19. According to the Ministry of Food and Agriculture of Ghana, see http://mofa.gov.gh/site/?page_id=6133.
20. See http://english.caixin.com/2010-08-11/100169742.html.
21. See http://www.forbes.com/sites/simonmontlake/2012/11/29/crossed-lines-zte-gets-tangled-in-u-s-china-telecom-gear-cold-war/, http://www.theregister.co.uk/2012/09/30/inside_huawei/.
22. The report is available at: http://www.gov.cn/english/official/2005-07/29/content_18334.htm (accessed 17 June 2013).
23. UNCTAD, *World Investment Report 2013*, available at: http://unctad.org/en/Pages/DIAE/World%20Investment%20Report/Annex-Tables.aspx (accessed 13 September 2013).

among Chinese companies that often have similar levels of technique and cost advantages. It is common for our interviewees to highlight the fierce competition among Chinese companies who are always the ultimate competitors of each other in the price-cutting race, for example when bidding for construction projects. For those arriving earlier than the official adoption of the policy, especially the provincial SOEs, with lesser financial resources and central state support in, for example, winning Chinese cooperation projects, they make use of their local experience to better tune their business strategy including a higher degree of workforce localization in competing in the open market.

Strategies of workforce localization

In this part of the article, we will analyse the workforce localization patterns especially in construction and telecommunication sectors, in which Chinese firms employ a large amount of workforce. Through examining the localization strategies we will see that the local context, among other factors, also plays a significant role on the usage of the local workforce in Chinese SOEs.

First, we will focus on the labour-intensive construction sector, where the degree of localization seems to closely link with the business duration of a SOE in Ghana. We will then focus on mid-level employees including supervisors, technicians and engineers. Even if a good part of them are still Chinese, SOEs both in construction and telecommunication sectors are making an effort in localizing much further. What's more, some Chinese SOEs have already localized management staff. It appears that those Ghanaian managers are in general in charge of the public relations of the company at the local level.

Building up the team: manual and skilled construction workers

Since the arrival of the China State Hualong Construction Engineering Co. Ltd in the 1980s, the manual construction workers have always been locals. Cost saving is no doubt a primary reason for using unskilled construction workers. In 2010, Ghanaian unskilled construction workers received around 100 Ghana cedi per month[24] (around US$70), while the salary in China is probably four time more—RMB 2,000 (around US$300), without taking in account transportation costs and living expenses.

For new Chinese construction companies in Ghana, it takes time to build up a team and train local workers. To overcome these difficulties, Chinese SOEs tend to bring more Chinese skilled workers at the beginning. The Sinohydro Corporation, building a hydroelectric dam since 2008, has not yet developed its own team—the proportion of Chinese staff is around 15%.[25] It is a much higher percentage compared to companies with a longer experience in Ghana. A pioneer SOE, for example, the China State Hualong Construction Engineering Co. Ltd, has only 6% that are Chinese in its total workforce—about 100 Chinese for 1,500 locals. The large majority of its

---

24. This was the general salary level of Chinese companies in early 2010, when 1US$ was equal to 1.45 Ghana cedi. The minimum wage set by the Ghanaian government was 2.5 Ghana cedi per day during the same period.

25. The company hires over 1,500 local workers (85% of the total workforce) and brings around 250 Chinese technicians and managerial staff.

local workforce were unskilled workers that have been trained for different tasks and to operate Chinese machines. Undoubtedly, those machines with instructions and labels written only with Chinese characters are impossible for non-Chinese workers to handle at the beginning. The Chinese construction SOEs usually provide on-the-job training to Ghanaian workers. At first, the Chinese skilled workers supervise or work together with a few local workers until the latter can work independently. Therefore, the degree of localization of construction workers tends to increase proportionally with the SOE's business duration in Ghana, as the Chinese firms gradually find or more often train local construction workers to replace the Chinese skilled workers.

Hire foreign workers or outsourcing. Apart from training local workers itself or bringing expensive Chinese skilled workers, an alternative is to hire cheaper foreign workers. This is the case for the Sinohydro Corporation as it is difficult to find local workers that can handle some specific tasks for their hydroelectric dam.[26] They employed 60 Pakistani technicians on the dam construction: 'We cannot find locals who know how to use our machines. They [Pakistani technicians] worked with us for our projects in Pakistan. They are good and cheaper than Chinese'.[27] The use of Pakistani technicians shows, once again, the pragmatism of Chinese companies. Without taking into account the political/social question of employment in China but rather considering in terms of labour cost and capacity that fit the company's needs, this central 'state-owned' enterprise chose to use technicians from another nationality. This reflects the fact that financial benefits are a much higher priority than national social issues in the reformed state-owned enterprises.

Another option to build a local team is to sub-contract to local companies. Even though the China International Water Electric group (CWE) entered the Ghanaian market in 1992, it still outsources part of the project tasks to local companies. The Chinese employees focus on coordinating and managing the project. The ratio of its Chinese staff to local workers and technicians is around 1:10, a ratio close to those SOEs that have been operating in Ghana for a long time.

The CWE subcontract policy is not only useful to avoid dealing directly with a large amount of local workers, but also the company can concentrate on developing its business advantage. As the Managing Director Mr X said 'the Ghanaian government thinks our project management skills are good', which is a good point for the SOE as it is mainly a contractor for Ghanaian ministry projects.

Similarly, in telecommunication sectors, Zhong Xing Telecommunication (ZTE) chooses not to hire construction workers directly. The telecommunication company focuses on their core technological competencies and outsources construction work to other companies, either Chinese or local. The subcontractors are in charge of the building of telecommunication infrastructure like laying underground cable. ZTE keeps in its own hands the more technological work. Moreover, ZTE's Chinese subcontractors may also further subcontract their work to local companies.

One of the advantages of subcontracting is to save training cost in a context where the turnover rate of local labour is high. The newly trained local technicians do not

---

26. The second hydroelectric dam in the country; the first one was built in the 1960s.
27. Interview with the Administrative Manager on 25 December 2009.

necessarily stay in the same company. 'Local staff know nothing when they start to work with us at the beginning. After training, they are qualified. Some of them move to work for other companies since the salary will be higher once they possess the technique' said Mr Z,[28] a Chinese project manager of an experienced provincial state-owned construction company. Therefore, Chinese companies need to provide incentives like promotions to keep those workers who are able to master the technology and train new workers, like in Mr Z's company: 'Some local workers have been working for us for more than ten years. They become the backbone of local staff and are promoted to supervise and train other locals'.[29]

The other main advantage is to avoid dealing with labour issues in an unfamiliar socio-political context. In Ghana, the labour law gives workers the right to set up a union when the number of employees is over 15.[30] It is also common that companies negotiate an annual collective contract with their labour union, listing clearly the remuneration and benefits details that are usually improved every year. Nevertheless, an annual contract does not guarantee that workers are always happy with the agreed conditions throughout the year. Labour unions sometimes organize strikes near project delivery deadlines, thus increasing their bargaining power as the employer does not have time to replace them with another team. The company is forced to negotiate and in most cases gives a wage increase. A labour strike took place in the Sinohydro Corporation (a company without subcontractors) because the workers were unsatisfied with the year-end bonus. 'The Chinese doesn't give bonus, but we worked well and there was no problem in the whole year so the labour union asked 1.5 time [of salary] for bonus' said the local union chairman.[31] Eventually, the company gave a bonus of 1.2 times in order to resume work. By outsourcing project tasks, the cost related to the labour conflicts is shifted to the subcontractors. Moreover, if the subcontractor is a local company, Chinese companies believe that negotiations or conflicts would be easier to be handled by Ghanaians.

Apart from cost saving. The localization of unskilled labour for Chinese companies in Ghana is nothing unique in the continent. A study of Chinese construction companies in Angola, Sierra Leone, Tanzania and Zambia shows that it is common for a Chinese company to employ locals for over 80% of its total workforce.[32] Cost minimization is a widely shared motivation in localizing workforce, but compliance to investment law also plays a role. According to the Ghana Investment Act, one of the requirements of setting up a business in the country is to employ at least ten Ghanaians.

Cost saving has been a justification in the debate over the large number of Chinese workers who were employed in Africa.[33] However, political incentives of the host countries and business reputation are more important reasons. As Tang

---

28. The name is fictitious as per the anonymity request of the interviewee.
29. Interview on 28 January 2010.
30. According to the International Labour Organization (ILO), available at: http://www.ilo.org/ifpdial/information-resources/national-labour-law-profiles/WCMS_158898/lang–en/index.htm (accessed 11 July 2012).
31. Interview on 25 December 2009.
32. Centre for Chinese Studies of the Stellenbosch University, *China's Interest and Activity in Africa's Construction and Infrastructure Sectors* (Stellenbosch University, November 2006).
33. Chris Alden and Martyn Davies, 'A profile of the operations of Chinese multinationals in Africa', *South African Journal of International Affairs* 13(1), (2006), pp. 83–96.

illustrates,[34] more Chinese technicians were recruited in cooperation projects in order to keep on schedule. Political interest is huge around those highly publicized projects which helps to image the African government's capacity in serving its people (while the work is outsourced). The infrastructure projects, usually widely publicized, serve as a landmark in the SOE's résumé which contributes to a good reputation in order to gain other contracts later. Chinese workers are certainly more expensive than Ghanaians but the SOEs consider the extra money well spent. Beside the deadline question, Chinese cooperation contracts usually offer a good profit margin.[35] Hence, bringing more expensive but better-qualified Chinese technicians is still affordable. Through these projects, the companies will be able to become more familiar with the Ghanaian business environment and progressively use local workers more extensively.

Ghanaian or Chinese foremen?

Many Chinese companies in Ghana consider that the relationship with local workers is important but not easy to master. In the absence of a common language, exchange between Chinese and locals—especially manual workers—is difficult. It often relies on body language, which sets a limit to communicate more complex instructions that potentially leads to cultural misunderstanding. For Chinese supervisors in construction sites, maintaining a special relationship with certain employees is essential. 'Occasionally, I give them a cigarette or a drink, and they help me transmit orders to others' said a Chinese chief technician. The Chinese bosses learn to react positively to various demands from their local employees:

> It is important to give them small gifts sometimes, not only on the occasion of marriages, births and funerals, but also when a child must go to hospital or if the price of certain commodities increases too much. After they are happy and work better.[36]

'Locals manage locals'. Mr R., the Managing Director of an experienced provincial construction company, considers that the lack of mutual understanding is a source of work inefficiencies and conflicts. After living for 20 years in the country, he explains why 'Chinese' encountered difficulties in handling local employees:

> First, Chinese look down on locals. Second, Chinese consider that locals are thieves. (...) Third, Chinese believe that if they treat locals well, like giving them gifts, locals should be grateful and work harder, but the locals don't keep it in mind. For them, we, the white, ought to give them presents.[37]

This is why Mr R. adopted the strategy of 'locals manage locals' which is believed to be a comparative advantage to newly arrived Chinese companies.

---

34. Xiaoyang Tang, 'Bulldozer or locomotive? The impact of Chinese enterprises on the local employment in Angola and the DRC', *Journal of Asian and African Studies* 45, (2010), p. 350.
35. For Chinese aid projects, payment is made directly by the Chinese government. According to interviews, delay and sometimes long delays in payment are common if it is done through local government.
36. Several Chinese interviewees have raised this point.
37. Interview on 21 January 2010.

Mr R. explained:

> To have a good local worker management system, we have promoted a group of local workers to supervisor positions. Today, some of them have been working for us for more than ten years. Although they have low education level, their operating technique is excellent [...] Those newly arrived Chinese construction companies have not yet developed a proper management system. It is why, labour conflicts occurred in Sinohydro.

The Corporation is part of a Chinese central SOE group that is also among the top 500 world companies according to *Fortune* magazine.[38] The SOE started its operation in Ghana by winning a hydroelectric dam project financed by a Chinese loan. However, it is so far the Chinese SOE which has had the most labour conflict issues reported in the media.

Respond to political pressure. Wider localization—hiring proportionally more locals—is also a response to political pressure. The Ghanaian authority requests that Chinese companies give as many jobs to Ghanaians as possible. To 'encourage' localization, the Ghanaian Immigration Department has tightened its measures on issuing working visas to foreigners in recent years. According to the Ghana Investment Act, an initial maximum immigration quota of four persons is set for an investment capital of US$500,000 or more.[39] Old Chinese companies said that this regulation was applied loosely before; however, now getting a visa has become one of the major obstacles to doing business in Ghana.

The Chinese Director of a state-owned fishery company that started operating in Ghana in the early 1990s told us how his company responded to the localization call of the Ghanaian authority:

> We have now around four–five Chinese staff and 20 Ghanaian staff per ship and we have eight ships. Before we had 13 Chinese per ship. However, the Immigration Department wanted us to leave the locals more jobs and they thought 13 Chinese per ship was too many. So now, jobs are left to locals, because after years, we had trained also Ghanaian workers to take up those jobs. Our Chinese captains have returned back home. Those staying are the very important crew members.[40]

'Chinese supervisors are always needed'. The strategy of 'locals manage locals' is adopted in some but not all Chinese SOEs and most of them do not want to implement it completely. Chinese staff may still occupy certain positions in the long run. The Chinese companies are ready to pay much more to keep some Chinese workers. A SOE subsidiary director put it more explicitly: 'if there is no Chinese supervising, the locals will do nothing, even though they have good techniques. They will work only when there is someone there to supervise'.

Moreover, 'unjustified and frequent absence' or 'locals often don't show up to work for a few days after receiving salary' are also among the common complaints.

---

38. Sinohydro belongs to the group China Power which ranks 390 on the list of Fortune 500 in 2012; see http://money.cnn.com/magazines/fortune/global500/2012/full_list/301_400.html.
39. Ghana Investment Act, Section 18, Part II, Article 30.
40. Interview on 6 March 2010.

Some Chinese managers seek to manage their staff in a paternalistic way. They learn to show their concern for the everyday problems confronted by their local employees. For instance, if a Ghanaian employee needs to borrow money, Chinese bosses will lend some and deduct it from the next month's salary of the worker or set up a repayment scheme deducting a fixed amount from their salary each month. A Chinese manager said:

> Personally I always lend money to local workers, but only to those who are obedient [to work order]. We deduct their debt bit by bit like 20 cedi from the salary per month. But you know they won't be able to pay them off before asking for the next credit again.[41]

Training local technicians and engineers

The Chinese SOEs find it difficult to find qualified and affordable technicians and engineers locally. Suitable ones usually have high salary expectations and prefer to work for multinational companies from developed countries or abroad. Most of the engineers in the construction and telecommunication companies are still Chinese. The ratio of Chinese staff in the Huawei Technologies Co. Ltd, which is 50%—only half of the 300 employees are local[42]—indicates the difficulties in hiring local technicians and engineers. Considering that the African market is increasingly important for them, Chinese SOEs started to formalize training programmes to guarantee their supply of cheap local technicians and engineers in the long run.

Nevertheless, Huawei's localization is going to expand in the next few years according to the Huawei human resources manager. When we met, he was sitting in front of his desk with piles of filled application forms and he explained, 'we are in the process of recruiting a few hundred graduates from universities in Ghana for our new projects. We will take people from all kinds of majors and train them'. The eagerness to replace Chinese staff is obvious due to the high expat cost, as the monthly salary of a mid-level Chinese staff member working in West Africa is around 30,000 RMB (US$4,500). The expat package includes three sets of return flight tickets to China per year, another set for leisure travel within Africa, insurance coverage, accommodation, meals and the expense of hosting family members. Hiring a local engineer reduces all expenses into a much lower salary. In order to retain trained local staff, Huawei has a training and loyalty programme for its local employees in Africa. After the first year working in a Huawei African subsidiary, the employees can have, for example, the opportunity to spend one month in its Shenzhen's headquarters.

Across the African continent, Chinese telecommunication firms have set up programmes for young or future IT professionals. Huawei in Tanzania, for example, has launched programmes to improve learning infrastructures in some primary and secondary schools and to provide free ICT training to excellent university students.[43] Another company, ZTE, has set up four training centres in the continent

---

41. Interview on 24 January 2010.
42. This ratio does not include the outsourcing part of a telecommunication project in which labour is mostly local. Chinese telecommunication companies subcontract construction work of a project to others while they themselves mainly take care of technology-related work.
43. 'Huawei launches "ICT Star" program for Tanzania education', *China Daily*, (16 November 2012).

(Algeria, Egypt, Ethiopia and Ghana) to train local ICT professionals.[44] All these efforts can help both train local IT professionals working for them in the future and build a socially responsible image. In addition, the two companies have recently started to offer internships to African students in China and may even provide scholarship to Africans in the future.

Apart from the fact that Chinese SOEs also take part in the selection of the African students that will receive scholarships to study in China,[45] some of them are now setting up special scholarships themselves for African students, like China Geo-Engineering Corp., an experienced SOE in Ghana for more than 15 years, which has regularly sent its Ghanaian staff to China for training. Last year, the construction subsidiary sent 13 Ghanaian managers and 22 technicians to follow management courses or professional construction training courses for between three and six months.[46] Similarly, the China Road and Bridge cooperation launched a five-year programme last year to send students from Congo Brazzaville to study civil engineering. The programme objective is to provide 100 engineers for the SOE working in the country.[47]

Toward management by Ghanaians?

During his speech at the company Christmas party, the Managing Director of Huawei told his staff: 'I'm happy with the business performance and revenue, but I'm not very happy to see so many Chinese here. There should be more Ghanaians here. It is a Ghanaian company'.[48] Even if the Chinese telecommunication companies in Ghana actively promote a socially responsible image,[49] recruit and train local technicians, they seem less in a hurry to localize their management level. It is true that many of their business partners are private and more internationalized. Having less political pressure to localize, they emphasized the difficulties and the cost of this process in our interview, as indicated by a ZTE manager:

> our company encourages localizing of our staff. But I believe it will be difficult to achieve in a short run. Local staff has lower technique and education level compared to the Chinese. [...] Project managers of our branches in India and Middle East countries are locals. In Ghana, management is still Chinese.[50]

We have been told that local managers with suitable profiles are usually schooled in developed countries and ask for a higher salary than a Chinese manager. It seems also that the Chinese telecommunication companies are not yet ready to pay an African manager more than a Chinese one. They prefer paying more to hire Europeans

---

44. See http://www.zte.com.cn/cn/about/corporate_citizenship/enriching_life/200812/t20081208_349647.html, http://www.zte.com.cn/cn/press_center/press_clipping/200903/t20090331_343720.html.
45. Interview in Xian with former ZTE employees in West Africa.
46. See http://africa.chinadaily.com.cn/weekly/2013-05/31/content_16550644.htm.
47. Interview in Xian with the person in charge of the programme.
48. As recited by an Huawei staff member.
49. See, for example, the website of Huawei on corporate citizenship: http://www.huawei.com/en/about-huawei/corporate-citizenship/index.htm.
50. We found that, for example, Ghanaians worked in the management level in the ZTE in Benin.

or Asians who speak English fluently to communicate with clients and are able to demonstrate the international character of the company. For instance, Juan,[51] an Asian who speaks fluent English but not a single word of Mandarin, has been working as a project manager in a Chinese telecommunication company in Ghana for several years.

In contrast, the Chinese construction SOEs with longer experience in Ghana have a clear strategy to localize their management staff. Are they more than the 'interpreter' or 'ambassador' of a Chinese company? As far as we know from our field studies, the exact role and decision-making power of those local managers remains limited as they essentially play a role in enhancing external and internal relations.

The Chinese managing directors we interviewed considered management localization generally as an essential strategy to make their business more successful. A construction SOE director said in an interview: 'Now our bidding team consists of Ghanaian staff only and no Chinese at all'. He was proud because not every Chinese company was able to develop its own local team. This would require, for instance, that Chinese management be able to communicate fluently in English and have enough trust in their Ghanaian colleagues, which is still uncommon. Most of the Chinese management in the newly arrived SOEs we met in Ghana have a very basic level of English.

In the same construction company, Sunny, the local general manager, has to accompany his Chinese colleagues to attend meetings with Ghanaian ministries to discuss the company project proposals. He said:

> After we had have meeting, the ministries people called me to ask if the Chinese company was OK or not. It is because Chinese and Ghanaians don't understand each other [...] I usually have to develop a good relationship with ministries in order to understand what they want on projects [to be bid for] and let ministries to better know our companies. Then, I pass information on to our bidding team.

The country director of that company was very happy to have localized its bidding team, 'in 2005, our total contract value was US$50 million. Now five years later, it is US$150 million'.[52]

Sunny received tertiary education in the United Kingdom where he also worked for several years upon graduation. Then, he got a job offer as the General Manager of his current Chinese employer. He admitted that he hesitated over whether to work for a Chinese company, though he was very interested in returning to work in his home country. Sunny's hesitation highlights the fact that Chinese companies are usually not the first choice for foreign-trained Ghanaian professionals. Emmanuel,[53] an ex-senior official in the Ghanaian government and now a businessman, received a scholarship to study in China 20 years ago and has never worked for Chinese companies since his graduation. He explained: 'like most of my fellow scholarship receivers [who also studied in China], we do not want to work for Chinese companies, because we don't want to become just an interpreter'.[54] Having a conservative image, the career

---

51. The name is fictitious as per the anonymity request of the interviewee.
52. Interview on 6 April 2010.
53. The name is fictitious as per the anonymity request of the interviewee.
54. Interview on 3 November 2010.

prospect in a Chinese company remains an important question. In the context of a shortage of high skilled professionals, Chinese companies have to compete with other Western multinationals.

To illustrate further, the same Chinese SOE in which Sunny is the Ghanaian General Manager, has two offices in Accra—one is 'official' and another is rather 'invisible'. The official office is situated in a busy commercial building in central Accra. Local management staff work and receive visitors in this office. The other one, located in an expat residential area, is an enormous house compound where all Chinese staff work and live. It has a big courtyard as a parking area with cars and drivers waiting for service and numerous domestic helpers and cooking staff to take care of the Chinese team. The Chinese Managing Director usually goes to the local office in the morning and spends the rest of the day in the house.[55] Apparently, the local staff office is a window to the public. Administration and communications with the headquarters in China all take place in the Chinese house and in the Chinese language. There is no sign outside indicating this big but invisible company office and only insiders know the address. Without investigating further the purposes behind these double management offices, their existence already indicates that local management is not yet totally integrated.

In general, Ghanaian management staff in Chinese companies often take up positions related to 'external affairs', such as accounting, sales, customs clearance, human resources that have to deal with authorities or local partners and employees. These Ghanaians definitely help the Chinese companies to build a more socially responsible image and convince others that the Chinese companies are reliable partners and employers. Until now, the degree of management localization and the function of local managers has been limited. Certain positions, such as in finance, will probably never be passed to locals. A Chinese staff member commented: 'like many foreigner companies, we don't want to leave certain sensitive works like finance, accounting, to others [non-Chinese]'. It is not unusual to have both Ghanaian and Chinese auditors in a Chinese company. The Ghanaian ones are responsible for dealing with the local authority and the Chinese ones are usually supervisors, responsible for internal management and communication with the Chinese headquarters.

In addition, it is also not surprising that experienced Chinese SOEs in the construction sectors are the most active in promoting management localization to improve their image, as one of the main channels to get construction project is through open bidding. The localization of management staff shows how more experienced Chinese SOE subsidiaries adapt to the changing local environment, fierce competition and public pressure aroused by the arrival of more Chinese companies in recent years.

SOE country directors: internationalization or personalization?

Paradoxically, those Chinese directors active in promoting the localization discourse may be themselves a barrier for further localization. Whereas most Chinese

---

55. All the Chinese staff, including the Chinese director, live in the same house compound.

interviewees complain about the 'terrible' living environment and suffer from separation from their family, particularly children who are being schooled in China, these Chinese directors and their senior management have been staying in Ghana or Africa for over 10–20 years. Evidently, the company headquarters prefer that the senior Chinese staff continue to stay abroad. Apart from the difficulty of getting qualified personnel to work in Africa, the local knowledge and network accumulated by these experienced branch managers are also hard to replace.

Mr X, one of these directors, said:

> I have been here for 18 years. I didn't imagine that I would stay for a long period of time. I was 28 years old when I came [1992]. In 1996, I was already promoted to Deputy Managing Director [in Ghanaian branch]. I felt that I could fully demonstrate and utilize my language skills and ability. Here there are a lot of development opportunities. It is what we say 'how big you want your business to be, it will be'.

He was then promoted to Managing Director in his Ghanaian branch. Under his management, the construction company in Ghana has diversified business in other sectors including manufacturing, real estate and hotels. He was recently promoted as one of the five vice-presidents of the headquarters of the Chinese provincial SOE. Nevertheless, he continues to stay in Ghana where he obtained his career success and in which he has extensive knowledge and a personal network.

Therefore, returning to China is less tempting for these Chinese directors. In their late 40s or 50s, after a prolonged career in the continent, they realize that not only their Ghanaian experience is not transferable to China, but also their loose personal network in China becomes much less resourceful. They will not have the same status and advantages. When working in Ghana, their salary is several times higher than in China, together with accommodation, cars, domestic helpers and drivers provided and all expenses covered by the company. Nonetheless, senior management in the state-owned Chinese enterprises in Ghana usually avoid giving an exact figure on their remuneration. The ownership of these Chinese branches has been transformed following the restructuring process of state-owned companies in China. Senior staff are now both employees and shareholders of many of the Ghanaian branches. As long as the branch can fulfil the business target set by the headquarters, they have a large autonomy and flexibility. The personal financial benefits obtained are definitely not for disclosure[56] but it could be significant enough to delay their return to China.[57] For example, the total responsibility of the company directors on financial performance also leaves them some flexibility in managing things such as gifts, commission and rebates that are useful in conducting business in Ghana, where many business-related transactions like salaries and bonuses are still in cash.

---

56. As Cyril Lin, 'Corporatisation and corporate governance in China's economic transition', *Economics of Planning* 34(1–2), (2001), pp. 5–35 said: 'many [Chinese] companies do not disclose annual salaries of their general managers. A significant proportion of total income accrues in non-monetary form [...] and other fringe benefits. Since such benefits are a hidden "black box", they are not publicly disclosed'.

57. Naughton, *The Chinese Economy*, p. 321, indicates that managers of state-owned companies in China have enjoyed 'an extraordinary degree of independence'. Overseas managers enjoy even an larger degree of business liberty.

Apart from economic advantages and professional opportunities, these Chinese branch directors without a privileged social background in China find themselves promoted to the elite group and are among the most influential Chinese businessmen in Ghana. On Sunday, in the bar of the golf club, they exchange business updates with other SOE managers and rich Chinese private entrepreneurs, sometime joined by the Chinese ambassador. From time to time, the discussion comes to rating the golf clubs of other African countries they have visited during holidays and they make jokes of themselves regarding how unaffordable it is to become a member of a golf club in China. It is in Ghana that they learn and play golf for the first time in their life. Many newly arrived Chinese SOEs' management staff learn to bring back a full set of golf equipment when they return from China the next time. While they are conscious that their status is limited to Ghana, if they manage to stay long enough, they could take early retirement in China. Financially speaking, they often say: 'one year in Ghana, a few years in China'. Chinese management staff in Ghana are university graduates, but from less prestigious universities, and they are originally from interior and poorer provinces. We seldom see Chinese staff coming from more affluent places like Beijing, Shanghai or other rich coastal cities. This background limits their opportunity to penetrate into elite/powerful circles in China that play an essential role in a decent career development. Without an influential network or rich family, working in Africa becomes a career opportunity although the continent is the least favourable destination. A few Chinese interviewees described:

> capable Chinese don't want to go to Africa. People want but cannot go to Africa because they don't have the capacity to work in Africa [e.g. education level, language ability]. Those really come to work in Africa are people who are capable but unable to have ways to succeed in China.

Probably, 'good' Chinese prefer working in China than in developed countries[58] let alone in 'terrible' Africa.

Mr Y, in his mid-30s is relatively young compared to directors of other SOEs. He told us that he hesitated over whether to pick up a job in Spain obtained via his uncle, who had been living there for many years, or stay with his current job in Ghana, but that returning to China was not on his list. Inspired by Western culture and lifestyle, the job opportunity in Spain was an interesting personal experience. What refrained him was that he would be earning much less than he does in Ghana, though he said he has already finished paying off the loan on his apartment in China after working for several years in Ghana. But he believed that should he take the Spanish job it would be difficult to support his current lifestyle. His favourite activities in Ghana are dining in high-end Western restaurants, spending weekend nights in bars and discos that are attended by well-off young locals and Western expats, and travelling in European countries during his annual vacation.

---

58. In an interview with the BBC *Business Daily Programme* on 10 May 2011, Nandani Lynton from China Europe International Business School in Shanghai said: '"Good" Chinese often don't want to go off as expatriates, because they also rely on their personal networks. They're afraid that if they get sent abroad for three or five years, their personal network will disintegrate. Many of them see it as possibly missing better opportunity at home and they don't want to go'.

The financial returns of working or doing business in Africa are attractive. However, generally speaking, Africa still remains a less favourable destination for Chinese as political instability raises safety concerns as highlighted by Zerba.[59]

Since 2011, the Chinese state and companies have finally become aware of the importance of promoting their localization efforts to demonstrate that they are operating in a socially responsible manner in Africa.[60] According to the *China Daily*, the overseas operations of Chinese companies have created over 800,000 local jobs in host countries.[61] In the African continent, the large Chinese MNCs, like Huawei and ZTE, employ around 37,000 Africans including in management positions.[62]

Localization versus 'Africanization'

In African studies, the term 'localization' of employment is not common but a similar practice does exist in the continent. During the decolonization process, for example, 'Africanization'[63] was a strong feature. In Ghana, the civil service has been Africanized much more quickly than in the business sector due to the fact that no penalty would result from non-Africanization and the lack of trust of local staff in Western companies. However, in responding to public pressure on Africanization, Western companies still recruited some local management staff just for 'window dressing', to improve their image.[64]

One of the early works by the French sociologist Bruno Latour[65] analysed the discourse of European managers and Ivorian workers regarding the local work competence in a context where Europeans still dominated in the company management in Africa in the 1970s. Though the study took place over 30 years earlier, similar racialized discourses were given by Europeans on Africans, and by Chinese on Africans, for example, about Africans being 'lazy',' dishonest' and 'incapable'. On the other hand, Africans complained that those European managers were not able to return to their home country to work and so had to make themselves irreplaceable and continued to propagate the stereotype. What we would like to stress here is that even if the encounter of China and Africa can often be unique, there are a lot of similarities when we compare with the previous experiences of other foreign companies and nationals in the continent.

---

59. Shaio H. Zerba, 'China's Libya evacuation operation: a new diplomatic imperative—overseas citizen protection', *Journal of Contemporary China* 23(90), (2014), doi: 10.1080/10670564.2014.898900.
60. Kingsley Edney, 'Soft power and the Chinese propaganda system', *Journal of Contemporary China* 21(78), (2012), pp. 899–914.
61. Ding Qingfen, 'Creating jobs seen as key responsibility', *China Daily*, (6 September 2013).
62. *Ibid.*
63. Mokubung O. Nkomo, 'A comparative study of Zambia and Mozambique: Africanization, professionalization, and bureaucracy in the African postcolonial state', *Journal of Black Studies* 16(3), (March 1986), pp. 319–342.
64. The public sector in Ghana took a few years after independence to be completely Africanized. However, there was still a significant expatriate presence in the private sector in Ghana in the 1980s. Stephanie Decker, 'Postcolonial transitions in Africa: decolonization in West Africa and present day South Africa', *Journal of Management Studies* 47, (July 2010), p. 5.
65. Bruno Latour, *Les ideologies de la compétence en milieu industriel à Abidjan*, Centre de Petit Bassam, Science Humaines, Série Etudes Industrielles, 9 (1974).

Since the arrival of the first Chinese construction SOEs in Ghana in the 1980s, local manual workers have formed the core workforce and, progressively, more Ghanaians were trained to replace the Chinese skilled workers or technicians. This first stage of localization was essentially driven by an economic motive, as bringing Chinese workers abroad is much more expensive.

The localization of mid-level employees takes more time and recently SOEs have begun to make more of an effort to intensify localization at this level. The economic factor is also important as the introduction of the 'going out' policy intensifies competition among Chinese SOE overseas subsidiaries. The African continent was initially the main overseas market for less privileged provincial SOEs in the 1980s and 1990s. However, turning to the new millennium, much larger SOE groups, including those central groups, have started to see Africa as a new strategic market. These groups are financially and politically much more powerful than their pioneers.

Facing more and stronger competition, Chinese SOEs established earlier (mainly from the construction sector) have had to adapt their business strategies including a higher degree of localization (localization of mid-level and management), which is one step forward in the internationalization process. With a longer experience in Ghana, they have obtained an organizational capacity in having trained technicians and local managers to improve their image and to better manage local workers. These older SOE subsidiaries in Ghana are proud of the fact that labour strikes occur only rarely in their enterprises but regularly in a newly arrived central group for example.[66] A closer relationship with the Chinese central state does not automatically produce the best master of micro-management of daily affairs and local public relationship in their Ghanaian branch.

Similar to the 'Africanization' during the decolonization period, the political context plays a role in shaping localization strategies. In a recent *China Daily* article, the journalist stresses the recent change in the legislation of many countries in Africa in order to promote localization, particularly in aid projects.[67] In the same article, the managing director of the Ghana branch of the China Geo-Engineering Corp, an advocate of workforce localization, advises, 'it means that Chinese companies need to move up a notch and start additional training programs for local workers, rather than depend on Chinese personnel for all technical and management problems'. The poling change indicates that the Ghanaian government, like many other governments in Africa, has to show publicly that they are actively taking action in this politically sensitive topic.

Since 2011, the Chinese Embassy in Ghana has started to publicly stress the importance of workforce localization. Nevertheless, directors of experienced Chinese SOEs in Ghana already shared their localization strategies and experiences during interviews conducted between 2009 and 2010. Therefore, workforce localization is not a top-down political initiative. This internationally-recognized and socially sound business strategy was first successfully adopted by certain SOE subsidiaries and then translated into an official discourse of 'best practice'. In the context of the controversial Chinese presence in Africa, the Chinese authority started to openly

---

66. See for example http://tain.ghanadistricts.gov.gh/?arrow=nws&read = 34415.
67. Zhong Nan, 'Mining for talent in Africa', *China Daily Africa Weekly*, (31 May 2013).

encourage its companies to create jobs to Africans. Hence, the increasing coverage in the Chinese state media (like the *China Daily* cited several times in the article) of the localization practice of Chinese companies in Africa serves as both encouragement and propaganda. As noted by Zhao, it can be seen as part of Beijing's recent efforts to readjust policy and 'insensitive business practices' in order to mitigate and improve the negative Chinese image in Africa.[68]

Nevertheless, complete localization of Chinese SOEs in Ghana is not going to take place in the near future. This is not only the case in Ghana, but also across the globe. Even though personnel localization is considered to be a key element for successful internationalization, it remains one of the major weaknesses of Chinese companies.[69] The ways in which the Chinese SOEs overseas organize, operate or communicate with the public still remain in Chinese and Chinese-oriented hands (Chinese headquarters, government and media). Their transformation into true multinationals will take time to achieve.

---

68. Suisheng Zhao, 'A neo-colonialist predator or development partner? China's engagement and rebalance in Africa', *Journal of Contemporary China* 23(90), (2014), doi: 10.1080/10670564.2014.898893.
69. Larçon, *Chinese Multinationals*.

Chinese State-owned Enterprises in Africa: ambassadors or freebooters?

XU YI-CHONG

The role of Chinese state-owned enterprises (SOEs) in Africa is puzzling: they pioneered China's inroads into Africa and shouldered the responsibilities of building and expanding cooperation with African countries, while these very activities and engagement, according to many scholars, often contradict or even undermine the political and diplomatic objectives adopted by the central government. To understand this puzzle, this article unpacks China's engagement in Africa, by examining large central SOEs in the resources and infrastructure sectors. It concludes that the commitment of large SOEs in Africa relies on small public and private contractors. The paradox therefore is, that while the central government encourages and supports the large SOEs to 'go global', it has limited capacity to control and regulate the small contractors.

Introduction

Africa has always been a chessboard for the manoeuvring of major powers. In the seventeenth and eighteenth centuries, European slave traders built posts along the African coast and started nibbling at the mysterious continent. In the 1880s, old and new European colonists grabbed territories in Africa, whether impenetrable jungle or waterless desert, to secure their trading routes and establish their prestige. By the end of the nineteenth century, 'Africa was sliced up like a cake, the pieces swallowed by five rival nations'.[1] As Africa became a battlefield of the Cold War worriers, its people suffered recurrent famines, persistent starvation, frequent civil wars and killing diseases, including a HIV/AIDS epidemic. At the turn of the twenty-first century, many saw a *Third Scramble for Africa*, with the world's old and new powers competing for control of strategic resources and markets in the continent.

---

* Xu Yi-chong is Professor, School of Government and International Relations, Griffith University, Australia; author of *The Politics of Nuclear Energy in China* (2010), *Electricity Reform in China, India and Russia* (2004) and *Powering China* (2002); co-author of *Inside the World Bank* (with Patrick Weller 2009) and *The Governance of World Trade* (with Patrick Weller 2004); editor of *The Political Economy of State-owned Enterprises in China and India* (2012) and *Nuclear Energy Development in Asia* (2011); and co-editor (with Gawdat Bahgat) of *The Political Economy of Sovereign Wealth Funds* (2010). All the projects were supported by research grants from the Social Sciences and Humanities Research Council of Canada and Australian Research Council (ARC).

1. Thomas Pakenham, *The Scramble for Africa* (New York: Perennial, 2003), p. xxi. Also see Martin Meredith, *The State of Africa* (London: Free Press, 2005).

In this new round of competition in Africa, China has become a noticeable player. Its initial focus on securing access to energy and natural resources quickly expanded to infrastructure construction, followed by activities in other areas: trade, agriculture, telecommunications, and financial and other services. Many of these activities are said to be conducted by state-owned enterprises (SOEs), backed by the government in Beijing in order to expand 'its influence in Africa to secure supplies of natural resources, to counter Western political and economic influence'.[2] SOEs are seen to be the agents in spreading 'bad behaviour' from China to Africa, whether corruption, human and/or labour rights violations, environmental pollution and even crimes,[3] and more importantly, spreading to African countries the practices of 'state capitalism'.[4]

The role of Chinese SOEs in Africa is puzzling. If, as some claim,[5] SOEs are the agents of the Chinese government and the Chinese Communist Party (CCP), pursuing the strategic goals set by the CCP, either by getting control of natural resources, inserting its power and influence, or undermining the values and practices of Western democracies, why and how are some of their activities undermining these objectives, as argued by some other scholars?[6] It is difficult to see how SOEs can simultaneously help their owner (the government) and work against its interests too. This article suggests that Chinese SOEs' engagement in Africa is not a simple either–or question. It is too easy to believe and then recycle as revealed truth the assumption that when Chinese SOEs operate in Africa, whether through investment, trade or foreign assistance, they are the handmaiden of the Party-state to 'sustain the existing authoritarian political system',[7] and/or a 'Trojan horse' to help China compete with the US and other democratic countries for influence around the world.[8]

Indeed, there is a contrary argument that Chinese SOEs have become profit-driven market privateers and powerful interest groups that can 'set policy agenda' in Beijing[9] and pursue their own interests when going overseas. This argument raises another set of questions: why has the Party-state permitted them to become semi-independent privateers, whose activities do not always coincide with the interests of the Party-state? More importantly, since they are, after all, state-owned, what are the mechanisms and levers of control the state still has? Is it able to exert control over its SOEs, their activities and behaviour in Africa?

---

2. Peter Brooks and J. H. Shin, 'China's influence in Africa: implications for the United States', *Background* no. 1916 (Heritage Foundation), (22 February 2006), p. 1; Horace Campbell, 'China in Africa: challenging US global hegemony', *Third World Quarterly* 29(1), (2008), pp. 89–105.

3. Yan Yang and Huanhuan Feng, '37 suspected Chinese gangsters extradited from Angola', (25 August 2012), available at: http://society.people.com.cn/n/2012/0825/c223276-18831787.html (accessed 26 August 2012).

4. Martyn Davies, *How China is Influencing Africa's Development* (OECD Development Centre, April 2010).

5. Carolyn Bartholomew, 'Testimony at hearing on "Assessing China's Role and Influence in Africa"', US House Committee on Foreign Affairs, 9 March 2012; Davies, *How China is Influencing Africa's Development*.

6. Chris Alden and Christopher R. Hughes, 'Harmony and discord in China's Africa strategy: some implications for foreign policy', *The China Quarterly* 199, (2009), pp. 563–584; Ian Taylor and Yuhua Xiao, 'A case of mistaken identity: "China Inc." and its "imperialism" in sub-Saharan Africa', *Asian Politics & Policy* 1(4), (2009), pp. 709–725.

7. Bruce J. Dickson, 'Integrating wealth and power in China', *China Quarterly* 192, (2007), p. 852.

8. Ian Bremmer, *The End of the Free Market* (New York: Penguin, 2010); Axel Berger, Deborah Brautigam and Philipp Baumgartner, 'Why are we so critical of China's engagement in Africa?', in Thomas Fues and Liu Youfa, eds, *Global Governance and Building a Harmonious World* (Bonn: Deutsches Institut fur Entwicklungspolitik, 2011), pp. 191–193.

9. Erica Downs, 'The fact and fiction of Sino–African energy relations', *China Security* 3(3), (2007), pp. 42–68.

SOEs are fewer in number than private enterprises in China and in Africa. Some are large and dominate their sectors, such as the China National Petroleum Corporation (CNPC) or Zhong Xing Telecommunication Equipment Company Limited (ZTE). Some are equally large, but have to compete with other SOEs, such as Sinohydro, China Civil Engineering Construction Corporation (CCECC), China Railway Construction Corporation (CRCC) or China Communication Construction Corporation (CCCC). While these large SOEs pioneered China's involvement in Africa,[10] their activities have brought to Africa a range of smaller and diverse SOEs and private players. Some scholars, with their sophisticated and detailed studies, have already refuted the argument that there is one master plan, one motive, one strategy and one coherent group of players to push forward China's inroads into Africa.[11] Some have gone so far as to argue that 'there are in fact many Chinas (in the sense of "China" as presented by the government)',[12] and that many SOEs compete among themselves and with others for agendas, strategies and operations.[13]

With these often contradictory narratives about Chinese SOEs in Africa, this article seeks to identify who is doing what in Africa, and why and how they operate there, to make sense of the relationship between SOEs and the government. It argues that large SOEs are not necessarily responsible for the problems reported in the media, which has caused headaches for Chinese diplomats in Africa and embarrassed politicians in Beijing. Large SOEs may dominate China's investment in Africa, but their activities often constitute only a small proportion of their global activities. To compete in mature economies, their reputation is important and depends on their adoption of internationally acceptable behaviour. A large group of small SOEs and private enterprises, however, have brought into Africa fierce and unregulated competition and practices which are not acceptable in Western democracies and African countries. Tension therefore often arises between local people and these poorly regulated small enterprises.

The following section provides a brief background of China's recent engagement in Africa, especially the rationale and players in this exercise. It is followed by a brief outlook of government agencies in charge of China's policy in Africa and state-owned policy banks that help finance these policies. In 1994, as part of the general economic reform, the Chinese government separated policy and commercial banking activities and formally established three policy banks. Two of them, the China Development Bank (CDB) and the Export–Import Bank (Exim Bank) actively support the 'going out' efforts by financing both state-owned and private enterprises. As it has been trying to gain a commercial bank status, CDB has been transformed 'from a bank created for the explicit purpose of undertaking policy-driven lending into one of the most dynamic and successful financial institutions'.[14] Its interests are

---

10. Jian-Ye Wang, *What Drives China's Growing Role in Africa?*, IMF Working Paper, WP/07/211 (2007).
11. Deborah Brautigam, *The Dragon's Gift: The Real Story of China in Africa* (New York: Oxford University Press, 2009); Lucy Corkin, 'Redefining foreign policy impulses toward Africa: the roles of the MFA, the MOFCOM and China Exim Bank', *Journal of Current Chinese Affairs* 4, (2011), pp. 61–90.
12. Taylor and Xiao, 'A case of mistaken identity', p. 714.
13. Jean-Paul Larçon, ed., *Chinese Multinationals* (Singapore: World Scientific Publishing Co, 2008); Alden and Hughes, 'Harmony and discord in China's Africa strategy'.
14. Erica Downs, *Inside China, Inc.: China Development Bank's Cross-Border Energy Deals*, John L. Thornton China Centre Monograph Series, No. 3 (The Brookings Institution, March 2011), p. 2.

no longer restricted to advancing the interests of the Chinese government but also its own. The nexus between state banks and SOEs is a key to our understanding of the operation of SOEs in Africa.

In the past decade or so, a new governing structure has emerged. Several ministries at 'the pinnacle of Party-state over the Chinese economy' were reorganised and consolidated.[15] New agencies, especially the State-owned Assets Supervision and Administration Commission (SASAC), were created to manage the new corporate players. Some see this development as a new political economy of regulation emerging in China;[16] some argue that it was merely old wine in new bottles; some see it as an emergence of a 'co-governance' structure;[17] others insist it is part of the efforts of the Party-state to reinvent its control.[18] To understand whether SOEs are handmaidens of the Chinese government, we need to examine the government agencies in charge of the 'going out' policy and whether and how they could work together to facilitate and support the revitalisation and global expansion of the Chinese SOEs. They include the Ministry of Foreign Affairs (MOFA), the National Development and Reform Commission (NDRC), SASAC, Ministry of Finance, Ministry of Industry and Information Technology (MIIT) and other line ministries, such as health, education, water resources, agriculture and railways.

The final section discusses several key SOEs, their motivations and activities in Africa, as they have attracted the most attention and are seen as the threat to the West's strategic positions and influence in Africa. They are divided into two groups—those in energy and natural resources, such as CNPC, and those engaging in infrastructure projects, such as Sinohydro, China Three Gorges Corporation, CCCC or CRCC. Each of these players has its political patrons in the central government—ministries and commissions. Those in resource industries are often financed by the policy or commercial banks with coordination between government agencies. They tend to be better organised and more closely monitored by the central government. Those in the infrastructure construction industry vary significantly: some have had a long history working on overseas projects while others are newcomers; some work on projects financed by the policy banks but many take on projects through international bidding and their projects therefore depend 'largely upon the development priorities of the recipient country'.[19]

It concludes that it is too simplistic, or simply wrong, to see China as a conventional aid donor and therefore the behaviour of its SOEs as 'disappointing', or to see China as a colonialist only interested in looting the land. Its interests are much broader and its SOEs and other players vary significantly in motivation and behaviour. For some, Africa 'is a stepping stone to a commercial presence around the

---

15. Margaret M. Pearson, 'The business of governing business in China', *World Politics* 57(2), (2005), p. 304.
16. Dali Yang, *Remaking the Chinese Leviathan* (Stanford, CA: Stanford University Press, 2004).
17. Bo Kong, *China's International Petroleum Policy* (Santa Barbara, CA: Praeger Security International, 2010), p. 27.
18. Christopher A. McNally, ed., *China's Emergent Political Economy* (New York: Routledge, 2008); Hon S. Chan and David H. Rosenbloom, 'Public enterprise reform in the United States and the People's Republic of China', *Public Administration Review*, (December 2009, special issue), pp. S38–45.
19. Charles Wolf, Jr, Xiao Wang and Eric Warner, *China's Foreign Aid and Government-Sponsored Investment Activities* (Rand Corporation, 2013), p. 31.

globe' and a place to build their reputation. For others, they work in Africa simply because 'it pays'.[20]

Background of China's engagement in Africa

China's involvement with Africa goes back to the early days of decolonisation. Its intentions were primarily to shore up votes for the eventual rejection of Taiwan's position at the United Nations and to compete with superpowers for influence. China's presence in Africa in those days could easily be identified with large projects, such as the railroad linking Lusaka and Dar es Salaam, or the Benin Friendship Stadium in 1982, and with technical experts, doctors, scholarships and other forms of aid. In the 1980s, the 'good-will' projects faded when China became preoccupied with its domestic economic reforms.[21] In the 1990s, the Chinese government consciously re-built its ties with Africa, first driven by political and diplomatic reasons and then economic ones. Any analysis of the recent Chinese activities in Africa needs to take into consideration four developments in the 1990s:

(a) after being isolated after 1989, the Chinese government resumed its efforts to expand its relations with those 'old friends who gave China the necessary sympathy and support';[22] this engagement was initiated for political and diplomatic reasons;

(b) as China became a net oil importer in 1993, its state-owned oil companies ventured into global oil equity markets, especially in countries which international oil companies had either abandoned, such as Sudan, or perceived as too risky to invest in;[23]

(c) following the Asian financial crisis of 1997–1998, many affected countries, China included, started building their foreign reserves, which the Chinese government was urged by experts to transfer to large SOEs to seek overseas assets;[24] and

(d) with the reforms in the late 1990s, SOEs were stripped of government, regulatory and public responsibilities but had to face domestic and international competition.[25]

After a nearly two-decade retreat, China resumed its activities with Africa, tentatively at first. In the 1990s, it was unwilling and unable to expand its ties with African countries aggressively. It was unwilling because of Deng's foreign policy: 'hide our capabilities, bide our time; keep low profile and never try to take the lead'.[26]

---

20. 'The Chinese in Africa: trying to pull together', *The Economist*, (20 April 2011).
21. Ian Taylor, 'China's foreign policy towards Africa in the 1990s', *The Journal of Modern African Studies* 36 (3), (1998), pp. 443–460; Ian Taylor, *China and Africa: Engagement and Compromise* (New York: Routledge, 2006).
22. Taylor, 'China's foreign policy towards Africa in the 1990s', p. 447.
23. Xu Xiaojie, *Petro-Dragon's Rise* (Florence: European Press Academic Publishing, 2002); Kong, *China's International Petroleum Policy*.
24. Joshua Aizenman and Jaewoo Lee, *International Reserves*, IMF Working Paper, WP/05/109 (2005).
25. Edward S. Steinfeld, *Forging Reform in China* (New York: Cambridge University Press, 1998); Yang, *Remaking the Chinese Leviathan*; Ross Garnaut, Ligang Song, Stoyan Tenev and Yang Yao, *China's Ownership Transformation* (Washington, DC: The World Bank, 2005).
26. Aaron Friedberg, *A Contest for Supremacy* (New York: W.W. Norton, 2011).

It was unable because the central government had seen its revenue steadily decline as a share of GDP while the country still faced a serious shortage of foreign reserves. With limited foreign reserves, the central government was unwilling to support outgoing investment, especially in highly-risky African oil fields, even though it had long realised that the country sooner or later would have to depend on offshore oil supplies. By 2000, China's outward FDI stock accounted for only 0.4% of the world's total with US$25.6 billion.[27]

A decade later, the situation was different: China's foreign reserves rose from US $22.4 billion in 1993 to US$286.4 billion in 2002, and then rose steadily with an average annual growth rate of 32% in 2002–2011. SOE reforms had not only led to the closing of thousands of SOEs but also released many large and more successful ones from state control.[28] These newly commercialised and corporatized large SOEs embraced the central government's 'going out' policy with such an enthusiasm as they were under pressure from domestic and international competition. A mix of direct investment, concession loans and foreign assistance helped these firms expand their presence in Africa quickly in the last decade. Chinese foreign direct investment (FDI) to Africa increased from US$7.48 million to US$2,112 million in 2003–2010,[29] yet Africa is not the major destination (taking in merely 3% of China's total FDI outflow in 2010 and 13.8% in 2005–2010). China's trade with Africa grew too but remained minute in its total trade, accounting for 4.1% and 4.3% of its total trade in 2009 and 2010. Meanwhile, nearly a quarter of its economic cooperation in 2009 and 2010 went to sub-Saharan Africa (SSA).[30] Comparative with these data, China's presence in Africa has attracted disproportionate attention. It is controversial partly because of the nature of its engagement and partly because of the players involved. Kaplinsky and Morris identify four groups of players from China in Africa—central SOEs, provincial SOEs, private firms incorporated in China and private firms incorporated in SSA only.[31]

This article focuses on the first group of players, central SOEs, which may be small in number, but account for 77% of China's FDI in Africa.[32] They are concentrated in two sectors—energy and natural resources and infrastructure construction, both of which are capital intensive. The presence and activities of these large SOEs brought to Africa many smaller SOEs and private players, which predominantly are in light industries, the retail sector or sub-contractors of the central SOEs. Their activities and problems in Africa mirror those in China where rapid economic development has not been matched by the same speed of development in regulation and improvement in

---

27. UNCTAD, *World Investment Report* (Geneva: United Nations, 2002), Table B.4.
28. Garnaut *et al.*, *China's Ownership Transformation*; Barry Naughton, *The Chinese Economy* (Cambridge, MA: MIT Press, 2007); Barry Naughton, 'A political economy of China's economic transition', in Loren Brandt and Thomas G. Rawski, eds, *China's Great Economic Transformation* (New York: Cambridge University Press, 2008), pp. 91–135; Pearson, 'The business of governing business in China', pp. 296–322; Margaret Pearson, 'Governing the Chinese economy', *Public Administration Review* 67(4), (2007), pp. 728–730.
29. Ministry of Commerce, China, *Statistical Report of Outflow Foreign Direct Investment, 2010* (Beijing: Ministry of Commerce, 2011), p. 43.
30. *China Statistical Yearbook 2011* (Beijing: China Statistical Press, 2011).
31. Raphael Kaplinsky and Mike Morris, 'Chinese FDI in sub-Saharan Africa', *European Journal of Development Research* 21(4), (2009), pp. 551–569.
32. Ministry of Commerce, China, *Statistical Report of Outflow Foreign Direct Investment, 2010*, p. 19.

labour conditions, especially in construction sectors. More importantly, 'Beijing has long had problems in controlling what companies do in China'[33] and has even less capacity for controlling and regulating the behaviour of an increasing number of players in Africa.[34] Business objectives and practices of large SOEs can differ significantly from those of small companies taking on contract jobs.[35] Lumping together all Chinese firms and their activities obscures the relationship between government and SOEs in Africa.

Government agencies

The 'state' is never a single entity. Several government agencies drive China's engagement in Africa and each has its own mission. Diplomats from the Ministry of Foreign Affairs in Africa are 'frontline' officers, identifying the 'national interest' and making connections with local governments. Their mission is political and diplomatic, whether competition with Taiwan for diplomatic recognition or with others for influence, building broader political support or ensuring China's access to natural resources and markets. Even though it is common to use foreign aid to promote national political, economic and commercial interests, in most developed countries, foreign aid agencies are either independent or affiliated with the external affairs ministry. In China, foreign assistance is under the auspice of the Ministry of Commerce whose main mission is to promote and protect Chinese industries through trade, investment and other forms of international cooperation. This unique combination of the two missions shapes Chinese engagement and activities in Africa.

When foreign aid is delivered as a bundle to allow Chinese SOEs to compete for resources and markets, the two objectives may be achieved simultaneously. This practice often draws international criticism and is controversial at home too when investment and assistance in Africa sometimes lead to competition with domestic firms for the Chinese markets. Coordination between the Ministry of Foreign Affairs and the Ministry of Commerce may be necessary as a Chinese diplomat explained it, 'we have to coordinate with the Ministry of Commerce because they have China's commercial interests in mind, but they also have to take our views into account because foreign aid is for policy objectives, not to make money'.[36] Their relationship, however, is not always easy, because foreign affairs officials 'do not have a seat at the table' when projects are negotiated between Chinese SOEs and African counterparts.[37]

---

33. Taylor and Xiao, 'A case of mistaken identity', p. 718.

34. Bates Gill and James Reilly, 'The tenuous hold of China Inc. in Africa', *The Washington Quarterly* 30(3), (2007), pp. 37–52; Linda Jakobson, 'China's diplomacy toward Africa: drivers and constraints', *International Relations of the Asia–Pacific* 9(3), (2009), pp. 403–433.

35. Chih-shian Liou, 'Bureaucratic politics and overseas investment by Chinese state-owned oil companies', *Asian Survey* 49(4), (2009), pp. 670–690; Lucy Corkin, *Chinese Construction Companies in Angola*, MMCP Discussion Paper No. 2 (March 2011), available at: http://commodities.open.ac.uk/8025750500453F86/(httpAssets)/ 9407B19720D08F308025787E0039E3FC/$file/Chinese%20Construction%20Companies%20in%20Angola.pdf (accessed 15 December 2012); Corkin, 'Redefining foreign policy impulses toward Africa'.

36. Brautigam, *The Dragon's Gift*, p. 110.

37. *Ibid.*, p. 111.

SASAC was created in 2003 as the owner of the central SOEs, responsible for maintaining and increasing the value of state assets. Even though SASAC's 'establishment did not mark any rupture with previous policy',[38] it did create a set of bureaucratic interests. SASAC sees its responsibility as being to promote large modern corporations and encourage and support central SOEs to 'go out' and compete with their international counterparts. 'If you cannot be on the top three in your sector', SASAC's former chairman used to say, 'get out'. It evaluates SOEs and their top managers based on their performance and profits. SOEs' profit-maximisation operations can occasionally create headaches for the Ministry of Foreign Affairs, for example, when their operation deviates from international norms, when disputes with the local workforce over wages or working conditions erupt, or when Chinese workers are kidnapped. While the Ministry of Foreign Affairs prefers cautious and orderly engagement in Africa, SASAC prefers a much more aggressive approach in encouraging its SOEs to expand their operations overseas. Foreign affairs officials are called in only when problems emerge.

Until the early 1990s, the banking system in China had worked on a credit plan which had two components—mandatory and indicative. The mandatory plan determined credit allocations for all enterprises; banks were told to finance specific projects, such as electricity, railways or other infrastructure projects that were the bottlenecks. Indicative credit plans covered those investments that were a priority for the central government.[39] Through the credit plans, government controlled money supply, influenced aggregate demand, and balanced the interests between SOEs and newly emerged non-state or semi-state enterprises, and among provinces. By the early 1990s, the system was no longer sustainable as the government increasingly relied on the banks to bail out loss-making SOEs.

In 1993, Zhu Rongji, the vice premier of the State Council, took over as the Governor of the People's Bank of China (PBC), the central bank, to strengthen the State Council's leadership in banking. The following financial reform led to the separation of 'policy' from commercial lending. Three policy banks were established to 'assume responsibility for the non-commercially oriented loans of the four large state banks'[40]—Agricultural Development Banks (ADB), China Development Bank (CDB) and Export–Import Bank (Exim). CDB and Exim Bank were asked to assist Chinese enterprises to expand overseas, but neither had sufficient resources to do so until the 2000s. Reforms in the 1990s created a financial market where CDB and Exim could raise capital by issuing bonds to assist SOEs. At the initial stage, the debts issued by the two policy banks in 1998 reached about 6.4% of GDP,[41] yet, in most cases, their loans were not recovered. This demanded further reforms, which brought a new generation of bankers to the central bank, the Ministry of Finance, CDB and Exim Bank in the early 2000s.

---

38. Barry Naughton, 'China's distinctive system: can it be a model for others?', *Journal of Contemporary China* 19(65), (2010), p. 441.
39. Christine Kessides, Timothy King, Marlo Nuti and Catherine Sokil, eds, *Financial Reform in Socialist Economies* (Washington, DC: World Bank, 1989).
40. Robert Cull and Lixin Colin Xu, 'Who gets credit?', *Journal of Development Economics* 71(2), (2003), p. 539.
41. 'When you borrow, you must pay', *China Economic Quarterly*, (2nd quarter, 2000).

Established in 1994, CDB is responsible for helping to finance infrastructure projects (e.g. national highways and rail networks, gas pipelines, water projects and power stations), key sectors of the economy (petroleum–chemical refining, telecommunications) and the development of the western provinces. It was also mandated to provide loans to Chinese firms as part of the national 'Going Global' strategy which was incorporated into the 11th Five-Year Plan (2006–10) in 2001. This combination of domestic and external objectives 'makes CDB China's most important bank'.[42] It has a large pool of capital allocated by the government budget for development and raised in financial markets by issuing bonds, which accounted for 20.8% of the total domestic debt markets. This allows CDB to support SOEs' expansion around the world.

As of the end of 2011, a little over 15% of its total loans (RMB5.5 trillion) went to overseas projects and its total outstanding foreign currency loans of US$187.3 billion were spread across all continents. African projects made up only a small share of this. In 2007, CDB created the China–Africa Development Fund (CAD Fund) as an independent subsidiary 'to support Chinese companies to develop cooperation with Africa and enter the African market'. Its initial capital of US$1 billion was later replenished with the Phase II capital of US$2 billion. It is designed to follow the international model of direct equity investment so that CDB could leverage its own lending by bringing in more investment to Africa. According to its 2011 annual report, 'it is estimated that the various CAD Fund projects will bring in an additional investment of over US$8 billion to Africa from Chinese enterprises, likely translating to increased annual African exports amounting to US$2 billion'.[43] In 2011, CDB set up 'representative offices with well-equipped financial service capabilities in Zambia and Ghana to manage its operations in Central and Western Africa' and it also has offices in 45 of the 53 countries in Africa.[44]

While acting 'as the link between government and market, coordinating relations between the engineering contracting companies involved in the infrastructure projects financed by the fund, the oil companies, the borrower, and the credit insurance agencies',[45] CDB is expected to generate positive returns from its investment. Even the CAD Fund is instructed to operate on market economy principles. Non-concessional financing is provided to both SOEs and private ones, for their development. CDB simply 'does not give official development aid',[46] as many have assumed. Its role to fill 'the gap under the traditional model of free aid and loans'[47] makes it difficult to categorise the loans made by CDB to Chinese firms in Africa as commercial loans or foreign aid, as both objectives are served.

Exim Bank is mandated to promote exports and foreign investment. Its export credits and investment loans particularly focus on similar infrastructure projects as CDB. Its main source of funding comes from the bond market too. Like the export credit agency in other countries, Exim Bank is asked by the central government to

---

42. Wang, *What Drives China's Growing Role in Africa?*, p. 15.
43. CAD Fund, 'The company overview', (2012), p. 56, available at: http://www.cadfund.com/en/NewsInfo.aspx?NId=48 (accessed 28 August 2012).
44. Downs, *Inside China, Inc.*, p. 29.
45. *Ibid.*, p. 58.
46. Brautigam, *The Dragon's Gift*, p. 115.
47. CAD Fund, 'The company overview'.

Table 1. Ratio of non-performing loans of CDG and Exim Bank, 2003–2010

	2003	2004	2005	2006	2007	2008	2009	2010
CDB	1.34	1.21	0.81	0.72	0.59	0.90	0.94	0.68
Exim	N/A	5.28	4.91	3.47	2.45	1.52	1.10	0.64
Large state-owned banks	19.6	14.9	10.5	9.22	8.05	6.29	2.0	1.43

Source: Annual report of CDB, Exim Bank, China Banking Regulatory Commission, available at: http://www.cbrc.gov.cn/chinese/home/docViewPage/110009&current=1.

help achieve its 'diplomacy, development, and business objectives'.[48] Unlike them, the Chinese government 'does not guarantee the bank's liabilities'.[49] This requires Exim Bank to be financially viable as it 'administers China's concessional foreign aid loan programme using subsidies from the foreign aid budget to soften the terms of its concessional loans'.[50] Its financial standing is improving. Both CDB and Exim Bank have the same credit rating as China as a sovereign country from three international agencies: Aa3 was given by Moody's, AA− by Standard & Poor's and A + by Fitch. Both also maintain a much lower ratio of non-performing loans than the commercial banks (Table 1).

Exim Bank financing primarily goes to large SOEs not only because they are considered low risk but also because the projects they are involved in tend to be large and capital intensive. How each project is financed differs significantly. In some cases, teams from CDB or Exim Bank identify investment opportunities. In some, SOEs go to CDB or Exim Bank for financing after their own team identified the opportunities. In others, the embassies in African countries are either approached by local governments with specific projects or notice the opportunities and then pass them on to large SOEs. Increasingly, either CAD Fund or Exim Bank would lead a consortium consisting of commercial banks from China in financing SOEs' projects in Africa. Sometimes, their financing targets the same SOEs in different operations. For example, in 2004, CDB identified six large SOEs and in the following two–three years, it assisted these SOEs in their 'going out' efforts—CNPC, China Chemical, CITIC, China Minmetals Corp, China Metallurgical Corporation, China State Construction Engineering Corporation (CSCEC). Again, most of these projects were in other continents, rather than in Africa.

Corporate players in Africa

The media often reports that China has gone to Africa mainly to exploit its natural resources. A World Bank report stated that 'most of the total commitments were directed at a natural resource development'.[51] Chinese activities in Africa present a

---

48. Deborah Brautigam, *China, Africa and the International Aid Architecture*, Working Paper No. 107 (African Development Bank, April 2010), p. 17.
49. Wang, *What Drives China's Growing Role in Africa?*, p. 14.
50. Brautigam, *China, Africa and the International Aid Architecture*, p. 7.
51. Vivien Foster, William Butterfield and Chuan Chen, 'Building bridges: China's growing role as infrastructure financier for sub-Saharan Africa', *Trends and Policy Options* no. 5 (The World Bank, 2009), p. 43.

much more complex picture. Chinese oil companies first moved into Sudan in the mid-1990s and they were followed by SOEs investing in oil and natural gas in Nigeria, cobalt in Angola, timber in Cameron and Gabon, bauxite and diamond in Guinea, copper in Zambia, etc. These projects led to a surge of studies on the 'new scramble of Africa' or 'China's colonisation of Africa'.[52] Meanwhile, investing in natural resources is one of the objectives of the 'going out' policy, but it is not the only one, and Africa is not the main destination of such investment (Table 2). Investments in natural resources accounted for 21.8% of China's total outward FDI in 2004–2010 and those to Africa constituted less than 4.5%. A large proportion of FDI, economic cooperation and foreign assistance to Africa went to infrastructure projects. This is especially the case since 2004. In 2008, for example, 90% of China's overseas infrastructure projects went to Africa.[53] Finally, SOEs investing in energy and natural resource seldom get involved in infrastructure projects.

Chinese national oil companies went overseas long before the central government adopted the 'going out' policy. As the International Energy Agency (IEA) explained more than a decade ago, with the reform, Chinese oil companies had 'tried hard to adopt standard international investment practices, and benefitted greatly from an array of joint-venture operations'.[54] They ventured into the global oil market in the early 1990s when China became a net oil importer. CNPC went to Sudan for a series of chain-reacted reasons: (a) domestic oil reserves were drying up, which contributed to the financial difficulties of CNPC; (b) CNPC was nearly insolvent and needed to sell more oil to overseas markets to service its borrowing of US$8.65 billion from domestic and foreign sources; (c) 'with its entry into the volatile global oil bazaar as an importer, China quickly learned the hazards of relying solely on purchase policies in open markets';[55] (d) much of the world oil fields were dominated by giant international oil companies; and (e) the production costs in offshore oil, especially in several 'un-occupied' places, were lower than in the new oil fields in western China.[56]

The initial move to overseas markets was tentative; 'Ironically, the government planners took little notice of the company's first forays into Peru, Sudan and Kazakhstan until the mid-1990s'.[57] Sudan became the first destination of CNPC's overseas venture when the Canadian oil company, Talisman, withdrew because of the civil war.[58] Along with the Indian and Malaysian national oil companies, CNPC acquired four blocks of Sudanese oilfields, sold by Talisman. To argue that CNPC's 'overseas ventures were mainly dictated by the state's interest in acquiring secure

---

52. Sandra T. Barnes, 'Global glows', *African Studies Review* 48(1), (2005), pp. 1–22; Esther Pan, 'China, Africa, and oil', *Council on Foreign Relations*, (12 January 2006); Ernest J. Wilson, III, 'China's Influence in Africa, Testimony before the Sub-Committee on Africa, Human Rights and International Relations', US House of Representatives, 28 July 2005.
53. Wolf, Jr et al., *China's Foreign Aid and Government-Sponsored Investment Activities*, p. 31.
54. IEA, *China's Worldwide Quest for Energy Security* (Paris: OECD, 2000), p. 58.
55. *Ibid.*, p. 61.
56. Kong, *China's International Petroleum Policy*.
57. Xiaojie Xu, *Chinese NOCs' Overseas Strategies: Background, Comparison and Remarks*, Working Paper (The James A. Baker III Institute for Public Policy, Rice University, March 2007).
58. Xu Yi-chong, 'China and the United States in Africa', *Australian Journal of International Affairs* 62(1), (2008), pp. 16–37.

Table 2. China outward foreign direct investment, 2003–2010 (in US$ million)

	2003	2004	2005	2006	2007	2008	2009	2010
World	2,855	5,498	12,261	17,634	26,506	55,907	56,529	68,811
Africa	75	317	392	520	1574	5491	1439	2112
Africa as %	2.6	5.8	3.2	2.9	5.9	9.8	2.5	3.1
Mining	N/A	1,800	1,675	8,540	4,063	5,824	13,343	5,715
Mining as %	N/A	32.7	13.7	48.4	15.3	10.4	23.6	8.6

Source: Ministry of Finance, China, *2010 Statistical Bulletin of China's Outward Foreign Direct Investment* (Beijing: Ministry of Commerce, National Bureau of Statistics and State Administration of Foreign Exchange, 2011), pp. 36–37, 48.

access to strategic natural resources'[59] much overlooked the opportunities available as the result of changing international politics at the time and neglected the reluctance of the central government to support such projects. Indeed, the Chinese government supported this project with great reluctance and its official endorsement did not come until 1997.

By the early 2000s, national oil companies, along with other SOEs, had been commercialised and corporatized. Economic reform provided further incentives for the national oil companies to seek opportunities around the world as they were 'driven to maximise profit'.[60] They had to compete among each other and with their international counterparts and competition intensified for CNPC as easy credits that were made available in China following the Asian financial crisis allowed other SOEs to follow suit. The second largest oil company, Sinopec, competed fiercely with CNPC for financing from the state policy banks, for global oil market shares and for their own continuing survival.[61]

One indication that national oil companies operate more like their international counterparts is the financing for their overseas investment. While attention has been placed on the so-called 'Angola model' of financing in Africa,[62] 'China's petroleum companies and state-owned mineral firms generally seem to shy away from these complicated packages, preferring to bid in auctions, obtain concessions directly, or purchasing shares of existing oil/mineral companies'.[63] By the first decade of the twenty-first century, these Chinese oil companies were still smaller than their international counterparts. They were new in global oil equity markets and learning to be global players. Nonetheless, they attracted a lot of attention not only because they are state-owned but also because of the fast speed of their overseas expansion.

China's SOEs have been building power stations, railways, highways, dams and water supply systems across the African continent. Some see these infrastructure projects as contributors to African development. Mwangi Kimenyi, director of the Africa Growth Initiative at the Brookings Institution, commented:

> One of our biggest problems is what you call the 'infrastructure deficit', yet few African countries have the capital to narrow this infrastructure deficit. The lack of infrastructure in some African countries such as roads, rails, ports and telecommunications facilities, hinders trade and slows development. China is actually helping bridge the gap.[64]

Others, however, see these projects as the expansion of China's influence in Africa. Many are more concerned about disputes emerging from these projects than the projects *per se*.

To understand the controversies over China's SOEs in infrastructure sectors, we need to emphasise that there are a lot more players than in energy or any natural

---

59. Chin-shian Liou, *The Politics of China's 'Going Out' Strategy*, Ph.D. thesis, The University of Texas at Austin, 2010.
60. Naughton, 'China's distinctive system', p. 440.
61. IEA, *China's Worldwide Quest for Energy Security*; Kong, *China's International Petroleum Policy*.
62. Harry G. Broadman, *Africa's Silk Road: China and India's New Economic Frontier* (Washington, DC: The World Bank, 2007); Foster *et al.*, 'Building bridges'.
63. Brautigam, *China, Africa and the International Aid Architecture*, p. 21.
64. 'US misread China's investment in Africa', *China Daily*, (15 August 2012), available at: http://www.chinadaily.com.cn/business/2012-08/15/content_15677358.htm (accessed 7 September 2012).

resource sector. Several large SOEs have had a long history of working in Africa, going back to the 1960s. This small group of SOEs includes, for example, China Machinery Engineering Corporation and China National Electric Engineering Co. Ltd (both of which are subsidiaries of Sinomach), Sinohydro and China Gezhouba Group Corporation (CGGC), both of which are descendants of China's now-defunct hydropower ministry, and China Railway Construction Corporation (CRCC) that originated from the army corps of engineers which was placed under the Ministry of Railways in 1984. They are not only large in size and operation, but have also built extensive networks with government agencies and other large SOEs. Their economic weight and political clout are often seen as the reason they are favoured in getting government projects.

Many of them went to Africa motivated by domestic developments as well as demands for infrastructure development in Africa. Chinese economic development has been driven by fixed capital investment. The construction industry may not contribute a large share to the national economy, but its corollary industries often drove about 40–50% of the total economic activities.[65] The infrastructure-related construction industry opened up for competition in China earlier than many other sectors. Competition in this sector was fierce as an abundant labour force significantly affected the cost of production and made it easy for new players to enter the field (see Table 3). Finally, several construction companies had a long history of working in Africa and other parts of the world. The new expansion of China's activities in Africa therefore should be considered as an extension of a long tradition of Chinese construction SOEs working overseas rather than something new, as is the case with the resources industries.

'China has developed one of the world's largest and most competitive construction industries, with particular expertise in the civil works critical for infrastructure development.'[66] This was the consequence of domestic development since 1980. In the following decades, China added more power stations and built more roads than anywhere else in the world; Chinese firms accumulated experience in undertaking and managing large infrastructure projects. Two sub-sectors particularly stood out: power generation (thermal and hydro) and transport (roads and railways). The total installed generation capacity in China rose from 66GW in 1980 to 316GW in 2000 and 966GW in 2010. Its hydro capacity increased from 20GW in 1980 to 216GW in 2010 and a dozen mega hydro stations, each with a capacity larger than the largest station of the Tennessee Valley Authority, Cumberland (1.3GW). Sinohydro and China Gezhouba Group (CGGC) have consequently acquired a large pool of professionals and experience in building and managing thermal and hydro projects, all of which have made them competitive in winning tenders on power projects around the world.

The same can be said about construction companies, CCECC, CRBC, CSCEC, China Harbour Engineering Co. Group (CHEC) and Sinomach. These companies have climbed steadily up the ENR international ranking. Long experience working on large projects and relatively low supply-chain and labour costs were the main reasons Chinese construction firms won about one-third of the projects funded by the World

---

65. Centre for Chinese Studies, *China's Interest and Activity in Africa's Construction and Infrastructure Sectors* (Stellenbosch University, November 2006), p. 9.
66. Foster *et al.*, 'Building bridges', p. xv.

Table 3. Employment and output of construction enterprises in China

	Employment			Gross output value		
	Total (million)	SOEs	SOEs as %	Total (billion RMB)	SOEs	SOEs as %
1980	6.5	4.8	73.8	28.7	22.1	77
1990	10.1	6.2	61.4	134.5	93.5	69.5
2000	19.9	6.4	32.2	1,249.8	505.4	40.4
2010	41.6	5.8	13.9	9,603.1	1,814.9	18.9

Source: National Bureau of Statistics of China, *China Statistical Yearbook 2011* (Beijing: China Statistics Press).

Bank and other multilateral institutions, such as the Asian Development Bank (ADB) and the African Development Bank in Africa. According to one study, 'in the area of civil works Chinese firms accounted for 31 per cent of total contract value over the period of 2004–06', far ahead of France which won around 12% of the World Bank civil works; no other country won more than a 5% share.[67]

About half of the projects that Chinese SOEs are undertaking in Africa were funded by non-Chinese multilateral or bilateral donors; 40% of them were financed by Chinese funds of either export credits or concessional loans; and the other 11% were funded by host governments.[68] In terms of getting projects, Chinese SOEs often behave no differently from other market players—that is, they prefer projects which are large with guaranteed funding from the Chinese government either as foreign assistance or as a combination of cooperation and aid. Projects funded by multilateral institutions fall into this category too. Chinese SOEs prefer not to bid for projects funded by host governments because 'if the host government funds the project, the date of payment is uncertain, especially when the ruling party changes and the new government may not honour the contract signed by the previous government'.[69]

Undoubtedly, SOEs in China have to take on some political and social responsibilities. They can seek, and often justify, government support for carrying out these responsibilities. They seldom do so by sacrificing their own financial gains. Indeed, this is part of the 'strange' politics in China today—large SOEs are frequently asked to absorb the costs for connecting remote areas with roads, bridges or electricity, subsidising clean energy development, or not laying off a redundant workforce. Yet, managers of large SOEs will lose their job or be demoted if these firms are loss making. This is also the case for their overseas operations. One CNPC manager explains, 'if we operate a loss-making project in one of the suggested countries, we will face troubles from the SASAC'.[70] In addition, their employees depend on the profits. Profit-maximisation behaviour by large SOEs is widely

---

67. Executive Research Associates, 'China in Africa: a strategic overview', (October 2009), available at: http://www.ide.go.jp/English/Data/Africa_file/Manualreport/pdf/china_all.pdf (accessed 27 August 2012).
68. Chuan Chen, Andrea Goldstein and Ryan J. Orr, 'Local operations of Chinese construction firms in Africa', *The International Journal of Construction Management* 9, (2009), pp. 75–89.
69. *Ibid.*, p. 81.
70. Quoted from Liou, *The Politics of China's 'Going Out' Strategy*, p. 144.

observed in many sectors.[71] A large number of Chinese firms and fierce competition among them in infrastructure sectors have created serious challenges for the Chinese government to regulate them at home and overseas. Also, while they have expanded in size and operation, the government's capacity to oversee, monitor and regulate their activities has been undermined by the competing interests and bickering among its agencies. These features of government–SOE relations, in addition to the poorly-educated, low-paid and low-quality labour force these firms hire to do the work in Africa, 'generate unintended consequences that do not chime well with the orthodox rhetoric' of China's foreign policy for Africa.[72] One study comments:

> To the Chinese government, CSCES [China State Construction Engineering Corporation] is by no means a faithful policy agent that pursues national goals regardless of potential negative impacts on the firm's financial performance. This is especially true for CSCES's cross-border project, where monitoring becomes even more difficult for China's fragmented regulatory system.[73]

Chinese SOEs have two advantages in the construction industry—low supply-chain costs and labour productivity. Both 'advantages' are highly contentious. Chinese SOEs tend to purchase equipment and materials from China. They argue that (a) it is cheaper to purchase equipment and other materials from China than from developed countries or even host countries; and (b) they have a network in China which makes it easier to get what they need at the price they would be willing to pay: 'A 50kg bag of Angolan-made cement would cost US$10, while that made in China costs US$4' and the suppliers were more reliable.[74]

> For example, for a 3-year project, we can import a foreign bulldozer, which costs US$300,000 and can work for more than 10 years. If we do not manage to win the next project, this bulldozer remains idle with high residual value. In contrast, although a Chinese bulldozer only lasts three years, it is very cheap (US$100,000) and completely depreciated by the end of the project.[75]

This practice is quite controversial in the continent, as Chinese SOEs out-compete market players from developed countries and pose a serious threat to the manufacturing industries in OECD countries.[76] They also foster resentment from local suppliers.

Labour is a highly contentious issue. Many complain that Chinese SOEs do not employ local people and their projects therefore do not really benefit local development. In construction industries, about half of the labour force used for African projects is from China. The Chinese government is aware of the problems. At the 5th Ministerial Conference of the Forum of China–Africa Cooperation in July

---

71. IEA, *China's Power Sector Reforms: Where to Next* (Paris: OECD, 2006); P. Nolan, *Transforming China: Globalisation, Transition and Development* (London: Anthem, 2004).
72. Alden and Hughes, 'Harmony and discord in China's Africa strategy', p. 572.
73. Liou, *The Politics of China's 'Going Out' Strategy*, p. 177.
74. Zhen Yu Zhao and Li Yin Shen, 'Are Chinese contractors competitive in international markets?', *Construction Management and Economics* 26(3), (2008), p. 230.
75. Chen *et al.*, 'Local operations of Chinese construction firms in Africa', p. 82.
76. Hany Besada, Yang Wang and John Whalley, *China's Growing Economic Activity in Africa*, NBER working paper series (May 2008); Davies, *How China is Influencing Africa's Development*; Foster *et al.*, 'Building bridges'.

2012, the Chinese President Hu Jintao promised that the Chinese firms in Africa would increase the usage of local labour. It is not clear whether the promise from the central government would be delivered because Beijing only has a 'tenuous' control over its SOEs in Africa.[77] Furthermore, according to the Ministry of Commerce in Beijing, large SOEs operating in Africa had more problems with China-based labour export agencies than with African local employees.[78]

Competition

Africa increasingly is seen as a large market where Chinese SOEs and private firms compete for a piece of the pie. It is reported that revenues of construction companies in central and southern Africa grew by 31.7% to US$27.52 billion in 2009. The share of Chinese enterprises in the African market rose from 26.9% in 2007 to 42.4% in 2008 and back to 36.6% in 2009. In the construction sector, overseas projects are more attractive to SOEs and others because 'foreign projects often pay ten to twenty-some per cent above expenses', explained one firm. 'This is very good for our cash flow.'[79] Yet they need to compete for this market with their domestic and international counterparts and also with small semi-public or private Chinese firms, which are much less regulated. When competition is not properly regulated, it undermines the very objective of the 'going out' policy.

'China's state-owned enterprises dominate the domestic construction market as well as almost all the share of China's enterprises in the international construction market.'[80] Yet, there are a lot more SOEs competing in the field than in resource industries. One survey of 35 Chinese construction SOEs shows all of them have operations in Africa. Some of them are central SOEs under the auspice of SASAC. These tend to be large conglomerates, formed following the model of the Japanese *keiretsu* and Korean *chaebol*. Some have been created in the past 10–20 years, encouraged and approved by the Ministry of Commerce to engage in overseas projects. Recently provincial or local SOEs have entered into competition for overseas projects. This competition among Chinese firms for overseas projects mirrors the competition in the domestic construction industry.

The domestic construction market has allowed these firms to accumulate experience in managing large projects and building a wide network of supplies. Meanwhile, to compete, they tried to limit their regular staff to reduce the cost. For example, China Geo-Engineering Corporation (CGC) set up a subsidiary, CGC-International, with the approval of the Ministry of Commerce to engage in overseas projects. CGC-International has only 200 employees. When it won large projects financed by the World Bank and other multilateral and bilateral donors, it would provide managers and other professionals only and rely on contract workers to

---

77. Gill and Reilly, 'The tenuous hold of China Inc. in Africa'; Jakobson, 'China's diplomacy toward Africa'; Liou, 'Bureaucratic politics and overseas investment by Chinese state-owned oil companies'.
78. Quoted from Chen *et al.*, 'Local operations of Chinese construction firms in Africa', p. 83.
79. Quoted from Chuan Chen and Ryan J. Orr, 'Chinese contractors in Africa', *Journal of Construction Engineering and Management* 135, (2009), p. 1206.
80. Low Sui Pheng and Hongbin Jiang, 'Internationalisation of Chinese construction enterprises', *Journal of Construction Engineering and Management* 129, (2003), p. 591.

operate in Africa. These contract workers are primarily from China's rural areas, with little or no education and no exposure to the outside world and unable to communicate with local people. Cultural differences often exacerbate the tension: 'locals assume the highly disciplined Chinese workers in identical boiler suits they see toiling day and night must be doing so under duress'.[81]

When provincial SOEs compete with large SOEs for projects, they 'can clash with Beijing's strategic objectives'.[82] The central government has limited capacity to regulate these non-central SOEs and private sub-contractors. The difference is not about the ownership—both public and private firms go after 'profits'. Their size, however, can decide how they operate. Large SOEs, especially central SOEs, tend to have operations in both developed and developing countries. Adopting practices of international standard is important for their reputation and business prospects. Meanwhile, they are required to integrate 'corporate social responsibilities' (CSRs) into their missions. No matter how incompletely these CSRs were implemented, the managements of these SOEs were held accountable in their evaluation by SASAC. This is not the case with smaller public or private firms; as one foreign affairs official explained, 'they are interested in quick money and not aspects such as CSR'.[83] Also, large SOEs are obligated to bring local people to China for training as part of the central government's policies to increase the number of local employees; small firms do not see it as their responsibility and the government does not have the means to enforce the policy. Small sized firms often pay their own employees and local workers less than the level suggested by Chinese governments; they ignore the working conditions or environmental impacts. When small sub-contractors hire poorly-educated and poorly-paid workers, who cannot integrate with local communities, it becomes a headache for the Chinese diplomats and an embarrassment for the politicians in Beijing. These problems are not unique in those operations in Africa; they are serious problems in China too. The real issue is to develop adequate governing capacities in China.

Conclusion

'On the whole, African nations welcome China's engagement, because increased trade, aid, investment, education, vocational training, and relief have all benefitted these societies.'[84] The problems of Chinese engagement in Africa are apparent too. Instead of discussing these problems, this article analyses the argument that all Chinese investment in Africa is controlled by the CCP and all activities are closely managed and monitored by the central government. It shows that: (a) players from China in Africa vary significantly; (b) their activities and behaviour differ too; and (c) the government in Beijing has limited capacity to direct and regulate SOEs and other economic players. Central SOEs and a myriad of smaller provincial SOEs and private enterprises have invested in Africa in natural resources, infrastructure and many other

---

81. 'The Chinese in Africa: trying to pull together', *The Economist*.
82. Corkin, *Chinese Construction Companies in Angola*, p. 30.
83. *Ibid.*, p. 37.
84. David Shambaugh, *China Goes Global* (New York: Oxford University Press, 2013), p. 111.

commercial activities. When investors are central SOEs in natural resources, they are better organised, better resourced and often better supported as well as regulated by the central government. That they do care about their international reputation as credible commercial players provides them with further incentives to adopt practices that can meet basic international standards. They are least likely to cause embarrassment, not because they are state-owned, but because they are or want to be international players.

In contrast, when small SOEs and private enterprises work as sub-contractors in the infrastructure sector in Africa, their unregulated competition and operation in Africa often work against the very principles of the Chinese foreign policy of engagement in Africa. The competitive nature of the industry shapes the rules of the game and their behaviour—that is, their labour force is less educated, less exposed, lower-paid and, using the words of *The Economist*, 'rough-and-tumble' and 'anything goes'.[85] Their game is short term. Corruption, bribery and embezzlement are all too common to the embarrassment of the Chinese government.

Finally, at best, the government in Beijing has limited capacity to monitor and regulate SOEs and their operations in Africa, let alone those small enterprises answerable to provincial or local governments or themselves. This is not an excuse for the 'bad' behaviour of Chinese firms in Africa but it does explain why increasingly Chinese involvement and activities in Africa are contradictory or even undermining the objectives of the central government. How to build adequate governance capacity in China where 'policies induce counter-policies' (*duice* 对策)[86] is a serious challenge for the government in Beijing as well as for those in political science interested in governance.

---

85. 'The Chinese in Africa: trying to pull together', *The Economist*.
86. Lynn T. White III, 'Temporal, spatial and functional governance of China's reform stability', *Journal of Contemporary China* 22(83), (2013), p. 791.

China–Africa Trade Patterns: causes and consequences

JOSHUA EISENMAN

China's trade patterns with African countries have made Beijing the focal point of new anti-Chinese resistance narratives in Africa. Unlike the Maoist era, when China's trade policies served its leaders' political goals, now they aim to access markets as part of China's larger domestic development strategy. China's state-run firms can channel China–Africa trade through extra-market decisions that influence flows, yet, ultimately, Beijing's ability to direct trade with Africa is constrained by market forces. Despite suggestions that shared illiberalism drives China–Africa trade the author concludes that five interrelated causal factors overwhelmingly determine China–Africa trade: China's comparative advantage in labor-intensive and capital-intensive production; Africa's abundant natural resource endowments; China's rapid economic growth; China's emphasis on infrastructure building at home and in Africa; and the emergence of economies of scale in China's shipping and light manufacturing sectors.

Introduction

China's trade patterns with African countries have made Beijing the focal point of new anti-Chinese resistance narratives in Africa.[1] Unlike the Maoist era, when China's trade policies served its leaders' political goals, now they aim to access markets as part of China's larger domestic development strategy. China's state-run firms can channel China–Africa trade through extra-market decisions that influence flows, yet, ultimately, Beijing's ability to direct trade with Africa is constrained by market forces. Five causal factors overwhelmingly determine China–Africa trade: China's comparative advantage in labor-intensive and capital-intensive production; Africa's abundant natural resource endowments; China's rapid economic growth; China's emphasis on infrastructure building at home and in Africa; and the emergence of economies of scale in China's shipping and light manufacturing sectors.

China trades extensively with African democracies and autocracies alike, making it doubtful that an African country's regime type explains much about its trade with

---

*Joshua Eisenman is Assistant Professor at the LBJ School of Public Affairs at the University of Texas at Austin, TX, USA, and Senior Fellow in China Studies at the American Foreign Policy Council in Washington, DC., USA. This chapter was written with support from the American Strategy Program of the New America Foundation.

1. The term 'trade patterns' refers to both the quantity of trade flows as well as the composition of goods traded.

China. Rather, China's trade with African countries is deeply rooted in powerful market dynamics that have begun to undermine China's image among average Africans. The spread of anti-Chinese narratives in Africa need not have coherent or predictable consequences. For instance, in December 2010 soccer fans chanting 'Chinese go home' rioted and attacked Chinese businesses in Lubumbashi, Democratic Republic of Congo (DRC) after their side lost to Inter Milan. Congolese fans mistook the Japanese referee for a Chinese, and were angered by some of his decisions in the match.[2] Nine months later in Zambia anti-China rhetoric helped populist Michael Sata win the presidency after three previously unsuccessful campaigns.[3] Incidents like these suggest that simmering anti-Chinese sentiment at the grassroots level is an increasingly powerful force that can be activated by either spontaneous, unforeseen events or ambitious politicians for their own purposes.

For a decade a dynamic expansion in the size and diversity of China's trade with Africa has been underway and along with it a growing number of anecdotal reports and interviews by scholars and government officials. Many, if not most of them, argue that China's trade is particularly friendly to repressive regimes. Beijing, the argument goes, prefers similarly autocratic African trading partners and its businesses assist in their efforts to resist calls for liberal democratic political reforms. Meanwhile, there are long standing economic theories of trade (i.e. factor abundance theory, the gravity model of trade, new trade theory, and structural economics) that also promise to explain China's trade with Africa. These explanations get more attention among economists than they do in the popular press. This paper seeks to reconcile these sometimes-conflicting explanations in an effort to better answer two related questions about China's trade with African countries: (1) what causal arguments help explain the patterns of China–Africa trade; and (2) what are the economic and political consequences of the patterns of trade between China and Africa?

Part I: Causes of China–Africa trade patterns

Overview: unanswered questions

Economic considerations determine trade patterns, but so do political ones, particularly between developing countries like China and African states. But why and how do political variables influence China's trade with Africa? Economies of scale and technological development, for instance, are often the product of a centrally administered national industrial policy born of a political process rather than an efficient market mechanism. This is particularly true in authoritarian developing countries like China, suggesting that nowhere is the role of politics in trade more prominent than between developing economies with autocratic political regimes. Many analysts have included political variables in their trade models yet none has developed a model that can predict the role they play, not only in determining the quantities of trade, but also in explaining which types of goods are traded and which

---

2. 'Unrest in DR Congo after TP Mazembe lose to Inter Milan', *BBC*, (18 December 2010), available at: http://www.bbc.co.uk/news/world-africa-12030051.

3. Howard W. French, 'In Africa, an election reveals skepticism of Chinese involvement', *The Atlantic*, (29 September 2011), available at: http://www.theatlantic.com/international/archive/2011/09/in-africa-an-election-reveals-skepticism-of-chinese-involvement/245832/.

are not. Yet, it is often the *types* of goods traded (raw materials vs. manufactured consumer products and capital equipment) that many observers believe determine the economic and political consequences we see between China and African countries.[4]

So what factors determine the patterns of China's trade with African countries (see Figure 1)? Amid the last decade or more of unprecedented growth in China's trade with Africa this question has elicited a myriad of publications describing its unique causes and implications. These works have revealed an assortment of arguments, each promising to explain the patterns of Chinese trade with Africa; many compare Chinese trade with other countries, such as India or the US. These explanations generally include elements of one or more of three prominent theories of trade: relative factor abundance theory, gravity trade theory, and political trade. Political trade is an amalgam of generally accepted views that some American, European, and African observers regularly use to help explain close economic ties between two liberal countries and two autocracies. President Bill Clinton broadly summarized the concept in his 1994 State of the Union address when he said: 'Democracies don't attack each other, they make better trading partners and partners in diplomacy'.

Differences between economic theories' predicted patterns of China–Africa trade and those observed in the real world have fueled political trade arguments. In 2005, for instance, a report for the United Kingdom's Department for International Development by Rhys Jenkins and Chris Edwards predicted a change in the structure of African exports and identified 'unexploited export opportunities which should be explored'. The report argued that rising per capita incomes in China would 'lead to a growing demand for food, particularly those with high income elasticity of demand such as meat products, fish, fruit and beverages'. The authors also identified 'new opportunities for other agricultural exports from Africa such as coffee and sugar'. They were optimistic that 'in the future for some [African] countries there may be opportunities for exporting labour intensive agricultural products such as fruit or coffee which could create more income or employment opportunities for poorer sections of society'.[5] By 2007, however, a World Bank report, authored by economist Harry Broadman, acknowledged that expected trade patterns had not emerged. Broadman was particularly puzzled by the slow growth in labor-intensive exports from African countries to China; he noted that only 'five oil- and mineral-exporting countries account for 85 percent of Africa's exports to China'[6] (see Figure 2). The report concluded that: 'Whether in terms of nationality, mode of entry, scale of investment, or geographic diversification, among other factors, one would expect to observe significant differences in the patterns of the exports and imports'.[7]

---

4. The early roots of this argument can be traced back to dependency theory. See Patrick J. McGowan, 'Economic dependence and economic performance in black Africa', *The Journal of Modern African Studies* 14(1), (March 1976); Michael B. Dolan and Brian W. Tomlin, 'First world–third world linkages: external relations and economic development', *International Organization* 34(1), (Winter 1980); And James A. Caporaso, 'Dependence, dependency, and power in the global system: a structural and behavioral analysis', *International Organization* 32(1), (1978).

5. Rhys Jenkins and Chris Edwards, *The Effect of China and India's Growth and Trade Liberalisation on Poverty in Africa*, DCP 70 (Department for International Development of the United Kingdom, May 2005), p. 18.

6. Harry G. Broadman, *Africa's Silk Road: China and India's New Economic Frontier* (Washington, DC: The World Bank, 2007), p. 10.

7. *Ibid.*, p. 296.

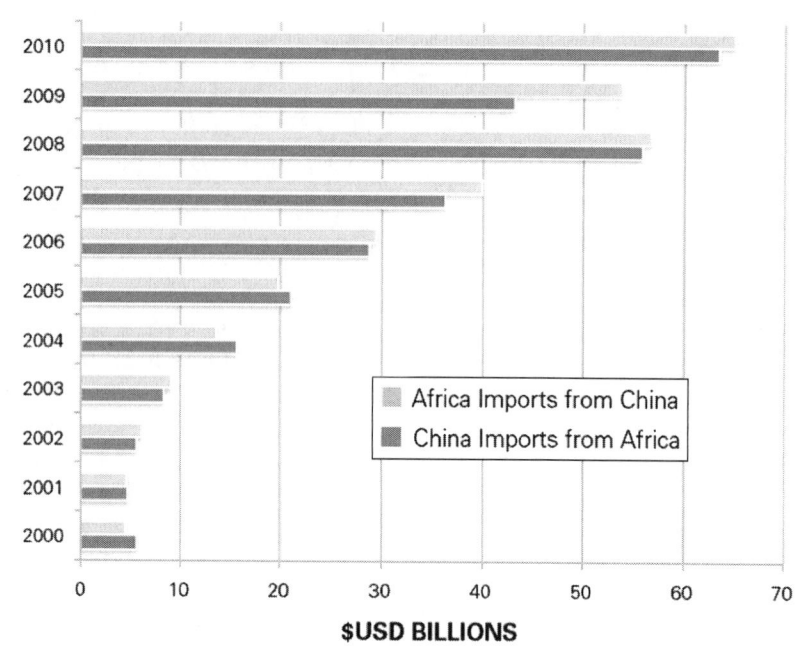

Figure 1. China–Africa trade, 2000–2010. *Source*: IMF Direction of Trade Statistics, China and Africa country pages, import data. David H. Shinn and Joshua Eisenman, *China and Africa: A Century of Engagement* (Philadelphia, PA: University of Pennsylvania Press, 2012), p. 115. Graphical displays reprinted with permission from the University of Pennsylvania Press.

Economists at the Common Market for Eastern and Southern Africa (COMESA) were similarly stumped by the slow growth in dozens of labor-intensive African exports to China. In July 2007, the East African regional trade bloc published a report on the effects of China's tariff-free entry program for products from its member countries. COMESA concluded that despite China's preferential tariff scheme on 454 African product lines, resource products including copper, cobalt, and marble consistently made up the bulk of member countries' tariff-free import value. Based on a comparative advantage index the report identified dozens of products in which member countries might have expected a comparative advantage — e.g. textiles, cotton, salt and sulfur, rawhides and skins, coffee and tea, and fish and crustaceans— yet could not gain a foothold in Chinese markets.[8] China's trade with Africa, at least, does not appear to function as relative factor abundance theory alone might predict. So what *else* determines China's trade with African countries?

Researchers remain divided among those who believe existing economic theories of international trade can satisfactorily explain the patterns of China's trade with Africa and those who believe political considerations are also highly salient. Even those who agree that China's trade patterns do not conform to expectations do not necessarily agree on why; some researchers have privileged economic considerations, while others have

---

8. Themba Munalula, 'China's special preferential tariff Africa', *Statistical Brief Issue No. 3*, Statistics Unit, Regional Integration Support Programme, Division of Trade, Customs and Monetary Affairs COMESA Secretariat, (July 2007).

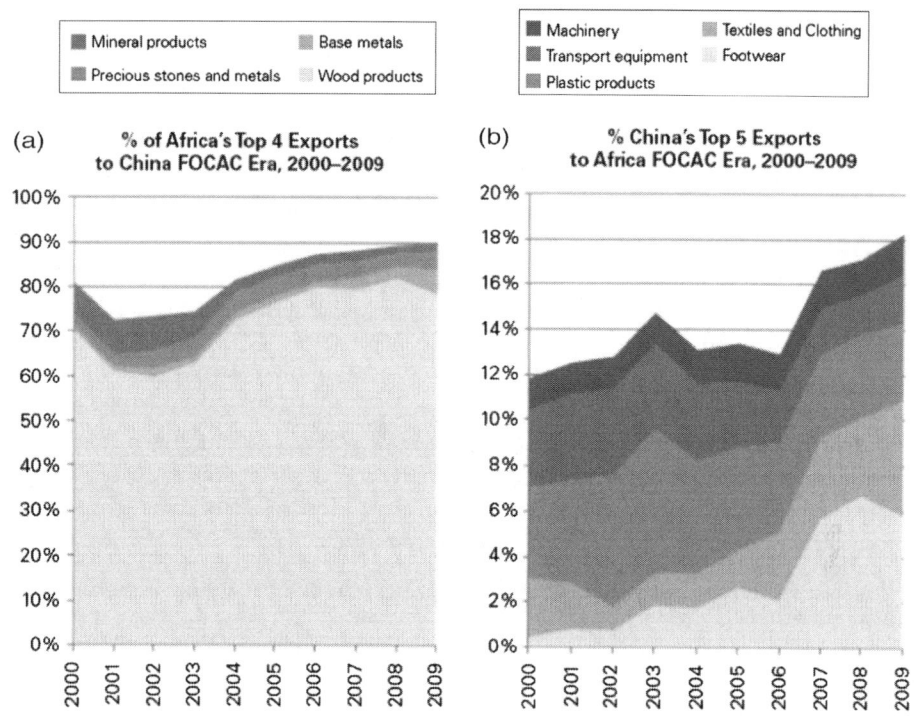

Figure 2. (a) Percentage of Africa's top four exports to China FOCAC era, 2000–2009. (b) Percentage of China's top five exports to Africa FOCAC era, 2000–2009. *Source*: Based on a study of detailed PRC trade data conducted by the Trade Law Centre for Southern Africa, TRALAC, Stellenbosch, South Africa (see Shinn and Eisenman, *China and Africa*, p. 119).

advocated a mix of economic and political ones. Those who argue that politics also help determine Chinese trade patterns with Africa are, broadly speaking, unsatisfied with traditional economic explanations. These researchers generally agree that economic theories of trade have failed to adequately predict the actual patterns of trade we see in the real world. Before reviewing political explanations for China's trade with Africa, however, the following section identifies the key elements and assumptions of leading trade theories and their relevant amendments to determine where and how politics influences China–Africa trade.

Relative factor abundance theory

Relative factor abundance theory, privileged by most trade economists, holds that differing national endowments of resources, labor, and capital should explain the composition of China's trade with Africa. The Heckscher–Ohlin (H–O) model, rooted in David Ricardo's theory of comparative advantage, is the economists' traditional explanation for bilateral trade flows between countries. H–O (also known as neoclassical trade theory) predicts that the relative abundance or scarcity of a country's fixed factor endowments (resources, labor, and capital) compared with those of its trade partner determines what it will sell and what it will buy. A country will export goods produced with its abundant endowments and import goods produced with inputs that are locally scarce. According to H–O, once trade barriers are removed fixed relative factor endowments will determine trade patterns.

Supporters of relative factor abundance theory, including economist Jian-Ye Wang, predict that China–Africa trade patterns 'will be shaped by shifts in comparative advantage and changes in global supply chains'.[9] Jeffrey Herbst and Greg Mills agree that in Africa 'the market, not grand strategy, is the Chinese motivation'.[10] They suggest that relative factor abundance theories have proven particularly valuable in explaining China–Africa trade because China's relatively scarce natural resource endowments mean it must 'lock up as many raw materials as possible'.[11] Indeed, many African countries' relatively abundant factor endowments are natural resources and they do tend to dominate those countries' exports to China. Wang's International Monetary Fund (IMF) working paper described the resulting China–Africa trade pattern:

> Strong growth of the Chinese and African economies, together with the complementary trade pattern—China imports fuel and other commodities, Africa purchases investment and manufactured products from China—largely explains their surging trade in recent years.[12]

Theories of relative factor abundance have come to dominate contemporary thinking on the patterns of international trade between countries, but researchers have also begun to recognize differences between traditional economic theories' predicted patterns of China's trade with Africa and those observed in the real world. To account for this researchers regularly relax one or another of the H–O theory's six basic assumptions.[13] When studying China the most relevant amendments are increasing returns to scale and differences in technology and transportation costs. As discussed below these modifications to H–O assumptions help explain China–Africa trade patterns and identify the role that political factors play.

Gravity trade theory

Some have turned to gravity trade theory (aka the gravity model) to fill in the gaps between the H–O theory's predictions and those patterns observed in the real world.

---

9. Jian-Ye Wang, *What Drives China's Growing Role in Africa?*, Working Paper No. 07/211 (International Monetary Fund, August 2007), p. 22.

10. Jeffrey Herbst and Greg Mills, 'Commodities, Africa and China', *S. Rajaratnam School of International Studies (RSIS) Commentaries* (Singapore: Nanyang Technological University, 9 January 2009), available at: http://www.rsis.edu.sg/publications/Perspective/RSIS0032009.pdf.

11. *Ibid.*

12. Wang, *What Drives China's Growing Role in Africa?*, p. 20.

13. The H–O model is based on a half-dozen assumptions, and they are: *Dimensionality*: the number of goods is equal to the number of productive factors; *Factor Immobility*: (a) the factors of production move freely among industries within a country but are completely immobile among countries, and (b) goods move internationally with no transport costs and there are no other impediments to trade; *Competition*: both goods and factor markets clear competitively, all agents act as if they could buy or sell unlimited quantities at the prevailing market price; *Technology*: the same technological knowledge about the production of goods is costlessly available to all countries; *Factor Endowment Similarity*: the variability of factor endowment ratios among countries is less than the variability of factor input intensities across industries; and *Demand Similarity*: individuals consume as if each were maximizing an identical utility function. See Edward E. Leamer, *Sources of International Comparative Advantage: Theory and Evidence* (Cambridge, MA: Massachusetts Institute of Technology, 1984), p. 2.

The gravity theory of trade predicts that the *distance* between any two countries (measured in shipping distance) and the *size* of their respective economies (measured in GDP) will be the principal determinants of the quantity of trade between them.[14] Advocates note that the gravity model can be made increasingly robust with the inclusion of social and political explanatory variables such as population size, common boarders, common language etc. This has made it attractive to many researchers investigating the link between politics and trade, while accounting for both economic and political drivers. In their examination of democracies, autocracies, and international trade, Mansfield *et al.*, for instance, found that 'the gravity framework is quite successful in explaining the flow of interstate commerce'.[15]

Increasing returns to scale

The concept of increasing returns to scale, first conceived by Bertil Olin and overlooked for decades, resurfaced in the early 1990s as a modification to relative factor abundance theory.[16] Paul Krugman describes this contribution, which came to be known as the new trade theory:

> In the new trade theory, the basic point was that increasing returns are a motive for specialization and trade over and above conventional comparative advantage, and can indeed cause trade even where comparative advantage is of negligible importance among industrial countries with similar resources and technology.[17]

Increasing returns to scale, according to Helena Marques, proved an important amendment to H–O and helped explain why we see both inter-industry trade, still governed by the factor endowment differences, and intra-industry trade, where developed countries produce different varieties of the same good and trade them.[18] New trade theory has proven most helpful in explaining trade between two developed countries with similar technological and factor endowments, conditions not found between China and Africa. Yet, the concept of increasing returns to scale is also very important for China since it has spent massive sums on government devised and funded industrial polices.

Perhaps the most well known of Beijing's initiatives designed to take advantage of increasing returns to scale was the selection of a handful of Special Economic Zones (SEZs) in southern China. Cheap labor, tax breaks, and other government-approved SEZ-specific incentives attracted foreign investment, promoted Chinese exports, and in a decade transformed the small fishing village of Shenzhen into a bustling

---

14. Although it is still unclear why and how it works, the gravity model, according to Jeffery Bergstrand, 'has long been recognized for its consistent empirical success in explaining many types of flows, such as migration, commuting, tourism, and commodity shipping'. See Jeffrey H. Bergstrand, 'The gravity equation in international trade: some microeconomic foundations and empirical evidence', *The Review of Economics and Statistics* 67(3), (August 1985), pp. 474–481.

15. Edward D. Mansfield, Helen V. Milner and B. Peter Rosendorff, 'Free to trade: democracies, autocracies, and international trade', *The American Political Science Review* 94(2), (June 2000), p. 312.

16. Paul Krugman, *Was it All in Olin?*, Massachusetts Institute of Technology website, (October 1999), available at: http://web.mit.edu/krugman/www/ohlin.html.

17. *Ibid.*

18. Helena Marques, *The 'New' Economic Theories*, FEP Working Papers 104 (Universidade do Porto, Faculdade de Economia do Porto), p. 8.

metropolis of over eight million people. Yet, it is important to remember that the comparative advantage conveyed to Chinese producers via the SEZs' economies of scale was not the result of an efficient market mechanism, but rather of bargaining among political elites in Beijing. Shanghai, for instance, was not included among the SEZs for political reasons—the personal power of party elder Chen Yun—and thus its development initially lagged behind those cities that were selected. The SEZs' success (and that of China's economic reforms more generally) was only ensured after Deng Xiaoping's 1992 'Southern Journey' silenced powerful political critics who sought to roll back his efforts to open China to foreign trade and investment.

Over the last decade China's state-built economies of scale have helped confer upon its producers huge trade advantage over their African competitors, which face towering barriers to entry. China has boomtowns specializing in only one or two consumer products: Datong, Zhejiang produces over nine billion pairs of socks per year; Shenzhou, Zhejiang is the world's necktie capital; and the province also boasts a Sweater City, Kids Clothing City, and an Underwear City. The same expansion of economies of scale has occurred in various consumer products in other Chinese cities and provinces—like sneakers in Quanzhou, Fujian. Jinfei Wang, the chairman of the Jiangsu Diao Garment factory in Nantong, Jiangsu, explained how economies of scale helped his firm gain trade advantage over African producers. He concluded: 'I've been to factories all over the world and we can compete with any of them. Without restrictions, certainly China is going to be No. 1'. On Christmas Eve 2004, *The New York Times* underscored the power of China's 'enclave' approach to consumer goods production:

> This remarkable specialization, one city for each drawer in your bureau, reflects the economies of scale and intense concentration that have helped turn China into a garment behemoth. Now, China is banking on its immense size and efficient operators to grab an even larger share of the world's clothing orders. China is not just becoming the leader of the pack. In many ways, it hopes to run away with as much of the market as possible.[19]

Technology and transportation costs

The introduction of technology and transport costs into relative factor endowment theory (a component that H–O theory considers to be either zero or prohibitive) has been another innovation relevant for the study of China–Africa trade patterns. Structural economists suggest that differences in technology and industrial upgrading play a critical role in countries' economic development. Large differences in technology and infrastructure between China and most African countries are important deviations from neoclassical trade theory's basic assumptions of costless domestic factor mobility and equal technology. 'Comparative advantage in the modern world is created not endowed', notes Thomas I. Palley; '[It] is driven by

---

19. David Barboza, 'In roaring China, sweaters are west of socks city', *The New York Times*, (24 December 2004), available at: http://www.nytimes.com/2004/12/24/business/worldbusiness/24china.html?pagewanted=print&position.

technology, and technology can be importantly influenced by human action and policy'. Palley, a structural economist, believes that 'differences in technology can confer an absolute advantage on one country'.[20]

Although their focus has typically been on understanding the conditions favorable to economic growth in Africa rather than trade flows, structural economists' insights strongly suggest political causes for China–Africa trade. A 2009 report issued by the World Bank's Senior Vice President and Chief Economist, Justin Yifu Lin, outlined the central tenants of structural economics and its relationship to neoclassical theories of trade. Lin explained that: 'Structural economics applies the neoclassical approach to study the mechanism of economic development', except it:

> ... suggests a different, larger conception of endowments, including the factor endowment and hard and soft infrastructure endowment. Both factor endowment and infrastructure endowment are given at any given time and changeable over time. Based on the factor endowment, the individual firms make their choices of industries and technologies as well as production decisions. The infrastructure endowment will affect individual firms' transaction costs and rate of return to their investments.[21]

As noted above, World Bank and COMESA reports have both acknowledged that relative factor abundance theory alone is insufficient to explain China's trade patterns with African countries. This is because China's government exerts significant control over corporations and national interest is factored into business strategy.[22] China's domestic industrial policies push its production down the cost curve allowing it to poach demand from other developing countries and become the low cost producer at their expense. Beijing's trade and industrial policies can weaken the explanatory power of relative factor abundance theory if it 're-roots corporations by realigning profit with the national interest' thus redistributing the gains of trade and, in turn, substantially determining its patterns.[23]

A country's infrastructure and large-scale technology investments kick-start a self-reinforcing and self-sustaining process of economic development, Lin notes. This idea is powerful in China where the government has channeled massive investment into highways, train systems, and shipping ports, etc. In 2009, for instance, China invested US$102.7 billion in its railways, with high-speed rail lines accounting for almost 60% of that total.[24] In 2011 and 2012, China invested an additional US$75 billion and US$65 billion, respectively, to maintain and expand its 91,000 km of railroads.[25] In 2011 the country had the world's largest high-speed rail network with 8,358 km of track, which is expected to exceed 13,000 km by the end of 2012 and

---

20. Thomas I. Palley, 'Institutionalism and new trade theory: rethinking comparative advantage and trade policy', *Journal of Economic Issues* 17(1), (March 2008), p. 199.
21. Justin Yifu Lin, *New Structural Economics: A Framework for Rethinking Development*, Policy Research Working Paper (The World Bank, February 2010), p. 28.
22. Palley, 'Institutionalism and new trade theory', p. 204.
23. *Ibid.*, p. 202.
24. 'High-speed railway accounts for over half of China's railway investment', *The People's Daily*, (27 April 2010), available at: http://english.peopledaily.com.cn/90001/90778/90860/6965110.html.
25. 'China blames 54 officials for bullet train crash', *The New York Times*, (28 December 2011), available at: http://www.nytimes.com/aponline/2011/12/28/world/asia/AP-AS-China-Bullet-Train-Crash.html?_r = 2&partner = rss&emc = rss.

16,000 km by the end of 2020.[26] Chinese firms are also building extensive rail networks throughout Africa linking resource-processing zones in DRC and Zambia to shipping hubs like Dar es Salaam, Luanda, Angola and Lagos, Nigeria. These rail projects serve to connect African raw material suppliers with Chinese buyers as well as Chinese goods manufacturers with African customers.

The growth of China's roads and highways has also outpaced African countries, yet China's proficiency in road building has also spread to Africa where its state-run firms have won bids to construct highways in Algeria, Angola, Ethiopia, and Zambia, among other countries. China has also invested heavily in its shipping sector and overtook South Korea to become the world's top shipbuilder in the first half of 2010. This success, however, came with ample political assistance. For instance, with the approval of the powerful National Development and Reform Commission in 2009 the Tianjin Shipbuilding Industry Fund was established. Cui Jindu, deputy mayor of Tianjin, announced that China's two major shipbuilders, China CSSC Holdings and the China Shipbuilding Industry Corp., donated to the fund, which by August 2010 had already financed 45 new ship orders for domestic shipbuilders and invested 15 billion yuan in shipbuilding projects. Revealing the close relationship between shipbuilders and the state, the website Tax-News.com reported that: 'Tianjin has the ear of central government with regard to providing the right tax incentives for ship financing packages and is spearheading developments in this direction'.[27] China has also expanded domestic port capacity and worked to improve the African ports of Lamu, Kenya, Luanda, Angola, and Dar es Salaam, Tanzania, among others. In Somaliland, even a lack of official diplomatic relations with Beijing did not stop Chinese firms from agreeing in August 2011 to build Berbera port into a regional hub for Chinese traders and a disembarkation port for East African oil shipments to China.[28]

Economists acknowledge that economies of scale, infrastructure and technology investment are born of a politics-driven process, yet they generally do not investigate the political causes of this cycle, i.e. the process that generates the preferential trade and industrial policies that governments use to pick winners. Instead they argue that a country's efforts to efficiently exploit its relative factor endowments will determine its industrial policies in accordance with efficient market outcomes. In short, while 'new trade' and structuralist critiques of factor endowment theory have succeeded in identifying *where* political considerations might influence trade patterns (e.g. in the development of economies of scale, technology, and reducing transport costs), they continue to shy away from explaining *how* and *why* they do.

---

26. 'Hu signals new high-speed railways approach', *South China Morning Post*, (17 April 2011), available at: http://www.onenewspage.com/news/Asia-Pacific/20110417/21489795/Hu-signals-new-high-speed-railways-approach.htm.

27. Mary Swire, 'China's shipping sector develops in Tianjin', *Tax-News.com* (Hong Kong), (5 August 2010), available at: http://www.tax-news.com/news/Chinas_Shipping_Sector_Develops_In_Tianjin____44636.html.

28. Mark T. Jones, 'Somaliland President stops over in Addis Ababa enroute to China', *Somaliland Press*, (9 August 2011), available at: http://somalilandpress.com/somaliland-president-stops-over-in-addis-ababa-enroute-china-23229.

Political trade

'Trade substantially follows patterns of comparative advantage. Nonetheless, economics is not everything', Harry Bliss and Bruce Russett argued in response to existing economic theories of trade.[29] They contend that although the influence of shared regime type on trade may be difficult to observe, it should not be discounted entirely. Naazeen Barma, Ely Ratner and Steven Weber believe that the expansion of trade between China and Africa is 'in excess of what standard economic models of trade would predict. This means that these patterns cannot be explained away by blistering economic growth'.[30]

The existing academic literature supporting political trade arguments is based on past studies that identified a connection between shared levels of liberalism and increased bilateral trade. This linkage appears to have emerged during the Cold War, when political loyalties in a bipolar world were substantially determined by a country's trade partners. In their landmark article 'Democratic trading partners: the liberal connection, (1962–1989)', Bliss and Russett found that democracy is significantly and positively related to trade volume: 'Trade between pairs of states with democratic polities', they conclude, 'is greater than that between states not sharing such a polity type'.[31] Bliss and Russett suggested that for democratic states the causal mechanism is rooted in a state's security concerns:

> States attempt to control trading patterns on behalf of private interests, and on behalf of perceived state and national interests. They promote trade with states deemed stable and reliable sources, and discourage, by various barriers, with adversaries and potential enemies. A democratic trading state will feel its security less threatened by other democratic state than by many autocracies. Democratic statesmen need to be less concerned that a democratic trading partner will use gains from trade to endanger their security than when their country trades with a nondemocracy. Their countries can enter into relationships of economic interdependence for absolute gains, without worrying as much about the hazard of relative gains as they might with nondemocratic partners.[32]

Mansfield *et al.* affirmed Bliss and Russett's results that two liberal countries will tend to trade more than a mixed pair. They conclude that: 'Holding constant various economic and political factors, democratic dyads tend to trade more freely than dyads composed of a democracy and an autocracy'.[33] Mansfield *et al.*'s research suggests that: 'On average, a democracy and an autocracy engage in roughly 15% to 20% less commerce than a dyad composed of two democracies'.[34] Unfortunately, however, their model does not yield determinate predictions about whether trade between

---

29. Harry Bliss and Bruce Russett, 'Democratic trading partners: the liberal connection, 1962–1989', *The Journal of Politics* 60(4), (November 1998), p. 1127.
30. Naazneen Barma, Ely Ratner and Steven Weber, 'A world without the West', *The National Interest* 90, (July/August 2007), p. 25.
31. Bliss and Russett, 'Democratic trading partners', p. 1127.
32. *Ibid.*, p. 1128.
33. Mansfield *et al.*, 'Free to trade', p. 306.
34. *Ibid.*, p. 314. In 2002, and again in 2005, Mansfield *et al.* reaffirmed their earlier work and concluded that two democracies are more than twice as likely to sign a trade agreement as are a mixed pair. See Edward D. Mansfield, Helen V. Milner and B. Peter Rosendorff, 'Why democracies cooperate more: electoral control and international trade agreements', *International Organization* 56(3), (Summer 2002), p. 505; B. Peter Rosendorff, 'Do democracies trade more freely?', unpublished manuscript, dated 29 September 2005.

autocratic pairs is more likely than between mixed pairs.[35] The question thus remains unanswered by the academic literature: do similarly autocratic states also prefer to trade with each other?

Barma and Ratner sought to answer this question using China as a qualitative case study. They conclude that China advocates 'illiberal capitalism ... where markets are free but politics are not'.[36] They argue that, 'through a wide array of bilateral and multilateral arrangements, the Chinese government has begun to build an alternative international structure anchored by illiberal norms'.[37] They contend that 'Chinese illiberalism presents the real long-term geopolitical challenge: it is easily exportable, and it is dangerously appealing to a disaffected world'.[38] Furthermore, Barma and Ratner link China's political illiberalism with its economic relations by arguing that Beijing leverages its 'mercantilist strength in the international system' to attain its national interests and argue that 'nowhere is this trend more evident than in Africa'.[39]

Two prominent Africanists, Denis Tull and Ian Taylor, also suggest that illiberalism drives China's economic relations in Africa, but are unsure whether to attribute this to China's indifference or its design. 'What is different in comparison to other countries' foreign policies is that Beijing legitimizes human rights abuses and undemocratic practices under the guise of state sovereignty', Taylor writes. 'China has no civil society worth talking about.'[40] Tull concurs, citing Taylor in support of his assessment that:

> There is virtually no way around the conclusion that China's massive return to Africa presents a negative political development that 'almost certainly does not contribute to the promotion of peace, prosperity and democracy on the continent'.[41]

In the 'Illiberal regimes' section of his 2007 book, Chris Alden explained why he believes China's demand for resource commodities instinctively increases the tendency to do business with autocracies:

> From the Chinese perspective, these economies are generally closely tied to African elites' interests, and there are fewer obstacles to rapid investment in the resource sector than they might experience in a state with stronger institutions and commitment to constitutional law.[42]

---

35. Mansfield *et al.*, 'Free to trade', p. 314. Not everyone agrees, however. According to Penubarti and Ward, Mansfield *et al.*'s 'results are both biased and inconsistent [and] produce not only inefficient, but also biased estimates of the parameters', thus causing 'the linkages virtually all other studies have uncovered between "joint democracy" and trade to evaporate'. They are also critical of using only a 'loose fitting gravity model' to determine bilateral trade flows and include economic variables (i.e. factor endowments, scale economies, and trade barriers) as well. These variables and falling trade barriers, not liberal politics, they conclude, are an increasingly important determinant of bilateral trade flows. See Mohan Penubarti and Michael D. Ward, 'Commerce and democracy', conference paper presented at *The Development and Application of Spatial Analysis for Political Methodology*, at the University of Colorado, Boulder, 10–12 March 2000.
36. Naazneen Barma and Ely Ratner, 'China's illiberal challenge', *Democracy: A Journal of Ideas* 2, (Fall 2006), p. 57.
37. *Ibid.*, p. 64.
38. *Ibid.*, p. 61.
39. *Ibid.*, pp. 61 and 64.
40. Ian Taylor, 'The "all weather friend"? Sino–African interaction in the 21st century', in Ian Taylor and Paul Williams, eds, *Africa in International Politics: External Involvement on the Continent* (New York: Routledge, 2004), p. 99.
41. Denis M. Tull, 'China's engagement in Africa: scope, significance and consequences', *Journal of Modern African Studies* 44(3), (2006), p. 476; also see Taylor, 'The "all weather friend"?', p. 99.
42. Chris Alden, *China in Africa* (New York: Zed Books, 2007), p. 70.

More hawkish critics, such as Peter Brookes and Ji Hye Shin, depict China's commercial competition as part of a zero-sum game that presents an inevitable threat to both liberalism and Western predominance. They contend that Chinese support for political and economic repression counters the liberalizing influences of traditional Western trading partners.[43] Others argue that as China expands economic relations with Africa its influence can actually encourage them to become more illiberal. This concern was echoed in a Council on Foreign Relations report issued in June 2008:

> The way China does business—particularly its willingness to pay bribes, as documented by Transparency International—undermines local efforts to increase good governance and international efforts at macroeconomic reform by institutions like the World Bank and the International Monetary Fund.[44]

Part II: Consequences of China–Africa trade patterns

Part I identified a number of causal arguments that purport to explain China–Africa trade patterns and which of them support the contention that China prefers to trade with authoritarian regimes in Africa. This section will address the economic and political consequences of the patterns of trade between China and Africa.

Resource trade growth

Relative factor endowments of labor, capital, and resources substantially determine China–Africa trade patterns as well as the unique patterns of each African country's trade with China. The result is a generally balanced China–Africa trade relationship when China's trade with all 54 African states is taken together, but important differences emerge when the data are disaggregated. On a country-by-country basis the balance of trade between China and resource exporters tends to favor the African country, meanwhile China's exports dominate its commerce with non-resource exporting African trade partners. Comparing China's trade with Egypt to its trade with Libya—both tightly controlled autocracies until 2011—helps to illustrate the leading role factor endowments play in determining China–Africa trade patterns while holding regime type constant. In both cases China's comparative advantage in consumer goods and capital equipment drives steep growth in its exports. Yet, the presence of an export commodity (oil) in Libya results in a moderate Libyan trade surplus, while the lack of resources in Egypt has led to a severely imbalanced trade relationship in China's favor. The displays in Figure 3 are indicative of the stark split in China's balance of trade generally observed between non-resource and resource exporting African countries.[45]

---

43. Peter Brookes and Ji Hye Shin, *China's Influence in Africa: Implications for the United States*, Backgrounder #1916 (The Heritage Foundation, 22 February 2006), available at: http://www.heritage.org/research/asiaandthepacific/bg1916.cfm.
44. Stephanie Hanson, *China, Africa, and Oil*, Backgrounder (Council on Foreign Relations, 6 June 2008), available at: http://www.cfr.org/publication/9557/.
45. In 2005, Jenkins and Edwards (*The Effect of China and India's Growth and Trade Liberalisation on Poverty in Africa*) disaggregated China's trade among 21 sub-Saharan African countries and identified the widespread nature of this pattern.

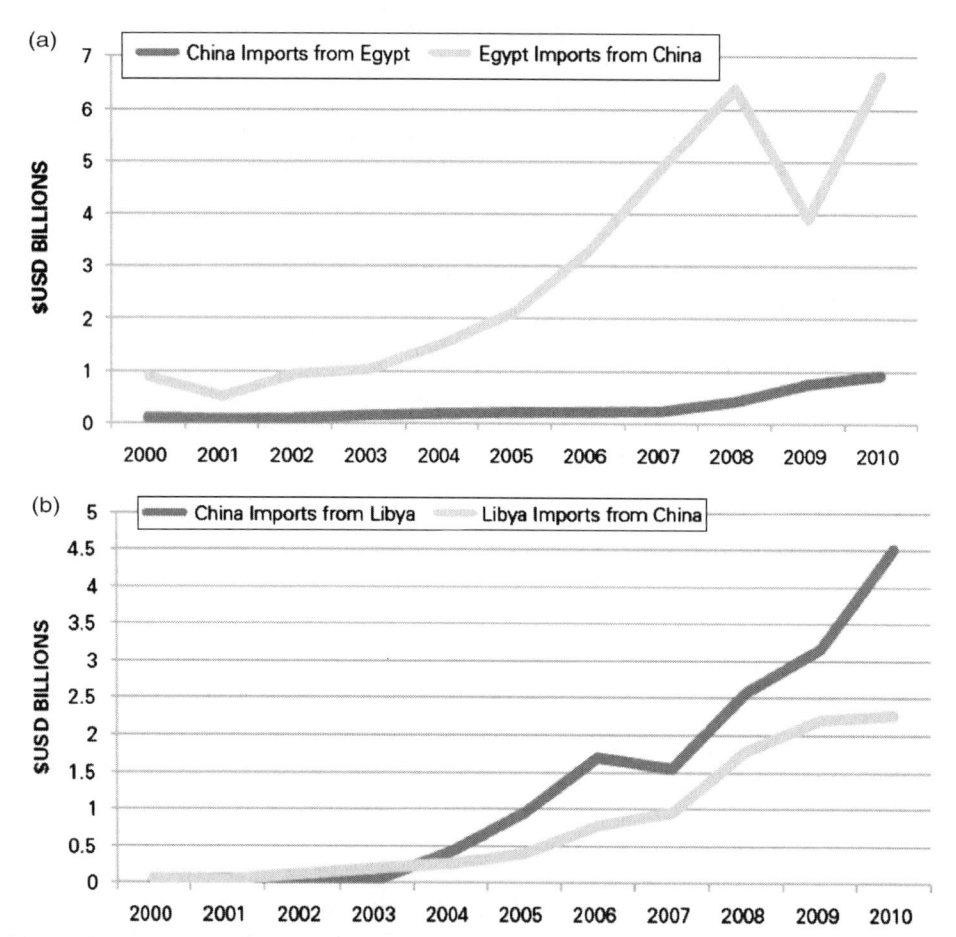

Figure 3. (a) China–Egypt trade, 2000–2010. (b) China–Libya trade, 2000–2010. *Source*: IMF Direction of Trade Statistics, China, Egypt and Libya country pages, import data (see Shinn and Eisenman, *China and Africa*, pp. 117–118).

African countries with few resources usually endure large trade deficits with China (e.g. Benin, Egypt, Ethiopia, Ghana, Liberia, and Morocco) while resource exporters, by contrast, enjoy surpluses (e.g. Equatorial Guinea, Republic of Congo, Angola, Libya, and Gabon). This pattern reflects the dichotomous nature of China–Africa trade: China's exports are diversified, i.e. each export category accounts for a small piece of total trade; by contrast, however, African exports to China remain concentrated in a narrow band of resource products. Since 2005 China's top five import categories from sub-Saharan Africa—mineral products, base metals (including oil), precious stones and metals, wood products, and textiles and clothing—alone make up 90% of its total purchases. In fact, by 2009, nearly 80% of China's exports from Africa were metals and petroleum products. Crude oil is Africa's top seller and since 2000 has made up over two-thirds of the total export value to China. Iron ore and platinum are also important African exports to China, although even they are swamped by the overwhelming value of China's oil imports (see Figure 2a).[46]

China's massive population and lack of labor rights protection have allowed for a largely cheap and generally compliant labor force that cannot be matched in Africa today. Meanwhile, China imports the large quantities of raw materials it needs to supply its domestic industries. Under such conditions, some African countries risk falling victim to the so-called Dutch disease whereby the economy becomes dominated by a single export commodity enriching a small group of elites who control natural resources at the expense of the larger workforce. By contrast, non-resource exporters risk accumulating large and sustained trade deficits with China that may have destabilizing political consequences over time.

The death of distance

As gravity trade theory predicts, the fast growing Chinese economy is attracting more trade from around the world, including Africa. This is particularly true of China's trade throughout the developing world, which has rapidly grown as a part of China's total trade portfolio. More interestingly, perhaps, is the decline of distance as an important determinant in China–Africa trade. This is largely due to China, which throughout the 1990s and 2000s invested heavily in shipping and port construction to expand the use of super-sized cargo shipping fleets. Thanks to policymakers in Beijing, distance, the single variable that hindered China–Africa trade for so long, has been rendered increasingly irrelevant.

One news story helps to illustrate why distance no longer hinders China–Africa trade. On 16 February 2011, Dar es Salaam became the first port of call for the world's largest car and truck megaship, which docked on its way from Xingang Port near Dalian, China. The 232-meter long vessel has a top speed of 24 knots per hour and can carry 8,000 vehicles per voyage. It can make the trip from China to East Africa in two weeks, shaving one third off the previous shipping time. Like countless other megaships this one now makes regular monthly voyages from China to and from African ports. This type of heretofore-unseen shipping capacity and speed has underpinned the growth of China–Africa trade over the last decade.[47]

Africans locked-out of labor-intensive sectors

China is building rail and road networks both at home and in Africa, allowing Beijing to exert influence over the destination of exports and the location of its suppliers. China's economies of scale, tariffs and subsidies to manufacturers for power, fuel, garbage collection, and other materials and social services have catalyzed its 'Go Global' strategy making it nearly impossible for African manufactured products to enter its markets and difficult for them to compete at home. By contrast labor-intensive African producers regularly pay hefty taxes and fees to political authorities for the

---

46. Data from 1995–2009 are available from the Trade and Law Centre for Southern Africa (TRALAC) in Stellenbosch, South Africa.
47. 'Largest cargo ship docks in Dar es Salaam', *Daily News* (Tanzania), (16 February 2011), available at: http://dailynews.co.tz/home/?n = 17346&cat = home.

same services while smugglers readily help Chinese products avoid African countries' tariffs. China's manufacturing capacity and economies of scale supply a steady stream of affordable Chinese consumer goods for the domestic market and for export to African (and other international) markets.[48] Together, China's increasing demand for raw materials and ability to produce affordable consumer goods has become the dual engine for the growth of Sino–African commerce. This looks set to continue as African consumers snap up Chinese consumer products and African governments prey upon, rather than protect, their labor-intensive domestic manufactures. These trade patterns have also begun to have a destabilizing effect on China's relations with some African countries as Jenkins and Edwards foresaw in 2005:

> Poverty reduction also depends on the type of growth generated by exports. There is a real danger that further expansion of mineral and petroleum exports to Asia will only reinforce an exclusionary model which will do little to reduce poverty and may exacerbate conflict and give rise to negative environmental impacts. Local producers may be displaced by competition from cheap imports from China and India and where these are in industries which employ significant numbers of unskilled workers, they may lose their jobs and be pushed into poverty.[49]

Anti-Chinese African resistance narratives

Although a boon for Chinese producers and African traders, the patterns of China–Africa trade also inhibit African countries from getting a foothold in labor-intensive manufacturing, the first rung of the development ladder. 'The potential danger, in terms of the relationship that could be constructed between China and the African continent, would indeed be a replication of that colonial relationship', South African President Thabo Mbeki said in December 2006. 'It is possible to build an unequal relationship, the kind of relationship that has developed between African countries as colonies. The African continent exports raw material and imports manufactured goods, condemning (it) to underdevelopment.'[50] Cautionary reports comparing China to colonial powers were news six years ago; today they have become a common refrain in conversations with average Africans and in press reports. One such article, published in August 2011 in Nigeria's *This Day* newspaper, was entitled 'A caution on China'. It said:

> It has become fashionable to welcome the rise and incursion of China into the African market as an unmitigated blessing. Chinese goods are cheap relative to what we get from the West. My attitude is that we need to take a long-term strategic view of China's coming if we have any degree of national self-interest left. At this point in time, the best attitude is one of great strategic caution and informed engagement. If we are a serious nation with an eye on the future of our children instead of our present greed, we need to have put in place a team of people who have a capability for strategic thinking to fashion out a China policy. There are lessons to be learnt and dangers that are clear and present. If

---

48. Although it is outside the scope of this publication African countries' inability to compete with China in third country markets like the US and EU is another troubling trend.
49. Jenkins and Edwards, *The Effect of China and India's Growth and Trade Liberalisation on Poverty in Africa*, pp. 18 and 39.
50. Victor Mallet, 'The Chinese in Africa: Beijing offers a new deal', *The Financial Times*, (23 January 2007).

we take out oil and gas, we are a very unproductive people because the funds we generate through real sectors and activities are hardly enough to keep this flag flying for longer than 30 days.[51]

Economically, many Africans welcome the benefits of rising energy and commodity prices driven by China's export industries' growing demand. Yet, China's willingness to aid autocratic regimes in some well-covered African countries like Zimbabwe, Equatorial Guinea, Guinea, Sudan, and Ethiopia risks tapping into a reservoir of anti-colonial resistance still widespread among Africans. On the one hand, China's officials use the old anti-imperialist rhetoric to differentiate themselves from the Western nations; on the other, they try to convince Africans that there are mutual gains from trade. This is a tough balancing act for China's policymakers, as Eric Kiss and Kate Zhou observe: 'China inadvertently follows the same pattern of other preceding great powers, spreading the seeds of discontent in a continent with diversified ethnicities and cultures'.[52] If China–Africa trade continues in accordance with existing patterns Beijing's interests will be increasingly pitted against an emerging narrative of grassroots African resistance.

China's trade has also made it the target of disaffected, unemployed Africans. This appears to be the case in Egypt, where an influx of Chinese goods over the last decade has swamped local production. Zambia and Zimbabwe have seen riots against Chinese merchants and products. Beijing provides some autocratic African governments with the weapons and censorship and monitoring equipment to keep social order. Meanwhile, its firms' comparative advantage in labor-intensive and capital-intensive production undermines the development of homegrown African industries that could underpin long-run stability. These two incompatible trends have placed China in the crosshairs. For instance, in an August 2010 article entitled 'Autocracy: China's unsolicited export', Ephrem Madebo, an Ethiopian commentator, accused 'China of mixing tyranny with its material exports to Africa'. The article, which appeared in *The Addis Voice*, suggested that Ethiopia's economic relationship with China was actively supporting its repressive political system. Madebo observed:

> The push of China into Ethiopia is driven by China's desperate need for raw materials and future market opportunities. China is wrecking the efforts of building democracy in Ethiopia by bankrolling the corrupt and repressive regime in Addis Ababa. We love your export of technology, investment, and pharmaceuticals, but please keep your unsolicited export of autocracy within China. The Ethiopian people do welcome mutually beneficial Chinese investment in Ethiopia, and the Chinese are encouraged to build dams in Ethiopia, but they should better build dams that contain water, not the flow of information.[53]

At the heart of this criticism is an argument that has become increasingly common among Africans: China's trade with African countries is supported by its close ties to

---

51. 'A caution on China', *This Day (Nigeria)*, (16 August 2011).
52. Eric Kiss and Kate Zhou, 'China's new burden in Africa', in Dennis Hickey and Baogang Guo, eds, *Dancing with the Dragon: China's Emergence in the Developing World* (Lanham, MD: Rowman & Littlefield Publishers, Inc., 2010), p. 156.
53. Ephrem Madebo, 'Autocracy: China's unsolicited export', *The Addis Voice*, (12 August 2010), available at: http://addisvoice.com/2010/08/autocracy-chinas-unsolicited-export/.

autocratic political forces and driven by its demand for natural resources and, hence, is exploitative.[54]

Conclusion

The largely market-driven disparities between China's trade with African resource producers and non-resource producers will continue to generate new political challenges. China's trade patterns with Africa are almost entirely determined by its comparative advantage in labor-intensive and capital-intensive production; Africa's natural resource endowments; the blistering growth of the Chinese economy; infrastructure expansion in both China and Africa; and the continuing emergence of government-supported economies of scale in China's shipping and manufacturing sectors. Shared autocracy appears unlikely to determine China's trade partners in Africa and is more likely a determinant of its political partners. China's state-run firms undoubtedly use extra-market decisions to influence trade flows; ultimately, however, Beijing's ability to direct trade is constrained by market demand and the location of Africa's resource supplies.

Given relative factor endowments of resources, labor, and capital there is little that can be done to reduce some African countries' overwhelming dependence on natural resource exports to China or African consumers' preference for low-cost, decent quality Chinese consumer goods. For African countries without resources and those that would like to diversify their exports, however, the continued inability of labor-intensive African exports, particularly agricultural goods, to freely access Chinese markets remain a source of frustration and concern. It is likely Africans will grow increasingly impatient with China's subsidies to its manufacturers, trade barriers against foreign goods, and political support for autocratic regimes. China will face increasing blowback from average Africans, many of whom will continue to watch their country's trade deficit with China grow over time. In sum, China's trade patterns with African countries are rooted in powerful market dynamics only partially created by government policies, but that has not stopped them from generating anti-Chinese resistance narratives among many Africans that threaten China's image as a long-term trading partner.

---

54. For additional African commentaries that reflect this emerging anti-Chinese narrative, see: Catherine Sasman, 'Chinese in Gobabis', *The Namibian* (Windhoek), (12 January 2012), available at: http://www.namibian.com.na/news/full-story/archive/2012/january/article/chinese-in-gobabis/; also Bisong Etabohen, 'Cameroon looks to fix Chinese employment blues', *Africa Review* (Nairobi), (15 January 2012), available at: http://www.africareview.com/Special+Reports/Cameroon+looks+to+fix+Chinese+employment+blues/-/979182/1305980/-/view/printVersion/-/69xk9i/-/index.html.

Index

Note: Page numbers in *italics* represent tables
Page numbers in **bold** represent figures

Accra (Ghana) 53–6, 63, 172
Acharya, A. 28; and Buzan, B. 29
actors 2, 65, 76, 85, 121, 125; African 62; Chinese 68, 75, 123, 132; Chinese state 62; ethical 31; external 63, 77, 121–2, 133; local 136; non-state 78; political 67; private 123; private economic 65; state 70; sub-state 130; Western 2, 23, 36, 74
Addis Ababa (Ethiopia) 14, 86
Addis Voice, The 213
Adebajo, A. 26
Aegean Sea 105
Afghanistan 83, 96, 101–3
Africa Development Fund 122
Africa Policy (China 2006) 64
Africa–Asia Confidential 69
African National Congress (ANC) 10; Mbete 69
African Union (AU) 67–8, 86
African Union Mission in Somalia (AMIMO) 89
Africanization 175–7; non- 175
agency: African 62–3
Agricultural Development Banks (ADB) 185
aid: foreign 184
Aid Policy (China 2011) 64
Aidoo, R.: and Hess, S. 117
AIDS (acquired immune deficiency syndrome) 178
Alden, C. 81, 208; and Large, D. 120–37
Alexandria (Egypt) 85
Algeria 85, 206
American dream 78
American exceptionalism 30
American hegemony 31–4
Angola 11, 44, 63–4, 72–5, 129, 166, 188, 206; Luanda 43, 73, 123, 206
Antananarivo (Madagascar) 56
anti-colonialism 22

anti-piracy 90, 114, 118
Apple (Consumer Electronics Company) 50
Arab Spring 17, 101, 104–6
Asahi, A. 88
Asahi Shimbun 103
Asia 16, 25, 32–3, 102, 119; financial crisis (1997–8) 182, 190
Asia Times 103
Asian Development Bank (ADB) 192
Australia 139
authoritarian capitalism 22, 25–6; state 50, 57
autocracy 197, 203, 207–8, 213–14
autonomy 130, 158–60, 173; enterprise 161; managerial 158–60
Awad al-Jaz 8

Ban Ki-moon 1, 89
Bandung (Indonesia) 66–7, 127
Barma, N.: and Ratner, E. 208; Ratner, E. and Weber, S. 207
Beijing Action Plan (FOCAC) 84
Beijing (China) 2–14, 40–53, 57–60, 73–8, 91–9, 100–11, 127–37, 179–80, 211–14; diplomacy 83; new diplomatic imperative 101–2; Olympic Games (2008) 18, 94, 97
Beijing Review 118
Beirut (Lebanon) 109–12
Benedict, R. 24
Bengal 24
Benghazi (Libya) 100, 108
bilateral diplomacy 86
Bliss, H.: and Russett, B. 207
Botswana 73; Khama 73
Bräutigam, D. 73
Brazil 76
Brazzaville (Democratic Republic of Congo) 56
British East India Company 24

INDEX

Broadman, H. 199
Brookes, P.: and Ji Hye Shin 209
bureaucracy 44, 54
Burundi 88–9
Busan (South Korea) 63
Bush, G.W. 31, 83
Buzan, B.: and Acharya, A. 29; and Little, R. 29

Cairo (Egypt) 124
Cameron, D. 22, 139
Cameroon 103, 188
Canada 56
Cape Town (South Africa) 10
capital-intensive production 197
capitalism 27; authoritarian 22, 25–6; authoritarian state 50, 57; market 132; state 8, 49; Western 59
Caritas 152
Central Intelligence Agency (CIA) 25, 32
Century of Humiliation (China 1839–1949) 104
Chad 88, 95, 129
challenges: of delivering difference in Africa 120–37
Chang, M. 58
Chaturvedi, S. 66
China Civil Engineering Construction Corporation (CCECC) 180
China Communication Construction Corporation (CCCC) 180–1
China Daily 105, 175–6
China Development Bank (CDB) 71, 123, 130, 180, 185–7, *187*
China Geo-Engineering Corporation (CGC) 170, 176, 194
China Gezhouba Group Corporation (CGGC) 191
China Harbour Engineering Co. Group (CHEC) 191
China International Water Electric group (CWE) 165
China Internet Network Information Center (CNNIC) 105
China National Fisheries Corporation (CNFC) 162
China National Petroleum Corporation (CNPC) 11, 15, 180
China Petrochemical Corp 11
China Railway Construction Corporation (CRCC) 180–1, 191
China State Hualong Construction Engineering Co. Ltd 161–4, 187, 193
China Water and Electric Corporation (CWE) 162
China's Libya Evacuation Operation 100–19; features 107–9; statistics *108*

Chinatown 56
China–Africa Business Conference 5, 124
China–Africa Business Council 5
China–Africa Cooperation 61–79, 116
China–Africa Development Fund 5, 49, 53, 82, 186
Chinese citizens: domestic pressure for protection overseas 103–7
Chinese Communist Party (CCP) 22, 104–5, 134–5, 159–61, 179, 195
Chinese People's Liberation Army (PLA) 87, 107–8
Chinese People's Political Consultative Conference (CPPCC) 111–13
Chinese State Owned Enterprise (SOE) 104
Chinese Young Volunteers Overseas Service Plan 87
Chinese–African People's Friendship Association 103
Christianity 34
Chrysanthemum and the Sword (Benedict) 24
Cissé, D. 71
citizens: Chinese overseas 103–7
Civil Aviation Administration of China 107
civil society 69, 74, 208
class: middle 119
Clinton, H. 1–2, 35, 139
Clinton, W. (Bill) 199
Coca Cola 50
Cold War 24–7, 45, 64, 84, 98
collective bargaining agreement (CBA) 150
Collum Coal Mine (CCM) (Zambia): shooting (2010) 104, 138–57
Collum Coal Mining Industries (Zambia) 13
colonialism 1, 10, 35, 51, 121, 126, 153; Chinese 52; neo- 33, 51, 75, 91–3, 124, 138–9; new 1
colonization 4
Coming War with Japan, The (Friedman and LeBard) 32
Common Market for Eastern and Southern Africa (COMESA) 86, 200, 205
communism 64
competition: strategic with West 91–8
Congo *see* Democratic Republic of Congo (DRC)
Congress of Unions ((COSATU) (South Africa) 76
Congressional Research Service (CRS) (USA) 25, 112
constitutive foreign policy rhetoric (China) 124–6
construction enterprises in China 191, *192*
cooperation: economic and trade (2010) 64, *see also* South–South cooperation (SSC)
corporate players: Africa 187–94
corporate social responsibility (CSR) 139, 195

INDEX

Cote d'Ivoire 88
counter-hegemony 36
Cresson, E. 32
Cummings, B. 24

Darfur (Sudan) 8, 18–19, 82, 89, 93–4
decolonization 7, 175–6, 182
democracy 7, 17, 31, 38, 197–9, 203, 207, 213; liberal 24, 30–2; non- 207; political 60; Western 179–80; Western liberal 21
Democratic Republic of Congo (DRC) 9, 17, 44, 81, 88, 198; Brazzaville 56
democratization 4, 93
Deng Xiaoping 22, 78, 115, 204
Denmark 162
dependency: African 62–3
development partners 1–20
diaspora 50, 77; Chinese 50
dichotomization 23
dichotomy 34, 50–2
difference: challenges of delivering 120–37
diplomacy 48, 80–4, 120, 162, 187, 199; African 80–1, 98; Beijing 83; bilateral 86; Chinese 82–7, 91, 94, 98; economic 80, 84, 91; energy 80–2, 91; foreign 103; frontier 95; Japanese 97; multilateral 86; new frontier 83, 91–2, 98; oil 80; pro-active 94
diplomatic capabilities: China 109–13
domestic pressure: for protection of Chinese citizens overseas 103–7
dynamism 122

East Timor (Timor-Leste) 101
Economic Community of West African States (ECOWAS) 86
economic cooperation: China-Africa (2010) 64
economic diplomacy 80, 84, 91
economic globalization 92, 130
economic growth 2–4, 22, 62, 80, 101–4, 197, 207
economic imperialism 56–8
economic reductionism 124
economic and trade cooperation (2010) 64
economic uncertainty: and expansive relations 122–4
Economist, The 196
economy: sociopolitical 48
Edwards, C.: and Jenkins, R. 199, 212
Egypt 14, 23, 43, 88, 104–13, 131, 209, 213; Alexandria 85; Cairo 124; Salum 107; Suez Canal 85; trade with China 209, **210**
eighteenth century 24, 178
Eisenman, J. 197–214
Elliot, E.A.: and Fei-Ling Wang 40–60
empowerment 106

energy diplomacy 80–2, 91
Engineering Institute of Zambia (EIZ) 143
English language 37, 52, 56, 141–3, 171
enslavement 4
enterprise autonomy 161
equality: political 120–1
Equatorial Guinea 8–9, 12, 17, 88, 129, 188, 213
Eritrea 88
ethical actors 31
Ethiopia 12–14, 49, 75, 88, 103, 132, 206, 213; Addis Ababa 14, 86
ethnicity 148, 154
Eurocentrism 23, 31, 36
Europe 1, 6–9, 18, 25, 29–30, 33, 114, 133
European Command (EUCOM) 109
European Parliament (EP) 97
European Union (EU) 14, 31–5, 97
EU–China relations 98
evacuation: Libya Operation 100–19, *108*; non-combatant operations (NEOs) (USA) 109–12, *110*
exceptionalism 120–37; American 30
expansive relations: amidst economic uncertainty 122–4
Expeditionary Strike Group (ESG) (USA) 109
exploitation: unsustainable 10–11
Export-Import Bank of China (EXIM Bank) 10, 65, 71, 85, 180, 185–7, *187*
external actors 63, 77, 121–2, 133
Extractive Industries Transparency Initiative (EITI) 9–10

factor endowment theory 206
Federal Bureau of Investigation (FBI) 25
Fei-Ling Wang: and Elliot, E.A. 40–60
feudalism 135
fifteenth century 90
Five Year Plan: Tenth (China 2001–5) 102
foreign aid 184
foreign diplomacy 103
foreign direct investment (FDI) 42, 183, 188, *189*
foreign policy: constitutive rhetoric (China) 124–6
Fortune (magazine) 168
Forum on China–Africa Cooperation (FOCAC) 1, 22, 43, 71, 86–7, 102, 122, 163, 193; *Beijing Action Plan* 84; top four exports 199, **201**
France 55, 96–7, 122, 192; Paris 63
French language 52, 55–6
Friedman, E. 15–16
friendship: Sino-African 47
frontier diplomacy 95

INDEX

Gabon 85, 188
Gadaffi, M. 17, 23, 68, 100
Gansu (China) 162
Gemstone and Allied Workers Union of Zambia (GAWUZ) 141–2
genocide 94
George, A.: Walt, S. and Nye, J. 28
Germany 30, 34, 45, 96, 162; Hitler 33
Ghana 12, 15, 44–8, 52–6, 85, 158–77, 185; Accra 53–6, 63, 172; Investment Act (1994) 166–8; relationship with China 52, 53; state-owned enterprises (SOEs) 158–77
Global Financial Crisis (2008–9) 6
global power 57, 80
Global Times 105
globalization 45, 77, 86; Chinese 61; economic 92, 130
Government of the Republic of Zambia (GRZ) 139–40, 143–7, 150–5
gravity trade theory 199, 202–3, 211
Great Britain 8, 139
Greece 107–10
Grimm, S. 61–79, 116
gross domestic product (GDP) 183–5
Group of 77 developing countries (G77) 67
growth: economic 2–4, 22, 62, 80, 101–4, 197, 207; resource trade 209–11
Grunberg, I. 31
Guangdong (China) 130
Guangzhou (China) 77
Gulf of Aden 90, 108, 113–14, 117–18

Han Fangming 113
Harare (Zimbabwe) 44, 48, 51
Hebei province (China) 12, 87
Heckscher–Ohlin (H–O) model 201–2
hegemony 22, 25, 29–31, 87; American 31–4; counter 36; non-Western 29; US 27; Western 23, 31
Herbst, J.: and Mills, G. 202
Hess, S.: and Aidoo, R. 117
Hitler, A. 33
HIV (human immunodeficiency virus) 178
Hong Kong (China) 107
Hoover, J.E. 25
Hu Fuming 73
Hu Jintao 5–6, 12–13, 17–20, 84–6, 92–8, 126
Huajian Shoes 14
Huang Ping 105, 110
Huang Zequan 103
Huawei Technologies Co. Ltd 42, 169–70, 175
human rights 4–7, 23–6, 44, 57, 70, 95, 125, 208; international organizations 18; practices 116; violations 8, 44

Huntington, S.P. 33–4
hydropower resources 11

identity: self- 31, 122
ideology 64, 68, 83–4, 122; revolutionary 84
illiberalism 197, 208
immigration 51
imperialism 24, 124; Chinese 45; economic 56–8; power 52
India 76, 96, 170, 212
information and communications technology (ICT) 169–70
instability: political 84, 103, 106
institutionalization 82
international community 22, 32, 82, 86, 90
International Energy Agency (IEA) 188
international human rights organizations 18
international media 7–8
International Monetary Fund (IMF) 6, 25, 139, 202, 209
International Relations (IR) 23, 29–36, 76, 97
international trade 200–3
International Trade Union Confederation (ITUC) 149
internationalization 159–64, 172–7
Internet 105–6, 119, 133, 156
investment: foreign direct (FDI) 42, 183, 188, *189*
Iraq 83, 96
Israel 108–9

Japan 16, 24, 30–6, 45, 77, 80, 95–8, 124; Junichiro Koizumi 95; relations with China 97; Tokyo International Conference on African Development (TICAD) 97
Japanese diplomacy 97
Jenkins, R.: and Edwards, C. 199, 212
Ji Hye Shin: and Brookes, P. 209
Jian-Ye Wang 202
Jiang Zemin 83
Jiangxi (China) 162
Jianwei Wang: and Jing Zou 80–99
Jones, W. 24
Jordan 107, 110
Juba University (Sudan) 15
Junichiro Koizumi 95

Kabonde, F. 144
Kagame, P. 7
Kaiser Kuo 105
Kaplinsky, R.: and Morris, M. 183
Kashimu, B. 143
Kazakhstan 188
Kenya 50, 81, 206; Nairobi 42, 47–9; University of Nairobi 50

INDEX

Kernen, A.: and Lam, K.N. 15, 158–77
KFC 50
Khama, I. 73
Khartoum (Sudan) 114
Kirkpatrick, J. 156
Kiss, E.: and Zhou, K. 213
Kitwe (Zambia) 123
Konkola Copper Mine (KCM) 144–5
Kordofan (Sudan) 129
Korea 45, 50, 77, 96
Korean War (1950–3) 156
Korea–Africa forum 96
Krugman, P. 203

labor rights 50
labor-intensive production 11, 91, 197, 200, 211–14; sectors 211–12
Lam, K.N.: and Kernen, A. 15, 158–77
Langfang (China) 87
Laos 118, 206
Large, D. 37; and Alden, C. 120–37
Latin America 25, 82, 102
Latour, B. 175
Layne, C. 32
Lebanon 108, 111–14; Beirut 109–12
legitimacy 2, 128
Leninism 126
Li Chengbao 113
Li Ruogu 10
liberal democracy 24, 30–2
liberalism 27, 207–9
liberalization 163
Liberia 42, 85, 88–9; Sirleaf 89
Libya 17, 23, 42, 68–9, 91, 94, 100–19, 209; Benghazi 100, 108; China's Evacuation Operation 100–19, *108*; Civil War (2011) 68; crisis (2011) 100; Gadaffi 17, 23, 68, 100; trade with China 209, **210**; Tripoli 100, 107
Lin, J.Y. 205
Little, R.: and Buzan, B. 29
local actors 136
localization 113, 159, 165–9, 172–7; management 171–2; workforce 158–77
Los Angeles Times 153
Luanda (Angola) 43, 73, 123, 206
Luanshy Copper Mine (LCM) 144–5
Luo Yuan 113
Lusaka (Zambia) 13, 138, 145, 182

Maamba Collieries Ltd (MCL) 154
Macau (China) 107
Madagascar 46, 52, 55–7, *56*; Antananarivo 56; Malagasy (national language) 55–6
Malawi 70
Mali 66, 70, 88

Malta 107–10
management: localization 171–2; micro- 176
managerial autonomy 158–60
Mansfield, E.D.: *et al.* 203, 207
Mao Era (China 1949–76) 41, 125, 161, 197
Mao Zedong 126; anti-imperialist struggle 3
Maputo (Mozambique) 11
marginalization 16, 163
Marine Corps (USA) 109; Marine Expedition Unit (MEU) 109
market capitalism 132
Marques, H. 203
Marxism 126
Mauritius 12–14, 49, 123
Mawdsley, E. 34–5
Mbeki,T. 71, 212
Mbete, B. 69
Mbulu, R. 149, 153
media: international 7–8; traditional 106
Mediterranean Sea 115
Mekong River 117–18
Member of Parliament (MP) 152
Memorandums of Agreement (MOAs) 112
mercantilism 49
Mhkondo Lungu 144
micro-management 176
microbloggers 106
middle class 119
Middle East 82–3, 90, 106, 114, 170
migrants 46, 55–7
migration 77, 133
military capabilities: development (China) 113–15
Military Central Command (CENTCOM) (USA) 109
Mills, G.: and Herbst, J. 202
Milner, H.V.: *et al.* 203, 207
Mine Safety Department (MSD) 142–3
Mineworkers Union of Zambia (MUZ) 146
Ministry of Commerce (MOFCOM) (China) 65, 100–1
Ministry of Foreign Affairs (MOFA) (China) 109–11, 181
Miwa Hirono: and Shogo Suzuki 21–39
mobilization 3–7, 120
modernity 131–3
modernization 90, 115, 135; neo- 121
modernization theory 135
Mopani Copper Mines (MCM) (Switzerland) 144
Morris, M.: and Kaplinsky, R. 183
Morrissey (singer/songwriter) 156
Mozambique 11, 18, 70, 88; Maputo 11
Mugabe, R. 7–8, 19
Mulafulafu, S. 152
multilateral diplomacy 86

INDEX

multilateralism 86, 123
Mulvenon, J. 115
Musyalike, M. 147
Mwangi Kimenyi 190
Myanmar 103, 118
myth-busting 21–39

Nairobi (Kenya) 42, 47–9; University of 50
Namibia 10, 43, 74
Nanjing University (China) 73
nation-building 126, 134–6
National Development and Reform Commission (NDRC) (China) 181
National People's Congress (NPC) 102
national security 90, 117; agenda 23–9
nationalism 31, 57–8, 68, 132, 135
nationality 199
neo-classical trade theory 204
neo-colonialism 33, 51, 75, 91–3, 124, 138–9
neo-colonialist predator 1–20
neo-liberalism: racial 155–7
neo-modernization 121
new colonialism 1
new diplomatic imperative 100–19; Beijing 101–2
new frontier diplomacy 83, 91–2, 98
New International Economic Order (1970s) 67
new trade theory 203
New York Times 152, 204
Newsweek 36
Nigeria 1, 6, 14–16, 49, 85, 103, 132, 188, 206; Niger Delta 16; Obasanjo 17
nineteenth century 24, 29, 46, 155, 178
non-Africanization 175
non-combatant evacuation operations (NEOs) (USA) 109–12, *110*
non-democracy 207
non-governmental organizations (NGOs) 15, 35, 94, 121, 143, 152; African 87; Western 67
non-OECD 72
non-state actors 78
non-West 29
non-Western hegemony 29
non-Western power 31, 31–6
Nye, J.: George, A. and Walt, S. 28
Nyerere, J. 129

Obasanjo, O. 17
oil diplomacy 80
Olin, B. 203
Olympic Games (Beijing 2008) 18, 94, 97
Organization for Economic Co-operation and Development (OECD) 5, 9, 65, 72, 126, 134, 193; non- 72
Orient 32

Orientalism 24, 38
orthodoxy 67
Ottawa Treaty (1997) 88
overseas citizen protection 100–19
ownership 63

Pakistan 101–3, 165
Palley, T.I. 204–5
Paris (France) 63
Patriotic Front (PF) (Zambia) 139, 155
Payne, D. 149
peace: building 87–93; keeping 89–90
Pehnelt, G. 25
Peking University (China) 106, 118
people-to-people relations 77–8
People's Bank of China (PBC) 185
People's Daily 18
personalization 172–5
Peru 188
PetroSA 15
Pew Global Attitudes Project (2002) 104
Philippines 105
piracy 114
pluralism 134, 137
political actors 67
political democracy 60
political equality 120–1
political instability 84, 103, 106
political relations 64, 129
political trade 199, 207–9
politically instable countries avoidance 16–17
postcolonial West 7
poverty 4, 15–16, 38, 85, 93, 131, 212
power: global 57, 80; imperialism 52; non-Western 31–6; relations 120, 129–31
pragmatism 64, 74, 165; Chinese engagement-economic 64–5
Prebitsch, R. 67
presence/perceptions and prospects 40–60
private actors 123
private economic actors 65
private sphere 155
privatization 160
pro-active diplomacy 94
pro-China 54
production: capital-intensive 197
protection: for Chinese citizens overseas 103–7; responsibility to protect (R2P) 67
public diplomacy 6
public sphere 155

Qinghai Province (China) 113

racial hierarchy 138
racial neo-liberalism 155–7

INDEX

racial superiority 132
racism 155
Ratner, E.: and Barma, N. 208; Weber, S. and Barma, N. 207
Reagan, R. 156
realism: structural 31
Red Sea 8
reductionism: economic 124
regionalism 80, 86
relations: China–Africa 19, 61–7, 73–4, 86, 121, 131; EU–China 98; expansive 122–4; international (IR) 23, 29–36, 76, 97; people-to-people 77–8; political 64, 129; Sino–African 2, 14, 21–39, 64, 71, 89, 99
relative factor abundance theory 199–205
repressive regimes 18–19
Research Excellence Framework (REF) 27
resource: trade growth 209–11; unsustainable trade pattern/exploitation 10–11
responsibility to protect (R2P) 67
revolutionary ideology 84
Rhodes University (South Africa) 38
Ricardo, D. 201
rights: human 4–7, 8, 23–6, 44, 57, 70, 95, 116, 125, 208; international organizations 18; labor 50, 211
Rohmer, S. 155
Rosendorff, B.P.: et al. 203, 207
Royal Dutch Shell 16
Russett, B.: and Bliss, H. 207
Russia 8, 94–6; Stalin 33
Rwanda 7, 16, 63, 67, 75; Kagame 7

Said, E. 24
Salum (Egypt) 107
Sampa, M. 2
Samuelson, R.J. 36
Sanusi, L. 1
Sata, M. 13–15, 59, 148, 153, 198
Sautman, B.: and Yan Hairong 138–57
Scott, G. 149, 153
security: national 23–9, 90, 117; Sino–African 86
Segou (Mali) 58
self-identity 31, 122
September 11 terrorist attack (9/11) (USA 2001) 82
seventeenth century 178
Shaanxi (China) 162
Shamenda, F. 151
Shanghai (China) 131, 174, 204
Shogo Suzuki: and Miwa Hirono 21–39
Sida International TV 47
Sierra Leone 88–9, 166
Simango, A. 141

Sinohydro Corporation 11, 164–6, 191
Sinopec 73
Sino–Africa Business and Investment Forum 5
Sino–African friendship 47
Sino–African relations 2, 14, 21–39, 64, 71, 89, 99
Sino–African security 86
Sino–African trade 41
Sino–American Strategic and Economic Dialogue 98
Sirleaf, E.J. 89
Snow, P. 132
social networking 105
socialism 133
socialization 29
sociopolitical economy 48
Somalia 59, 90, 118
Somaliland 206
Song Xiaojun 114
South Africa 10–11, 17–19, 46, 75–6, 129, 132–3, 156; Cape Town 10; Congress of Unions (COSATU) 76; Mbeki 71, 212; PetroSA 15; Rhodes University 38; University of Johannesburg 38; Zuma 1, 71
South Korea 95, 206
Southern African Development Community (SADC) 86
South–South cooperation (SSC) 61, 65–7, 70–2, 76–9; roots and core elements 65–70
sovereignty 29, 66–8, 93, 116, 120–2, 127, 135; state 67, 208
Soviet Union 24, 36, 98–9, 125
Spain 174
special economic zones (SEZs) 14, 49, 203–4
Spielberg, S. 18
Stalin, J. 33
state actors 70; Chinese 62
state capitalism 8, 49
state sovereignty 67
State-owned Assets Supervision and Administration Commission (SASAC) 160, 181, 185, 192–5
state-owned enterprises (SOEs) 178–96; Ghana 158–77
state-ownership 54, 64, 159
stereotypes 77, 175
strategic competition: with West 91–8
structural realism 31
sub-Saharan Africa (SSA) 183, 210
sub-state actors 130
Sudan 7–8, 17–19, 88–9, 93–7, 110–14, 128–9, 133, 188, 213; Awad al-Jaz 8; Darfur 8, 18–19, 82, 89, 93–4; Juba University 15; Khartoum 114; Kordofan 129; National Congress Party 128; Taha 18

INDEX

Suez Canal (Egypt) 85
Suisheng Zhao 1–20

Taha, A. 18
Taiwan 3, 41, 83, 107, 115, 125, 182–4; Taipei 41
Tang Xiaoyang 166
Tanzania 3, 19, 43, 59, 85, 129–31, 166, 169, 206; Nyerere 129; Tazara 41
Taylor, I. 208
Tazara (Tanzania-Zambia Railroad) 41
telecommunications 179
Telegraph, The 153
terrorism 103
Thailand 103, 117–18
Third World 22, 65–6, 93
This Day (Nigerian newspaper) 212
Tiananmen Square massacre (China 1989) 3, 22, 64
Tibet 66
Tokyo International Conference on African Development (TICAD) 97
Tonga 140–1
trade: China–Africa cooperation (2010) 64; China–Africa patterns 197–214; China–Egypt 209, **210**; gravity trade theory 199, 202–3, 211; international 200–3; new 203; patterns 197–214; political 199, 207–9; resource growth 209–11; Sino–African 41; unsustainable pattern/resource exploitation 10–11
trade unions 76
traditional media 106
Transportation Command (TRANSCOM) 109
Tripoli (Libya) 100, 107
Tsvangirai, M. 19
Tull, D. 208
Tunisia 107, 110
Turkey 107, 110
twentieth century 27, 66, 77, 155
twenty-first century 3, 74, 80, 87, 96, 178, 190
Twitter 105, **106**

Uganda 89
unification 83
United Arab Emirates (UAE) 110
United Kingdom (UK) 27, 96, 156, 171; Cameron 22, 139; Konkola Copper Mine (KCM) 144–5
United Nations Commission on Human Rights (UNCHR) 83
United Nations Conference on Trade and Development (UNCTAD) 67
United Nations (UN) 3, 41, 68, 82, 87–9, 94, 156, 182; Ban Ki-moon 1, 89; Organization Mission in Democratic Republic of Congo (MONUC) 88; Security Council 43–4, 82; Transition Assistance Group 87; Transitional Authority in Cambodia (TAC) 88
UN–Africa Union 18
United Party for National Development (UPND) 148
United States of America (USA) 6–9, 24–7, 30–7, 45–7, 55–7, 81–3, 95–9, 108–9, 139, 199; Bush 31, 83; Central Intelligence Agency (CIA) 25, 32; Clinton (Bill) 199; Clinton (Hilary) 1–2, 35, 139; Congressional Research Service (CRS) 25, 112; Department of Defense 27; Expeditionary Strike Group (ESG) 109; Federal Bureau of Investigation (FBI) 25; hegemony 27; Joint Planning Doctrine 112; Marine Corps 109; Marine Expedition Unit (MEU) 109; Military Central Command (CENTCOM) 109; Navy's Sea, Air and Land Teams (SEALs) 59; non-combatant evacuation operations (NEOs) 109–12, *110*; Payne 149; Reagan 156; September 11 terrorist attack (9/11) 82; Washington 83, 93, 135
University of Johannesburg 38
University of Nairobi 50
unsustainable trade pattern/resource exploitation 10–11
urbanizing 81

Vietnam 156

Wall Street Journal 153
Walt, S.: Nye, J. and George, A. 28
Wang Chengan 102
Wang Yi 97
Wang Yizhou 106, 118
Washington (USA) 83, 93, 135
Weber, S.: Barma, N. and Ratner, E. 207
Weibo 105–6
Wen Jiabao 5, 107
West 38–9, 40–1, 44–5, 50–2, 57–60, 125–8, 131–2, 155–7, 212; actors 2, 23, 36, 74; aid 66–7, 116; capitalism 59; democracy 179–80; exceptionalism 29–36; hegemony 23, 31; liberal democracies 21; NGOs 67; non- 29; postcolonial 7; strategic competition with 91–8
Western Sahara 88
Westmoreland, W. 156
Wheeler, T. 75
workforce localization 158–77
World Bank 6, 14, 25, 71, 139, 162, 191–4, 205, 209
World Trade Organization (WTO) 163
World War II (1939–45) 24, 28–30
Wu Bangguo 124
Wu Jiuhua 138

INDEX

xenophobia 50, 58, 77, 149
Xi Jinping 5, 19–20, 44, 47, 80, 98–9
Xiao Lishan 138
Xu Jianxue 140–3, 146, 150
Xu Yi-Chong 178–96
Xuzhou (China) 113–14

Yan Hairong: and Sautman, B. 138–57
Yellow Peril thesis 23, 32–6
Yemen 103
Yiwu (China) 77
Yu, G.T. 131
Yu Hong Wei 58

Zambezi Natural Stone Enterprise (ZNS) 154
Zambia 1–3, 11–15, 43, 75, 104, 188, 206, 213; Collum Coal Mine (CCM) shooting (2010) 104, 138–57; Collum Coal Mining Industries 13, 104; Engineering Institute of Zambia (EIZ) 143; Federation of Employers (ZFE) 148; Gemstone and Allied Workers Union (GAWUZ) 141–2; Government (GRZ) 139–40, 143–7, 150–5; Kitwe 123; Lusaka 13, 138, 145, 182; Mineworkers Union (MUZ) 146; neoliberal reform and CCM's troubled marginality 139–43; Patriotic Front (PF) 139, 155; Sampa 2; Sata 13–15, 59, 148, 153, 198; Tazara 41
Zambia-China Mulungushi Textiles Ltd 13
Zerba, S.H. 100–19, 175
Zhai Jun 2
Zhang Dejiang 107, 112
Zhao, S. 177
Zhejiang (China) 130
Zheng He (Admiral) 90, 133
ZhongXing Telecommunication Equipment Corporation (ZTE) 163–5, 169–70, 175, 180
Zhou, K.: and Kiss, E. 213
Zhu Rongji 102, 185
Zhu Weilie 116
Zimbabwe 7–8, 18–19, 25, 37, 44, 47, 51, 59, 75, 128, 157, 213; Harare 44, 48, 51; Mugabe 7–8, 19; visa fees 51, *51*; ZANU 128
Zoellick, R. 14
Zuma, J. 1, 71